Learning Guide

Kathleen Schmidt Prezbindowski

College of Mount St. Joseph

With Contributions by Laurie Prezbindowski

Introduction to the Human Body
The Essentials of Anatomy and Physiology
Seventh Edition

Gerard J. Tortora

Bergen Community College

Bryan Derrickson

Valencia Community College

WILEY

JOHN WILEY & SONS, INC.

Cover Photo: ©2005 Lois Greenfield

ISBN-13 978-0-471-76105-1
ISBN-10 0-471-76105-2

Printed in the United States of America

10 9 8 7 6 5 4 3 2 1

Printed and bound by Bind-Rite Graphics, Inc.

Contents

Preface

This *Learning Guide* accompanies Tortora and Derrickson's *Introduction to the Human Body,* Seventh Edition. It emphasizes active learning, not passive reading, and prompts students to examine concepts by means of a variety of activities and exercises. By approaching each concept several times from different points of view, the student sees how the ideas of the text apply to real, clinical situations. In this way, the students do not simply study but *learn.*

The 24 chapters of the *Learning Guide* parallel those of the Tortora and Derrickson text. Each chapter begins with an **Overview** that introduces content succinctly and a **Framework** that visually organizes the relationships among key concepts and terms. Objectives from the text promote convenient previewing and reviewing. **Objectives** appear according to major chapter divisions in the **Topic Outline. Wordbytes** present word roots, prefixes, and suffixes that facilitate understanding and recollection of key terms. **Checkpoints** offer a variety of learning activities that follow the exact sequence of topics in the chapter, with cross-references to pages in the text. This format makes the *Learning Guide* "student friendly" and enhances the effectiveness of this excellent textbook. **Answers to Selected Checkpoints** provide feedback to students. A **Critical Thinking** section provides questions that challenge students to practice writing skills. A 15-question **Mastery Test** includes both objective and subjective questions as a final tool for student self-assessment of learning. **Answers to the Mastery Test** immediately follow the Mastery Test.

The *Learning Guide* fits the needs of people with different learning styles. Visual learners will identify the Frameworks as assets in organizing concepts and key terms. Visual and tactile learners will welcome the coloring and labeling exercises with color code ovals. Further variety in Checkpoints is offered by short definitions, fill-ins, tables, matching, multiple-choice, and arrange-in-sequence exercises. *Clinical Challenge* introduce applications especially relevant for students in allied health. These are based on my own experiences as an R.N. and on discussions with colleagues. These challenging exercises enable students to apply abstract concepts to actual clinical settings.

This guide also provides readily available feedback for self-assessment. Answers to most Checkpoints are included at the end of the chapter. The placement of Answers to the Mastery Test immediately following the Mastery Test offers readily accessible feedback.

Many people have contributed to the development of this book. I would like to thank my colleagues and students at the College of Mount St. Joseph for their helpful comments and suggestions, and Laurie Prezbindowski, M.Ac., for contributions and editorial assistance, Gala Erland, and Bonnie Roesch, Editor, and Alicia Romano, Editorial Assistant at Wiley for their assistance in the preparation of this edition .

Kathleen Schmidt Prezbindowski

To the Student

This *Learning Guide* is designed to help you do exactly what its title indicates: learn. It will serve as a step-by-step aid to help you bridge the gap between goals (objectives) and accomplishment (learning). The 24 chapters of the *Learning Guide* parallel the 24 chapters of Tortora and Derrickson *Introduction to the Human Body,* Seventh Edition. Each chapter consists of the following parts: *Overview, Framework, Topic Outline with Objectives, Wordbytes, Checkpoints, Answers to Selected Checkpoints, Critical Thinking,* and *Mastery Test with Answers*. Take a moment to turn to Chapter 1 and look at the major sections in this *Learning Guide*. Now consider each one.

Overview. As you begin your study of each chapter, read this brief introduction, which presents the topics of the chapter. The Overview is a look at the "forest" (or the "big picture") before the examination of the individual "trees" (or details).

Framework. The Framework arranges key concepts and terms in an organizational chart that demonstrates interrelationships and lists key terms. Refer back to the Framework frequently to keep sight of interrelationships.

Topic Outline and Objectives. Objectives that are identical to those in the text organize the chapter into "bite-sized" pieces that will help you identify the principal subtopics you must understand. Preview objectives as you start the chapter, and refer back to them to review. Objectives are arranged according to major chapter divisions in the Topic Outline. Check off Objectives as you complete each one.

Wordbytes present the word roots, prefixes, and suffixes that will help you master terminology in the chapter and facilitate understanding and remembering key terms. Study these, then check your knowledge (cover meanings and write them in the margins). Try to think of additional terms in which these important bits of words are used.

Checkpoints offer a variety of learning activities arranged according to the sequence of topics in the chapter. At the start of each new Checkpoints section, note the related pages in the Tortora and Derrickson text. Read these pages carefully and then complete this group of Checkpoints. Checkpoints are designed to help you handle each new concept. Through a variety of exercises and activities, you can achieve a depth of understanding that comes with active participation, not simply passive reading. You can challenge and verify your knowledge of key concepts by completing the Checkpoints along your path of learning.

Diverse learning activities are included in the Checkpoints. You will focus on specific facts by doing exercises in which you provide definitions and comparisons, and fill in blanks in paragraphs with key terms. You will also color and label figures. Fine color felt-tipped pens or pencils will be helpful so that you can fill in related color code ovals easily. Your understanding will also be tested with matching exercises, multiple-choice questions, and arrange-in-correct-sequence activities. Applications for enrichment and interest in areas particularly relevant to nursing or other health-sciences students are found in selected *Clinical Challenges* exercises.

When you have difficulty with an exercise, refer to the related pages (listed with each Checkpoint of the *Learning Guide*) of *Introduction to the Human Body* before proceeding. You will find specific references, by page number, for many text figures and exhibits.

Answers to Selected Checkpoints. As you glance through the activities, you will notice that a number of questions are marked with solid boxes (■). Answers to these Checkpoints are given at the end of each chapter, in order to provide you with some immediate feedback about your progress. Incorrect answers alert you to review related objectives in the *Learning Guide* and corresponding text pages. Each *Learning Guide* (LG) figure included in *Answers to Selected Checkpoints* is identified with an "A," such as Figure LG 8.1A.

Checkpoints without answers are included intentionally in this book. This encourages you to verify some answers independently by consulting your textbook and to initiate discussion with students or instructors.

Critical Thinking questions near the end of each chapter offer you the opportunity to practice your writing skills in essays related to chapter content.

Mastery Test. This 15-question self-test provides an opportunity for a final review of the chapter as a whole. Its format will assist you in preparing for standardized tests or course exams that are objective in nature, since the first 10 questions are multiple-choice, true-or-false, and arrangement questions. The final 5 questions of each test help you evaluate your learning with fill-in and short-answer questions. Answers for each mastery test are placed at the end of each chapter.

I wish you success and enjoyment in learning concepts and relevant applications of your own anatomy and physiology, as well as applying this information in clinical settings.

K. S. P.

About the Author

Kathleen Schmidt Prezbindowski, Ph.D., M.S.N., is a professor of biology at the College of Mount St. Joseph, where she was recently named Distinguished Teacher of the Year. She has taught anatomy and physiology, pathophysiology, and biology of aging for 35 years. Dr. Prezbindowski's Ph.D. is in biology, and she has earned M.S.N.s in both Gerontological and Psychiatric/Mental Health Nursing. Through a grant from the U.S. Administration on Aging, she produced the first video series on the human body ever designed especially for older people. She is also the author of four other learning guides and a wellness manual for older adults.

Laurie Prezbindowski, Contributor, has earned a Masters in Acupuncture, and has taught Anatomy at the New England School of Acupuncture, Waterton, MA.

Learning Guide

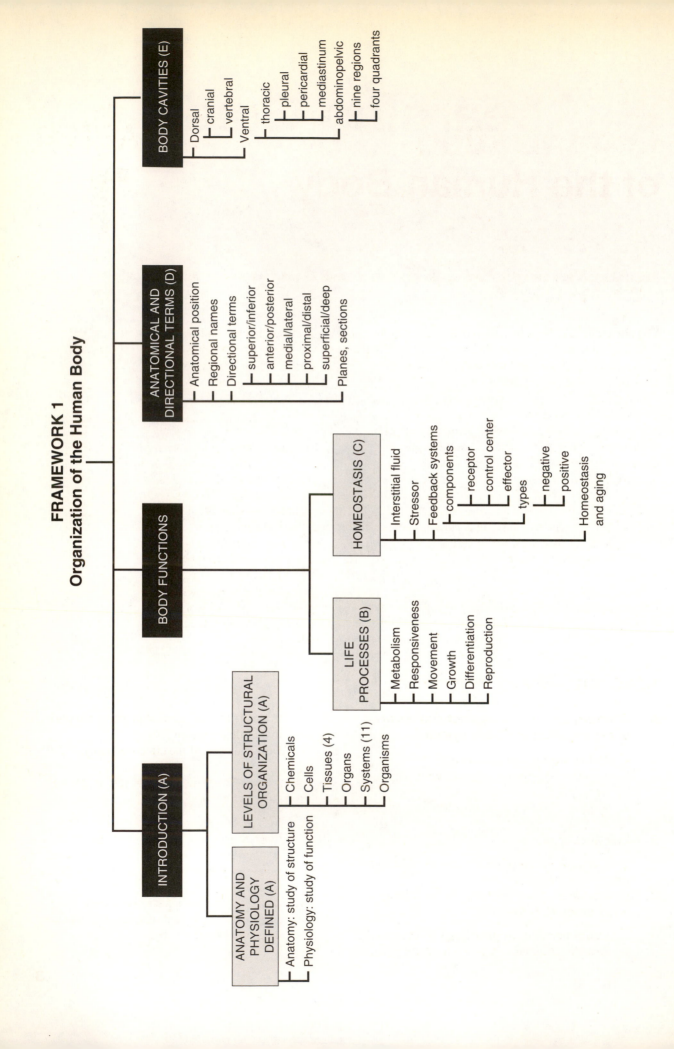

FRAMEWORK 1
Organization of the Human Body

INTRODUCTION (A)

ANATOMY AND PHYSIOLOGY DEFINED (A)
- Anatomy: study of structure
- Physiology: study of function

LEVELS OF STRUCTURAL ORGANIZATION (A)
- Chemicals
- Cells
- Tissues (4)
- Organs
- Systems (11)
- Organisms

BODY FUNCTIONS

LIFE PROCESSES (B)
- Metabolism
- Responsiveness
- Movement
- Growth
- Differentiation
- Reproduction

HOMEOSTASIS (C)
- Interstitial fluid
- Stressor
- Feedback systems
  - components
    - receptor
    - control center
    - effector
  - types
    - negative
    - positive
- Homeostasis and aging

ANATOMICAL AND DIRECTIONAL TERMS (D)
- Anatomical position
- Regional names
- Directional terms
  - superior/inferior
  - anterior/posterior
  - medial/lateral
  - proximal/distal
  - superficial/deep
- Planes, sections

BODY CAVITIES (E)
- Dorsal
  - cranial
  - vertebral
- Ventral
  - thoracic
    - pleural
    - pericardial
    - mediastinum
  - abdominopelvic
    - nine regions
    - four quadrants

Organization of the Human Body

CHAPTER 1

You are about to embark on a tour, one you will enjoy for at least several months, and hopefully for many years. Your destination: the far-reaching corners and fascinating treasures of the human body. Imagine the human body as an enormous, vibrant world, abundant in architectural wonders and bustling with activity. You are the traveler, the visitor, the learner on this tour.

Just as the world is composed of individual buildings, cities, countries, and continents, so the human body is organized into cells, tissues, organs, and systems. In fact, a close-up look at a cell (comparable to a detailed study of a building) will reveal structural intricacies—the array of chemicals that make up cells. Any tour requires a sense of direction and knowledge of the language: directional and anatomical terms that guide the traveler through the body. A tour is sure to enlighten the visitor about the culture in each region—in the human body, the functions of each body part or system. By the time you leave each area (body part or system), you will have gained a sense of how that area contributes to overall stability and well-being (homeostasis). You will also learn about how feedback systems (loops) contribute to homeostasis.

In planning a trip, an itinerary serves as a useful guide. A kind of itinerary is found below in the objectives (same as in your text), listed according to a topic outline that can help you to organize your study. It can also give you a well-deserved feeling of accomplishment as you check off the box next to each objective after you complete it. Then, at the end of the trip/chapter, look back at this itinerary/outline to put all aspects of the entire chapter in context. Also look at the Framework for Chapter 1, study it carefully, and refer back to it as often as you wish. It contains the key terms of the chapter and serves as your map and itinerary. Bon voyage!

TOPIC OUTLINE AND OBJECTIVES

A. Anatomy and physiology defined; levels of organization and body systems

☐ 1. Define anatomy and physiology.
☐ 2. Describe the structural organization of the human body.
☐ 3. Explain how body systems relate to one another.

B. Life processes

☐ 4. Define the important life processes of humans.

C. Homeostasis: maintaining limits; aging and homeostasis

☐ 5. Define homeostasis and explain its importance.
☐ 6. Describe the components of a feedback system.

☐ 7. Compare the operation of negative and positive feedback systems.
☐ 8. Describe some of the effects of aging.

D. Anatomical and directional terms, planes, and sections

☐ 9. Describe the anatomical position.
☐ 10. Identify the major regions of the body and relate the common names to the corresponding anatomical terms for the various parts of the human body.
☐ 11. Describe the directional terms and the anatomical planes and sections used to locate parts of the human body.

☐ 12. Describe the principal body cavities and the organs they contain.

☐ 13. Explain why the abdominopelvic cavity is divided into regions and quadrants.

WORDBYTES

Now become familiar with the language of this chapter by studying each wordbyte (word root, prefix, or suffix), its meaning, and an example of its use in a term. After you study the entire list, self-check your understanding by jotting the meaning of each wordbyte in the margin. As you continue through the text and *Learning Guide,* identify other examples of terms that contain these wordbytes.

Wordbyte	Meaning	Example(s)	Wordbyte	Meaning	Example(s)
ana-	up	*ana*tomy	-logy	study of	physio*logy*
ante-	before	*ante*rior	physio-	nature	*physio*logy
homeo-	same	*homeo*stasis	post-	after	*post*erior
inter-	between	*inter*cellular	-stasis, stat-	stand, stay	homeo*stasis*
intra-	within	*intra*cellular	-tomo	cutting	ana*tomy*

CHECKPOINTS

A. Anatomy and physiology defined; levels of organization and body systems (pages 2–6)

■ **A1.** Anatomy is the study of _____ and physiology is the study of

_____ .

A2. An intimate relationship exists between structure and function: one determines the other. Explain how the structure of each of the following determines the functions for which it can be used.

a. Chair/bed

c. Hand/foot

b. Spoon/fork

d. Incisors (front teeth)/molars (back teeth)

■ **A3.** Your text book authors, Tortora & Derricksen, compare levels of organization in the human body to levels of organization in language, for example, language that makes up this book. Fill in the blanks below to complete that analogy: _____

    a. Letters → _____ → sentences → paragraphs → _____ → the entire book

    b. Chemicals → cells → _____ → organs → _____ → the entire _____

■ **A4.** Match answers in the box with correct descriptions below. One is done for you.

a. One living individual made of many systems: _____

b. Group of related organs with common function: __**system**_____

4

c. Structure composed of two or more different types of tissues: _____

d. Group of similar cells that perform a particular function: _____

e. The basic unit of structure and function of an organism: _____

f. Composed of atoms; lowest level of organization: _____

■ **A5.** Muscle is one type of tissue. Name the other three types of tissue.

■ **A6.** Refer to Table 1.1 (pages 4–5 of your text) and identify which systems in the box match descriptions listed below.

D. Digestive	LI. Lymphatic and immune	S. Skeletal
E. Endocrine	R. Reproductive	U. Urinary
I. Integumentary		

_____ a. Produces hormones

_____ b. Includes kidneys, ureters, bladder, and urethra

_____ c. Includes ovaries and uterus (females), prostate and penis (males)

_____ d. Returns proteins to blood; protects against disease

_____ e. Supports the body, stores calcium, and serves as site of blood cell formation

_____ f. Consists of skin, hair, nails, and some glands

_____ g. Includes esophagus, stomach, intestine, liver, gallbladder, and pancreas

■ **A7.** Several systems listed in Table 1.1 were not included in Checkpoint A5. Name them.

A8. Suppose you decide to run a quarter mile at your top speed. Tell how your body would respond to this stress created by exercise. List according to system the changes your body would probably make in order to run the quarter mile.

a. Cardiovascular: _____

b. Respiratory: _____

c. Muscular: _____

d. Integumentary: _____

B. Life processes   (page 6)

B1. Define six life processes of humans. One is done for you.

a. _____

b. _____

c. _____

d. __Growth: increase in size and complexity__

e. _____

f. _____

■ **B2.**   Refer to the list of life processes in the box. Demonstrate your understanding of these processes by selecting the answer that best fits each of the activities in your own body, described below.

D.	Differentiation	Move.	Movement
G.	Growth	Repro.	Reproduction
Meta.	Metabolism	Resp.	Responsiveness

_____ a. Your hunger at 8:00 A.M. prompts you to head towards breakfast.

_____ b. During breakfast you chew your toast with jelly and an egg; your stomach and intestine then contract to help break apart the food and propel it through the digestive tract.

_____ c. Your body utilizes the starch, sugars, and proteins in your breakfast foods to provide building blocks for more muscle protein and to provide energy to your eyes and brain for studying anatomy and physiology.

_____ d. After you work out four days a week for a month, you note that your arm and thigh muscles (biceps and quadriceps) are larger (two answers).

_____ e. As you work out, "stem cells" in your bone marrow are stimulated to undergo changes to become mature red blood cells (RBCs) so you experience a healthy increase in your RBC count.

C.   Homeostasis: maintaining limits; aging and homeostasis   (pages 6–9)

■ **C1.**   Fill in the blanks in this definition of homeostasis.

a. *Homeo* means _____ and *stasis* means _____ .

b. Homeostasis is the condition in which the fluid around body cells, called _____ fluid or the

body's _____ environment, remains relatively stable so that it stays within certain limits.

c. List at least five qualities of your interstitial fluid that are maintained under the optimal conditions when your body is in homeostasis.

d. Consider how your internal environment is affected by your external environment. Which of the qualities of your interstitial fluid listed above would be altered if your body were invaded by infectious microorganisms that cause fever and diarrhea over a prolonged period of time? Circle answers in C1c.

■ **C2.**   Are you seated right now? For safety, hold on to the chair or desk, and then stand up quickly. Describe the homeostatic mechanism that follows as an effort to maintain your blood pressure (BP) at a normal level.

a. The effect of gravity as you stand up causes blood to flow to the *(upper? lower?)* parts of your body, _____ -creasing BP in your upper body. This change is called a(n) *(stimulus? output?)*, which causes *(effectors? receptors?)* in blood vessels of your neck to sense the decreased BP. As a result, nerve impulses, called *(input?*

output?), are sent to your brain, known as the _____ center.

b. Your brain then conveys nerve impulses, called *(effectors? output? response?)*, to your heart and blood vessels. These organs serve as *(effectors? output? response?)* and cause the desired *(effectors? output? response?)*: an elevation of your BP back to normal.

c. The homeostatic mechanism just described is a *(positive? negative?)* (or opposite) feedback mechanism because the slight decrease in BP (upon your standing) triggered mechanisms to _____ -crease BP back to normal. If mechanisms had caused your blood pressure to drop even further as you stood, this would have been a *(positive? negative?)* (or same direction) feedback mechanism.

d. Most of the body's homeostatic mechanisms are *(positive? negative?)*. Write an example of one that is positive.

C3. Write two or three health practices that you believe would improve your life. Explain how these would better maintain your homeostasis and contribute to disease prevention.

■ **C4.** Identify signs and symptoms in this list by writing correct answers on lines provided.

Decreased red blood count	Pain of a stomach ulcer
Fever	Rash
Joint pain	Vomiting

a. Signs: _____

b. Symptoms: _____

■ **C5.** Contrast terms within the following pairs that are related to illnesses and diagnoses.

a. Disease/disorder: _____

b. Inspection/palpation: _____

c. Auscultation/Percussion: _____

■ **C6.** Aging is considered a(n) *(normal? abnormal?)* process. List five of more natural aging changes.

D. Anatomical and directional terms, planes, and sections (pages 9–15)

D1. Assume *anatomical position* yourself. Now write a description of that position.

■ **D2.** Complete the table relating common names to anatomical terms. (See Figure 1.4, page 11, in your text to check your answers.) For extra practice use common names and anatomical terms to identify each region of your own body and that of a study partner.

Common Name	Anatomical Term
a.	Axillary
b. Fingers	
c. Arm	
d.	Gluteal
e.	Cephalic
f. Mouth	
g.	Inguinal
h. Chest	
i.	Cervical

■ **D3.** Using your own body, a skeleton, a torso, Exhibit 1.1 and Figures 1.5–1.8 (pages 13–15 in your text), determine relationships among body parts. Write the correct directional term(s) to complete each of these statements.

a. The liver is _____ to the diaphragm.

b. Fingers (phalanges) are located _____ to wrist bones (carpals).

c. The skin on the dorsal surface of your body can also be said to be located on your _____ surface.

d. The great (big) toe is _____ to the little toe.

e. The skin on your leg is _____ to muscle tissue in your leg.

f. When you float facedown in a pool, you are lying on your _____ surface.

g. The lungs and heart are located _____ to the abdominal organs.

h. A frontal plane divides the body into _____ and _____ portions.

i. A _____ plane divides the body into equal right and left halves.

E. Body cavities (pages 15–17)

■ **E1.** After you have studied Figures 1.8–1.11 (pages 15–17 in your text), complete this exercise about body cavities. Circle the correct answer in each statement.

a. The *(dorsal? ventral?)* cavity consists of the cranial cavity and vertebral canal.

b. Heart, lungs, and intestine are all located in the *(dorsal? ventral?)* cavity.

c. The *(dorsal? ventral?)* cavity appears to be better protected by bone.

d. *Pleural, mediastinal,* and *pericardial* are terms that refer to regions of the *(thorax? abdominopelvis?)*.

e. The *(heart? lungs? esophagus and trachea?)* are located in the pleural cavities.

f. The division between the abdomen and the *(thorax? pelvis?)* is marked by the diaphragm.

g. The stomach, pancreas, small intestine, and most of the large intestine are located in the *(abdomen? pelvis?)*.

h. The urinary bladder, rectum, and internal reproductive organs are located in the *(abdominal? pelvic?)* cavity.

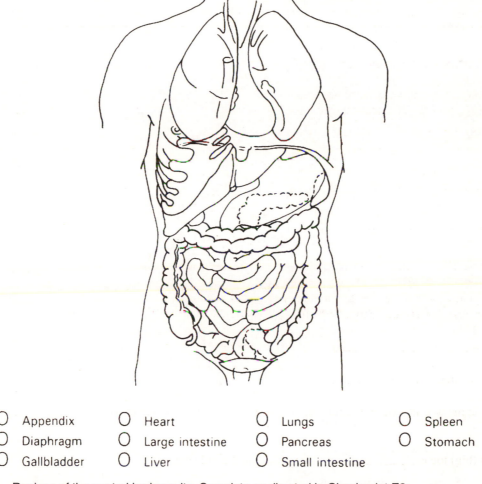

○ Appendix	○ Heart	○ Lungs	○ Spleen
○ Diaphragm	○ Large intestine	○ Pancreas	○ Stomach
○ Gallbladder	○ Liver	○ Small intestine	

Figure LG 1.1   Regions of the ventral body cavity. Complete as directed in Checkpoint E2.

■ **E2.** Complete Figure LG 1.1 according to the directions below. Then check your answers by referring to Figures 1.9 and 1.10, page 16 in the text.

a. Color each of the organs listed on the figure. Select different colors for each organ, and be sure to use the same color for the related color code oval (○).

b. Next, label each organ on the figure.

c. The liver and gallbladder are located in the _____ quadrant of the abdomen.

d. *For extra review.* Draw lines dividing the abdomen into the nine regions, and note which organs are in each region.

9

ANSWERS TO SELECTED CHECKPOINTS: CHAPTER 1

A1. Structure, function.

A3. (a) Letters ➔ (b) **words** ➔ (c) sentences ➔ (d) paragraphs ➔ (e) **chapters** ➔ (f) the entire book
(b) Chemicals ➔ (b) cells ➔ (c) **tissues** ➔ (d) organs ➔ (e) **systems** ➔ (f) the entire **organism (human body)**

A4. (a) Organism. (b) System. (c) Organ. (d) Tissue. (e) Cell. (f) Chemical.

A5. Connective, epithelial, and nervous.

A6. (a) E. (b) U. (c) R. (d) LI. (e) S. (f) I. (g) D.

A7. Muscular, nervous and sensory, cardiovascular, and respiratory.

B2. (a) Resp. (b) Move. (c) Meta. (d) G, Repro. (e) D.

C1. (a) Same, stays. (b) Interstitial, internal. (c) Gases (such as oxygen and carbon dioxide), nutrients (such as proteins and carbohydrates), electrolytes (such as sodium and potassium), fluids, temperature, and pressure. (d) Temperature, fluids, and electrolytes.

C2. (a) Lower, de; stimulus, receptors; input, control. (b) Output; effectors, response. (c) Negative, in; positive.

(d) Negative; labor contractions or blood clotting.

C4. (a) Decreased red blood count, fever, rash, vomiting, (b) Joint pain, pain of an ulcer.

C5. (a) More specific/less specific. (b) Involves sense of vision/involves senses of touch and pressure. (c) Involves listening, for example, with a stethoscope/involves tapping on body surface (such as over lungs or liver) and listening for the quality of echoes.

C6. Normal. See page 9.

D2. (a) Armpit. (b) Digital. (c) Brachial. (d) Buttock. (e) Head. (f) Oral. (g) Groin. (h) Thoracic. (i) Neck.

D3. (a) Inferior or caudal. (b) Distal. (c) Posterior. (d) Medial. (e) Superficial. (f) Anterior or ventral. (g) Superior. (h) Anterior, posterior. (i) Midsagittal.

E1. (a) Dorsal. (b) Ventral. (c) Dorsal. (d) Thorax. (e) Lungs. (f) Thorax. (g) Abdomen. (h) Pelvic.

E2. See Figure LG 1.1A.

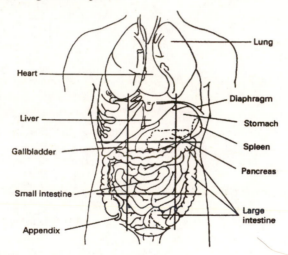

Figure LG 1.1A   Regions of the ventral body cavity.

CRITICAL THINKING: CHAPTER 1

1. Write two short paragraphs, each incorporating one of these lists of terms.
   a. cardiovascular, heart, homeostasis, ventral, system, thoracic
   b. abdominal, digestive, liver, quadrant, superior, transverse

2. Think of a person you know who has a chronic (ongoing) illness. Explain how changes in anatomy (abnormal body structure) in that person are associated with changes in physiology (inadequate or altered body function).

MASTERY TEST: ■ CHAPTER 1

Questions 1–5: Circle the letter preceding the one best answer to each question.

1. A negative feedback mechanism results in a change:
   A. Even farther in the same direction as the stimulus, making the problem worse

   B. In the opposite direction, tending toward correction of the problem

2. The following structures are all located in the ventral cavity *except:*
   A. Spinal cord   D. Gallbladder
   B. Urinary bladder   E. Esophagus
   C. Heart

3. Which pair of common/anatomical terms is mismatched?
   A. Eye/orbital   D. Neck/cervical
   B. Head/cephalic   E. Buttock/gluteal
   C. Armpit/brachial

4. The spleen, tonsils, and thymus are all organs in which system?
   A. Nervous   D. Digestive
   B. Lymphatic   E. Endocrine
   C. Cardiovascular

5. Which of the following structures is located totally outside the upper right quadrant of the abdomen?
   A. Liver   D. Spleen
   B. Gallbladder   E. Pancreas
   C. Transverse colon

Questions 6–7: Arrange the answers in correct sequence.

_____ _____ _____ 6. From most proximal to most distal:
   A. Ankle
   B. Hip
   C. Knee

_____ _____ _____ 7. From most superior to most inferior in location:
   A. Abdominal cavity
   B. Axillary region
   C. Pelvic cavity

Questions 8–10: Circle T (true) or F (false). If the statement is false, change the underlined word or phrase so that the statement is correct.

T F 8. In order to assume the anatomical position, you should lie down with arms at your sides and palms turned backward.

T F 9. The appendix is usually located in the left iliac region of the abdomen.

T F 10. Anatomy is the study of how structures function.

Questions 11–15: Fill-ins. Write the word or phrase that best fits the description.

_____ 11. If a response reverses the original stimulus, the system is a _____ feedback system.

_____ 12. A _____ is a group of similar cells that perform a particular function.

_____ 13. A _____ plane divides the body into equal right and left halves.

_____ 14. The _____ is a region of the thorax located between the lungs; it contains the heart, esophagus, thymus, some large blood vessels, and part of the trachea.

_____ 15. _____ is a life process in which unspecialized cells (such as those in an early embryo) change into specialized cells such as nerve or bone cells.

ANSWERS TO MASTERY TEST: ■ CHAPTER 1

Multiple Choice
1. B
2. A
3. C
4. B
5. D

Arrange
6. B C A
7. B A C

True or False
8. F. Stand erect facing observer, arms at sides, palms forward
9. F. Right iliac
10. F. Physiology

Fill-ins
11. Negative
12. Tissue
13. Midsagittal
14. Mediastinum
15. Differentiation

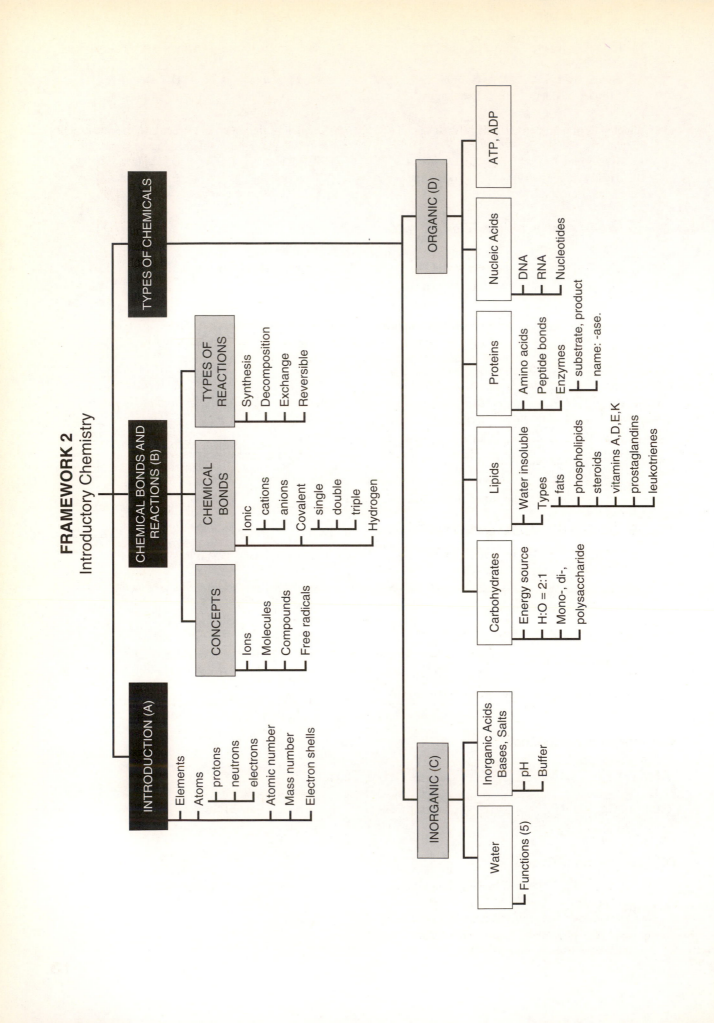

FRAMEWORK 2
Introductory Chemistry

TYPES OF CHEMICALS

INTRODUCTION (A)

Elements
Atoms
├ protons
├ neutrons
└ electrons
Atomic number
Mass number
Electron shells

CHEMICAL BONDS AND REACTIONS (B)

CONCEPTS
├ Ions
├ Molecules
├ Compounds
└ Free radicals

CHEMICAL BONDS
├ Ionic
│ ├ cations
│ └ anions
├ Covalent
│ ├ single
│ ├ double
│ └ triple
└ Hydrogen

TYPES OF REACTIONS
├ Synthesis
├ Decomposition
├ Exchange
└ Reversible

INORGANIC (C)

Water
└ Functions (5)

Inorganic Acids
Bases, Salts
├ pH
└ Buffer

ORGANIC (D)

Carbohydrates
├ Energy source
├ H:O = 2:1
└ Mono-, di-,
  polysaccharide

Lipids
├ Water insoluble
└ Types
  ├ fats
  ├ phospholipids
  ├ steroids
  ├ vitamins A,D,E,K
  ├ prostaglandins
  └ leukotrienes

Proteins
├ Amino acids
├ Peptide bonds
└ Enzymes
  ├ substrate, product
  └ name: -ase.

Nucleic Acids
├ DNA
├ RNA
└ Nucleotides

ATP, ADP

Introductory Chemistry CHAPTER

Chemicals make up the ultrastructure of the human body. They are the submicroscopic particles of which cells, tissues, organs, and systems are constructed. Chemicals are the minute entities that participate in all of the reactions that underlie bodily functions. Every complex compound that molds a living organism is composed of relatively simple atoms held together by chemical bonds. Two classes of chemicals contribute to structure and function: inorganic compounds (primarily water but also inorganic acids, bases, and salts) and organic compounds (including carbohydrates, lipids, proteins, nucleic acids, and the energy-storing compound ATP).

As you begin your study of the chemistry of life, carefully examine the Chapter 2 Framework and note the key terms associated with each section. Also refer to the Chapter 2 Topic Outline and check off objectives as you meet each one.

TOPIC OUTLINE AND OBJECTIVES

A. Introduction to basic chemistry

- [ ] 1. Define a chemical element, atom, ion, molecule, and compound.

B. Chemical bonds and reactions

- [ ] 2. Explain how chemical bonds form.
- [ ] 3. Describe what happens in a chemical reaction and explain why it is important to the human body.

C. Inorganic compounds

- [ ] 4. Discuss the functions of water and inorganic acids, bases, and salts.
- [ ] 5. Define pH and explain how the body attempts to keep pH within the limits of homeostasis.

D. Organic compounds

- [ ] 6. Discuss the functions of carbohydrates, lipids, and proteins.
- [ ] 7. Explain the importance of deoxyribonucleic acid (DNA), ribonucleic acid (RNA), and adenosine triphosphate (ATP).

WORDBYTES

Now become familiar with the language of this chapter by studying each wordbyte, its meaning, and an example of its use in a term. Check your understanding by jotting the meaning of each wordbyte in the margin. Identify other examples of terms that contain these wordbytes as you continue through the text and *Learning Guide*.

Wordbyte	Meaning	Example(s)	Wordbyte	Meaning	Example(s)
-ase	enzyme	prote*ase*	-philic	loving	hydro*philic*
de-	remove	*de*hydrogenase	-phobic	fearing	hydro*phobic*
di-	two	*di*phosphate	poly-	many	*poly*saccharide
-lysis	to break apart	hydro*lysis*	-saccharide	sugar	di*saccharide*
mono-	one	*mono*saccharide	tri-	three	*tri*phosphate

CHECKPOINTS

A.   Introduction to basic chemistry   (pages 23–25)

■ **A1.**   Complete this exercise about matter and chemical elements.

a. Matter exists in three states: solid, _____ , and _____ .

b. All matter is made up of building units called chemical _____ .

c. Of the total 112 different chemical elements found in nature, four elements make up 96 percent of the mass of the

   human body. These elements are _____ (O), _____ (C),

   hydrogen (_____), and _____ (_____).

d. Circle the chemical symbols of elements listed here that together make up 3.8% of human body mass:

   Ca   Cl   Co   Cu   Fe   I   K   Mg   Mn   Na   P   S   Si   Sn   Zn

e. Define this term: *trace elements.*

   Now write on the line the symbol of a trace element from the list of elements above (in d) that plays a key role in the body because it forms part of the thyroid hormone.

   _____

■ **A2.** Refer to Table 2.1 (page 23) in your text. Match the elements listed in the box with descriptions of their functions in the body. Use each answer only once. The first one has been done for you.

C. Carbon	Cl. Chlorine	N. Nitrogen	O. Oxygen
Ca. Calcium	Fe. Iron	Na. Sodium	P. Phosphorus

___C___ a. Found in every organic molecule

_____ b. Component of all protein molecules, as well as DNA and RNA

_____ c. Component of ATP; found in bone and teeth

_____ d. Essential component of hemoglobin, which gives red color to blood

_____ e. Composes part of every water molecule; functions in cellular respiration

_____ f. Contributes hardness to bone and teeth; required for blood clotting and for muscle contraction

_____ g. Anion (negatively charged ion) in table salt (NaCl)

_____ h. Cation (positively charged ion) in table salt (NaCl); helps control water balance in the body; important in nerve and muscle function

■ **A3.** Complete this exercise about atomic structure.

a. An atom consists of two main parts: _____ and _____ .

b. Within the nucleus are positively charged particles called _____ and uncharged (neutral)

particles called _____ .

c. Electrons bear (*positive? negative?*) charges. The number of electrons forming shells around the nucleus is (*greater than? equal to? smaller than?*) the number of protons in the nucleus of the atom.

■ **A4.** Refer to Figure 2.2 in your text (page 24), and complete this exercise about the structure of atoms.

a. The atomic number of potassium (K) is _____ . This means that an atom of potassium has _____ protons

and _____ electrons.

b. The mass number of potassium (K) is _____ . Mass number is the total number of (*protons + electrons? protons + neutrons?*) in the atom.

c. Calculate how many neutrons are found in potassium atoms. An average of _____ neutrons.

B. Chemical bonds and reactions (pages 25–29)

■ **B1.** The type of chemical reacting (or bonding) that occurs between atoms depends on the number of electrons of the bonding atoms. Refer to Figure 2.2 in your text (page 24) and do this exercise about electrons and bonding.

a. Electrons are arranged in _____ shells. The electron shell closest to the nucleus has a

maximum of _____ electrons, whereas the second electron shell is complete when it holds _____ electrons.

b. The combining capacity is the number of extra or deficient electrons in an atom's *outermost electron shell.* (Visualize that this is the electron shell atom A presents to nearby atom B, which may be a possible "bonding partner.")

In other words, the combining capacity of an atom tells the number of _____ an atom attempts to gain, lose, or share in an effort to have its outermost electron shell complete.

c. Now find potassium (K) in the same figure. Potassium has one *(extra? missing?)* electron in its outermost energy level. It is therefore more likely to become an electron *(donor? acceptor?)*. When potassium gives up the electron (which is a negative entity), potassium becomes more *(positive? negative?)*. In other words, potassium forms K^+, which is a(n) *(anion? cation?)*.

d. Now look at chlorine (Cl) in the figure. Chlorine has _____ electrons in its outermost energy level, so Cl has one *(extra? missing?)* electron in that energy level. As potassium chloride (KCl) is formed, chlorine *(accepts? gives up?)* one electron, and so becomes the *(anion? cation?)* Cl^-.

e. In summary, ions are atoms that have *(gained or lost? shared?)* one or more electrons into or from their outermost energy levels. A cation, such as K^+, Na^+, or H^+, is an atom that has *(gained? lost?)* an electron (a negative entity) by ionic bonding. A(n) *(anion? cation?)*, such as Cl^-, has gained an electron from another atom. Ions in

solution, such as table salt (Na^+Cl^-) in water, are called _____ .

■ **B2.** Write all of the answers that describe each chemical listed below.

C. Compound	F. Free radical	I. Ion	M. Molecule

_____ a. Glucose ($C_6H_{12}O_6$)

_____ b. Hydrogen (H_2)

_____ c. Water (H_2O)

_____ d. Superoxide which has an unpaired electron in its outer shell

_____ e. Ca^{2+}

_____ f. Carbon dioxide (CO_2)

■ **B3.** Match the description with the types of bonds listed in the box.

C. Covalent	H. Hydrogen	I. Ionic

_____ a. Atoms lose electrons if they have just one or two electrons in their outer orbits; they gain electrons if they need just one or two electrons to complete the outer electron shell, as in K^+Cl^- (see Activity B1).

_____ b. This type of bond is most easily formed by carbon (C) because its outer orbit is half-filled.

_____ c. A pair of electrons is *shared* in this type of bond, for example, in a C—H bond.

_____ d. These bonds are only about 5 percent as strong as covalent bonds and so are easily formed and broken. They hold large molecules such as protein and DNA in proper configurations.

_____ e. Double bonds, such as O══O (O_2), or triple bonds, such as N≡≡N (N_2).

_____ f. Types of bonds broken when electrolytes are dissolved.

_____ g. Bonds may be classified as polar or nonpolar.

■ **B4.** Identify the kind of chemical reaction described in each statement below.

D. Decomposition reaction	R. Reversible reaction
E. Exchange reaction	S. Synthesis reaction

_____ a. Two or more reactants combine to form an end product; for example, many glucose molecules bond to form glycogen.

_____ b. All chemical reactions involve making or breaking of bonds. This type of reaction involves only breaking of bonds.

_____ c. This type of reaction is utilized in hydrolysis of foods, for example, digestion of starch to glucoses.

_____ d. Reactions that can go in either direction (synthesis or decompositions), depending on conditions.

_____ e. In this type of reaction chemicals undergo decomposition, and the products form new chemicals.

■ **B5.** Energy released from chemical bonds via decomposition reactions is temporarily stored in high-energy bonds of *(ATP? H_2O? DNA?)*. Write out the full name of ATP:

A _____ T _____ P _____

C. Inorganic compounds: water, inorganic acids, bases, and salts; pH and buffers (pages 29–30)

■ **C1.** Write *I* next to phrases that describe inorganic compounds, and write *O* next to phrases that describe organic compounds.

_____ a. Always contain carbon and hydrogen; held together almost entirely by covalent bonds

_____ b. Tend to be very large molecules that serve as good building blocks for body structures

_____ c. The class that includes carbohydrates, proteins, fats, and nucleic acids

_____ d. The class that includes water, the most abundant compound in the body

C2. Use key words here to describe several parts of the "job description" of water in the human body:

a. Solvent

b. Chemical reactions

c. Heat

d. Lubricant

■ **C3.** Write the term *hydrophilic* or *hydrophobic* next to related descriptions.

a. Sugar and salt are _____ .

b. Most lipids such as peanut oil and corn oil are _____ .

c. Chemicals that are "water-loving" because they are soluble in water: _____ .

■ **C4.** Fill in the blanks in this exercise about acid, bases, and salts.

a. NaOH (sodium hydroxide) is an example of a(n) _____ , that is, a chemical that breaks apart into hydroxide ions (OH^-) and one or more cations.

b. H_2SO_4 (hydrogen sulfate) is an example of a(n) _____ , that is, a chemical that dissociates into hydrogen ions (H^+) and one or more anions.

c. Salts, such as NaCl (sodium chloride), are chemicals that dissolve in water to form cations and ions, neither of which is _____ or _____ .

d. Write a term that means "to break apart into ions": _____ .

■ **C5.** After you study Figure 2.7 and Table 2.2, page 31 in your text, circle the correct answer in each question about pH.

a. Which pH is most alkaline?   4   7   10

b. Which pH has the highest concentration of H^+ ions?   4   7   10

c. Which of these solutions has its pH closest to neutral (7.0)?

Blood   Distilled (pure) water   Lemon juice   Milk of magnesia   Stomach (gastric) juice   Vaginal fluids

List five other fluids in the human body that are classified as alkaline (pH >7).

d. A solution with pH 6 has *(10 times more? 10 times fewer?)* H^+ ions than a solution with pH 7.

e. A solution with pH 5 has *(10 times more? 20 times more? 100 times more?)* H^+ ions than a solution with pH 7.

17

C6. Write a sentence explaining each of the following:

a. How pH is related to homeostasis

b. The role of buffers in maintaining homeostasis

D. Organic compounds (pages 31–40)

■ **D1.** Do this exercise on carbohydrates.

a. Name two types of carbohydrates that provide energy for the body.

b. The ratio of hydrogen (H) to oxygen (O) in carbohydrates is _____ : _____ . For example, the chemical

formula for glucose is _____ , indicating that for every one carbon (C), there are

_____ hydrogens (H) and _____ oxygen (O).

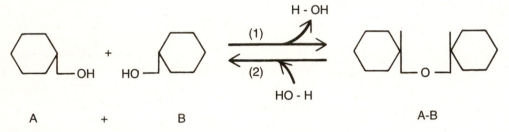

Figure LG 2.1 Dehydration synthesis and hydrolysis reactions. Refer to Checkpoint D1.

c. When two glucose molecules ($C_6H_{12}O_6$) combine, a disaccharide is formed. Its formula is

_____ , indicating that in a synthesis reaction such as this, a molecule of water is *(added? removed?)*. Refer to Figure LG 2.1. Identify which reaction *(1? 2?)* on that figure demonstrates synthesis.

d. Continued dehydration synthesis leads to enormous carbohydrates called _____ . One

such carbohydrate is _____ .

e. When a disaccharide such as sucrose is broken, water is introduced to split the bond linking the two mono-

saccharides. Such a decomposition reaction is called _____ , which literally means "splitting using water." (See Figure 2.8, page 32 in your text.) On Figure LG 2.1, hydrolysis is shown by reaction *(1? 2?)*.

■ **D2.** Identify the correct class of carbohydrates of each chemical listed below.

| D. Disaccharide | Mono. Monosaccharide | Poly. Polysaccharide |

_____ a. Cellulose

_____ b. Lactose and maltose

_____ c. Galactose

_____ d. Ribose and deoxyribose

D3. Defend or dispute this statement: "Lipids or fats are bad for us (humans)." Be sure to include these key words in your writing: *dissolve, insulate, energy, plasma membranes, hormones, vitamins, saturated,* and *cardiovascular disease.*

■ **D4.** Circle the answer that correctly completes each statement.

a. A triglyceride consists of:
  A. Three glucoses bonded together
  B. Three fatty acids bonded to a glycerol

b. A typical fatty acid contains (see Figure 2.10, page 33 in your text):
  A. About 3 carbons
  B. About 16 carbons

c. A polyunsaturated fat contains _____ than a saturated fat.
  A. More hydrogen atoms
  B. Fewer hydrogen atoms

d. Saturated fats are more likely to be derived from:
  A. Plants, as in corn oil or cottonseed oil
  B. Animals, as in beef or cheese

e. Oils that are classified as saturated fats include:
  A. Tropical oils such as coconut oil and palm oil
  B. Vegetable oils such as safflower oil, sunflower oil, and corn oil

f. A diet designed to reduce the risk for cardiovascular disease is a diet low in:
  A. Saturated fats
  B. Polyunsaturated fats

g. The part of a phospholipid that is polar is the:
  A. Fatty acid
  B. Phosphate

h. The portion of a phospholipid that faces outward in a cell membrane is the:
  A. Polar "head"
  B. Nonpolar "tail"

i. The steroid from which all other steroids are formed is:
  A. Estradiol
  B. Cholesterol

j. Essential fatty acids (EFAs), such as omego-3 and omega-6 fatty acids, that may protect against heart disease and otherwise improve health, _____ be synthesized by the human body.
  A. Can
  B. Cannot

k. The type of monounsaturated fatty acids that are unhealthy because they are hydrogenated are:
  A. *Cis*-fatty acids
  B. *Trans*-fatty acids

■ **D5.** Circle the four words from this list that are the best "job descriptions" of proteins.

| antibodies | cholesterol | contraction | enzymes |
| hormones | phospholipids | table sugar | vitamins E, K |

■ **D6.** Contrast proteins with the other organic compounds you have studied so far by completing this exercise.

a. Carbohydrates, lipids, and proteins all contain carbon, hydrogen, and oxygen. A fourth element,

_____ , makes up a substantial portion of proteins.

b. Just as large carbohydrates are composed of repeating units (the _____), proteins are

composed of building blocks called _____ . At least *(12? 20? 32?)* different types of amino acids form human proteins. They differ from one another based on the structure of the *(carboxyl group? side chain?)*. An amino group is the *(-COOH? -NH$_2$?)* portion of an amino acid.

c. As in the synthesis of carbohydrates or fats, when two amino acids bond together, a water molecule must be *(added? removed?)*. This is another example of *(hydrolysis? dehydration?)* synthesis.

d. The product of such a reaction is called a _____ -peptide. (See Figure 2.13, page 35 in your text.) When many amino acids are linked together in this way, a _____ -peptide results. One or more polypeptide chains form a _____ .

e. In order for a protein to have the correct structure and function, its polypeptide chains must each have amino acids arranged in correct sequence. Name one hereditary blood disorder that involves substance of just one amino acid for another one in the blood protein, hemoglobin. _____

■ **D7.** Do this exercise about roles of enzymes related to chemical reactions.

a. Explain why it is necessary for chemical reactions to take place rapidly within the human body.

b. Explain how enzymes act as "catalysts" to increase chemical reaction rates without requiring an elevated body temperature.

c. Chemically, enzymes consist mostly of *(carbohydrates? lipids? proteins?)*. Each enzyme reacts with a specific molecule called a _____ .

d. The substrate interacts (in lock-and-key fashion) with a specific region on the enzyme called the

_____ , producing an intermediate known as an _____ complex. Very quickly the *(enzyme? substrate?)* is transformed in some way (for example, is broken down or transferred to another substrate), while the *(enzyme? substrate?)* is recycled for further use.

e. Names of most enzymes end in the letters _____ . Enzyme names also give clues about their functions. For example, enzymes that remove hydrogen are called _____ .

Anhydrases remove _____ from substrates. Lipases are enzymes that break down _____ . Lactose intolerance involves a deficiency of the enzyme _____ .

f. Cofactors are *(ions? proteins? vitamins?)*, whereas coenzymes are _____. How do cofactors affect enzymes?

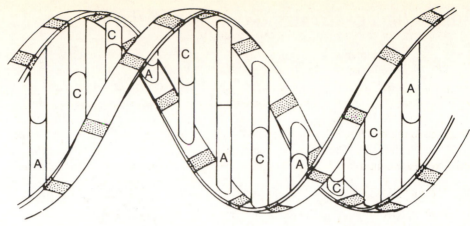

Figure LG 2.2   Structure of DNA. Complete as directed in Checkpoint D8 using the letters P, phosphate; S sugar (deoxyribose); A, adenine; T, thymine; G, guanine; C, cytosine.

■ **D8.**   Complete Figure LG 2.2 and fill in the blanks in this exercise about DNA.

a. The repeating units in nucleic acids such as DNA and RNA are called _____ . Circle one nucleotide in Figure LG 2.2.

b. Write in the complementary set of nucleotides in the figure of double-stranded structure of DNA.

Bases are paired _____ to _____ and

_____ to _____ . Also label sugars (S) and phosphates (P).

c. DNA molecules differ from one another by their sequence of _____ . Note that bases form the "rungs" of the DNA ladder. One gene is formed from about *(10? 100? 1,000? 100,000?)* rungs. Humans have about *(40? 400? 4,000? 40,000?)* functional genes, such as genes controlling eye color or blood type. Genes regulate sequencing of amono acids that form body proteins. Note the dire consequences of an omission or substitution (mutation) in the "four-letter genetic alphabet" A C G T. (See Checkpoint D6e.)

d. How does a strand of RNA differ from DNA?

■ **D9.**   Describe the structure and significance of ATP by completing these statements.

a. *ATP* stands for *adenosine triphosphate.* Adenosine consists of a base that is a component of DNA, that is,

_____ , along with the five-carbon sugar named _____ .

b. The *TP* of *ATP* stands for _____ . The final two phosphates are bonded to the molecule by high-energy bonds.

c. When the terminal phosphate is broken, a great deal of energy is released as ATP is split into

_____ + _____ .

d. ATP, the body's primary energy-storing molecule, is constantly reformed by the reverse of this reaction as energy is made available from foods you eat. Write this reversible reaction.

e. The anaerobic phase of cellular respiration yields _____ molecules of ATP, whereas the aerobic phase yields

_____ molecules of ATP.

A1. (a) Liquid, gas. (b) Elements. (c) Oxygen (O), carbon (C), hydrogen (H), nitrogen (N). (d) Ca, Cl, Fe, K, Mg, Na, P, S. (e) Fourteen elements found in very small amounts, together composing only 0.2% of the total body mass; I.

A2. (b) N. (c) P. (d) Fe. (e) O. (f) Ca. (g) Cl. (h) Na.

A3. (a) Nucleus, electrons. (b) Protons (p^+), neutrons (n^0). (c) Negative, equal to.

A4. (a) 19; 19, 19. (b) 39; protons + neutrons. (c) 20 (mass number − atomic number = 39 − 19).

B1. (a) Electron; 2, 8. (b) Electrons. (c) Extra; donor; positive; cation. (d) 7, missing; accepts, anion. (e) Gained or lost; lost; anion; electrolytes.

B2. (a) C, M. (b) M. (c) C, M. (d) F. (e) I. (f) C, M.

B3. (a) I. (b) C. (c) C. (d) H. (e) C. (f) I. (g) C.

B4. (a) S. (b) D. (c) D. (d) R. (e) E.

B5. ATP; adenosine triphosphate.

C1. (a–c) O. (d) I.

C3. (a) Hydrophilic. (b) Hydrophobic. (c) Hydrophilic.

C4. (a) Base. (b) Acid. (c) H^+ or OH^-. (d) Ionize (or dissociate).

C5. (a) 10. (b) 4. (c) Blood; semen, cerebrospinal fluid, pancreatic juice, bile. (d) 10 times more. (e) 100 times more.

D1. (a) Sugars (such as sucrose and table sugar) and starch; also lactose (milk sugar). (b) 2 : 1; $C_6H_{12}O_6$, 2, 1. (c) $C_{12}H_{22}O_{11}$, removed; 1. (d) Polysaccharides; starch or glycogen. (e) Hydrolysis; 2.

D2. (a) Poly. (b) Di. (c) Mono. (d) Mono.

D4. (a–d) B. (e, f) A. (g) B. (h) A. (i–k) B.

D5. Antibodies, contraction (of muscles), enzymes, hormones.

D6. (a) Nitrogen. (b) Monosaccharides, amino acids; 20; side chain; NH_2. (c) Removed; dehydration. (d) Di; poly; protein. (e) Sickle cell anemia.

D7. (a) Thousands of different types of chemical reactions are necessary daily to maintain homeostasis. (b) Enzymes orient molecules so that they are more likely to react, and increase frequency of collisions. (c) Proteins; substrate. (d) Active site; enzyme-substrate; substrate, enzyme. (e) *-ase*; dehydrogenase; water; fats (lipids); lactase. (f) Ions, vitamins (or portions of vitamins); they facilitate enzyme function.

D8. (a) Nucleotides; phosphate–sugar–base; see Figure LG 2.2A. (b) See Figure LG 2.2A; adenine, thymine, cytosine, guanine. (c) Bases (A C T G) within nucleotides; 1,000; 40,000. (d) RNA is single-stranded, its sugar is ribose, and it contains the base uracil (U), instead of thymine (T).

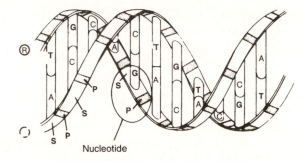

Figure LG 2.2A Structure of DNA. P, phosphate; S, sugar (deoxyribose); A, adenine; T, thymine; G, guanine; C, cytosine.

D9. (a) Adenine, ribose. (b) Triphosphate. (c) ADP, phosphate (P). (d) ATP $\rightleftharpoons$ ADP + P + energy. (e) 2, 36–38.

CRITICAL THINKING: CHAPTER 2

1. Write several functions of water that explain why this chemical is vital to living organisms.
2. Contrast the roles of carbohydrates and proteins in the human body.
3. Describe three factors that affect enzyme action: specificity, efficiency, and control.
4. Ahmed, age 6 months, measures at 5% on the height and weight scale for his age group. His mother tells the pediatric nurse that Ahmed "has no interest at all in the bottle, and just screams and wails after he does drink some milk." Ahmed is diagnosed with lactose intolerance. Explain how Ahmed's diagnosis is related to his size and his crying after he drinks his bottle. Suggest clinical interventions that may help Ahmed.

MASTERY TEST: ■ CHAPTER 2

Questions 1–6: Circle the letter preceding the one best answer to each question.

1. Choose the one *true* statement.
    A. A pH of 7.5 is more acidic than a pH of 6.5.
    B. Anabolism consists of a variety of decomposition reactions.

C. An atom such as chlorine (Cl), with seven electrons in its outer orbit, is likely to be an electron donor (rather than electron receptor) in ionic bond formation.

D. Polyunsaturated fats are more likely to reduce cholesterol level than saturated fats are.

2. Which of the following describes the structure of a nucleotide?
   A. Base—base
   B. Phosphate—sugar—base
   C. Enzyme
   D. Dipeptide
   E. Adenine—ribose

3. Which of the following groups of chemicals includes only polysaccharides?
   A. Glycogen, starch
   B. Glycogen, glucose, galactose
   C. Glucose, fructose

D. RNA, DNA
E. Sucrose, polypeptide

4. $C_6H_{12}O_6$ is most likely the chemical formula for:
   A. Amino acid
   B. Fatty acid
   C. Glucose
   D. Polysaccharide
   E. Ribose

5. All of the following consist of organic chemicals correctly paired with examples of those chemicals *except:*
   A. Nucleic acids: DNA and RNA
   B. Lipids: triglycerides and phospholipids
   C. Carbohydrates: glucose and sucrose
   D. Proteins: starch and ATP

6. Which is a component of DNA but not of RNA?
   A. Adenine   D. Ribose
   B. Phosphate   E. Thymine
   C. Guanine

Questions 7–10: Circle T (true) or F (false). If the statement is false, change the underlined word or phrase so that the statement is correct.

T   F   7. K^+ and Cl^- are both cations.

T   F   8.   ATP is a molecule that contains more energy than ADP.

T   F   9.   Gastric juice, lemon juice, grapefruit juice, tomato juice, and coffee all have pH less than 7.0.

T   F   10. The number of protons in an atom always equals the number of neutrons in that atom.

Questions 11–15: Fill-ins. Write the word or phrase that best fits the description.

_____ 11. A reaction in which a disaccharide is digested to form two monosaccharides is known as a _____ reaction.

_____ 12. Oxygen, water, carbon dioxide (CO_2), and NaCl are all _____ chemicals.

_____ 13. Human proteins are formed of _____ different amino acids.

_____ 14. The atomic number is always equal to the number of _____ in an atom.

_____ 15. Fill in the names for the elements whose symbols are C, Ca, K, N, Na, and P.

ANSWERS TO MASTERY TEST: ■ CHAPTER 2

Multiple Choice
1. D
2. B
3. A
4. C
5. D
6. E

True or False
7. F. K^+ is a cation and Cl^- is an anion.
8. T
9. T
10. F. Electrons

Fill-ins
11. Decomposition, catabolic, or hydrolysis
12. Inorganic
13. 20
14. Protons or electrons
15. C: carbon; Ca: calcium; K: potassium; N: nitrogen; Na: sodium; and P: phosphorus

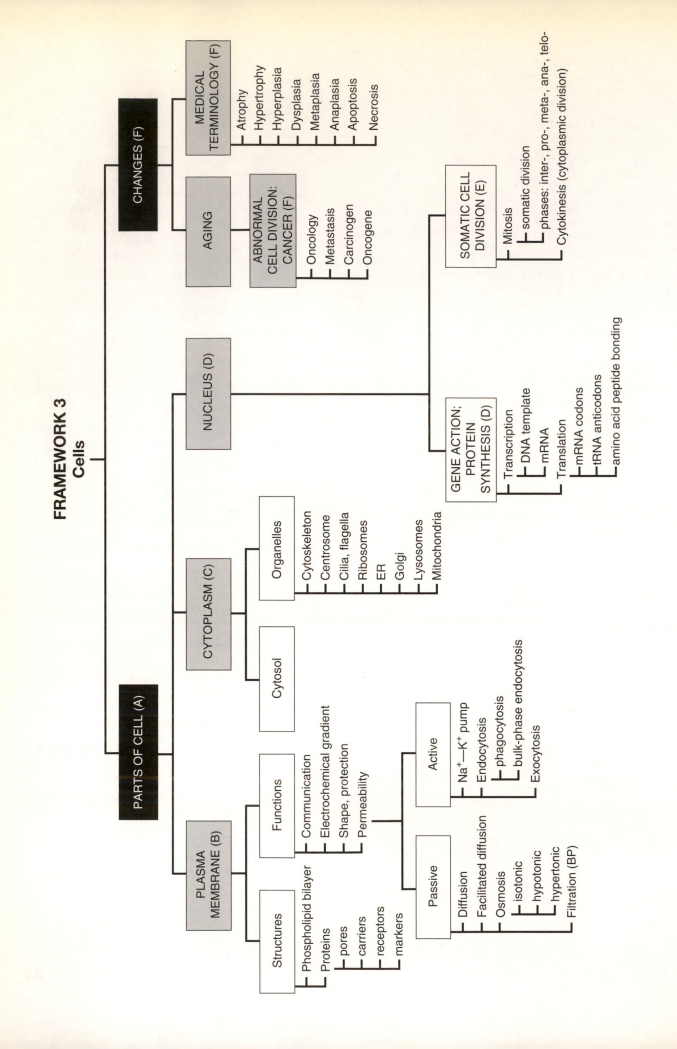

FRAMEWORK 3
Cells

Cells

Cells are the basic structural units of the human body, much as buildings make up the cities of the world. Every cell possesses certain characteristics similar to those in all other cells of the body. For example, membranes surround and compartmentalize cells, just as walls support virtually all buildings. Yet individual cell types exhibit variations that make them uniquely designed to meet specific body needs. In an architectural tour, the structural components of a cathedral, mansion, lighthouse, or fort can be recognized as specific for activities of each type of building. So too the number and type of organelles differ in muscle, nerve, cartilage, and bone cells. Cells carry out significant functions such as movement of substances into, out of, and throughout the cell; synthesis of the chemicals that cells need (such as proteins, lipids, and carbohydrates); and cell division for growth, maintenance, repair, and formation of offspring. Just as buildings change with time and destructive forces, cells exhibit aging changes. In fact, sometimes their existence is haphazard and harmful, as in cancer.

As you begin your study of cells, carefully examine the Chapter 3 Topic Outline and Objectives, noting relationships among concepts and key terms in the Framework.

TOPIC OUTLINE AND OBJECTIVES

A. Generalized view of the cell

☐ 1. Name and describe the three main parts of a cell.

B. Plasma membrane; transport across the plasma membrane

☐ 2. Describe the structure and functions of the plasma membrane.

☐ 3. Describe the processes that transport substances across the plasma membrane.

C. Cytoplasm

☐ 4. Describe the structure and functions of cytoplasm, cytosol, and organelles.

D. Nucleus; gene action: protein synthesis

☐ 5. Describe the structure and functions of the nucleus.

☐ 6. Outline the sequence of events involved in protein synthesis.

E. Somatic cell division

☐ 7. Discuss the stages, events, and significance of somatic cell division.

F. Aging and cells; wellness; disorders; medical terminology

☐ 8. Describe the cellular changes that occur with aging.

WORDBYTES

Study each wordbyte, its meaning, and an example of its use in a term. Check your understanding by jotting meanings of wordbytes in margins. Identify other examples of terms that contain these wordbytes as you continue through the text and *Learning Guide*.

Wordbyte	Meaning	Example(s)	Wordbyte	Meaning	Example(s)
a-	without	*a*trophy	meta-	beyond	*meta*stasis
auto-	self	*auto*phagy	neo-	new	*neo*plasm
cyto-	cell	*cyto*sol	-oma	tumor	carcin*oma*
-elle	small	organ*elle*	-philic	loving	hydro*philic*
homo-	same	*homo*logous	-phobic	fearing	hydro*phobic*
hydro-	water	*hydro*static	-plasm	formed, molded	cyto*plasm*
hyper-	above	*hyper*tonic	-some	body	lyso*some*
hypo-	below	*hypo*tonic	-stasis, -static	stand, stay	meta*stasis*
iso-	equal	*iso*tonic	-tonic	pressure	hyper*tonic*
lyso-	dissolving	*lyso*some	-trophy	nourish	hyper*trophy*

CHECKPOINTS

A.   Generalized view of the cell   (pages 45)

■ **A1.**   List the three principal parts of a generalized cell.

A2.   Contrast *cytoplasm* with *cytosol*.

B.   Plasma membrane; transport across the plasma membrane   (pages 45–52)

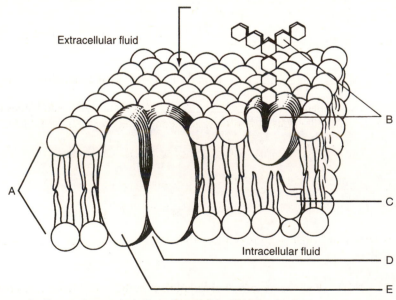

Extracellular fluid

Intracellular fluid

A
B
C
D
E

Figure LG 3.1   Plasma (cell) membrane. Label as directed in Checkpoint B1.

■ **B1.** Refer to Figure LG 3.1 and complete this exercise about plasma (cell) membrane structure.

a. The basic framework of the membrane consists of a bilayer of *(protein? lipid?)* composed mainly of *(phospholipids? cholesterol? glycolipids?)*.

b. The other major component of membranes is *(carbohydrate? protein?)*. Most of these molecules are *(glycoproteins? lipoproteins?)*.

c. Label parts A–E of the membrane in Figure LG 3.1.

d. Now write on the figure (at D) the chemicals listed below that are likely to pass through region D of the membrane:

steroids   vitamin A or D   large proteins   water   ions O_2 CO_2

Write next to the arrow on the figure the chemicals in the above list that are likely to pass through the membrane at the site of the arrow.

■ **B2.** List four or more specific functions of proteins found in plasma membranes.

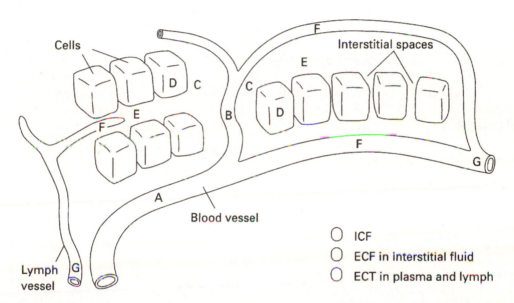

Figure LG 3.2   Internal environment of the body: types of fluid. Color as directed in Checkpoint B3. Areas labeled A to G refer to Checkpoint B4.

■ **B3.** Describe the types of fluid in the body and their relationship to homeostasis by completing these statements and Figure LG 3.2.

a. Fluid inside cells is known as _____ fluid (ICF). Color areas containing this type of fluid yellow.

b. Fluid in spaces between cells is called _____ . It surrounds and bathes cells and is one form of *(intracellular? extracellular?)* fluid. Color the spaces containing this fluid light green. The condition of

maintaining this fluid in relative constancy is known as _____ .

c. Another form of extracellular fluid is that located in _____ vessels and

_____ vessels. Color these areas dark green.

d. The body's "internal environment" (that is, surrounding cells) is _____-cellular fluid (ECF) (all green areas in your figure).

■ **B4.** Refer again to Figure LG 3.2. Show the pattern of circulation of body fluids by drawing arrows connecting letters in the figure in alphabetical order (A → B → C, etc.). Identify the lettered areas.

■ **B5.** The energy of molecular motion is known as _____ energy. Passive transport processes depend on *(ATP? kinetic energy?)*, whereas active transport processes depend on *(ATP? kinetic energy?)*. Diffusion and osmosis are *(active? passive?)* processes, whereas the Na^+/K^+ pump is a(n) *(active? passive?)* process.

B6. Briefly describe each of these words or phrases:

a. Net diffusion

b. Down the concentration gradient

c. Equilibrium

■ **B7.** Gases such as O_2 and CO_2 are *(water? lipid?)*-soluble, and they therefore diffuse across the *(lipid bilayer? pores in channels?)* of plasma membranes. Ions such as K^+ diffuse primarily across the *(lipid bilayer? pores in channels?)* of plasma membranes. There are more channels in plasma membranes for passage of *(K^+? Na^+?)*. Ion channels *(are never? may be?)* gated. Review Checkpoint B1 for more details on sites of diffusion across membranes.

■ **B8.** Select from the following list of terms to identify passive transport processes described below.

BFE. Bulk-phase endocytosis	O. Osmosis
E. Exocytosis	P. Phagocytosis
FD. Facilitated diffusion	SD. Simple diffusion

_____ a. Net movement of any substance (such as cocoa powder in hot milk) from region of higher concentration to region of lower concentration; membrane not required

_____ b. Same as (a) except movement across a semipermeable membrane with help of a transporter; ATP not required

_____ c. Net movement of water from region of high water concentration (such as 2 percent NaCl) to region of lower water concentration (such as 10 percent NaCl) across semipermeable membrane; important in maintenance of normal cell size and shape

_____ d. "Cell drinking"

_____ e. Process of cellular ingestion of food by endocytosis

_____ f. Secretion of neurotransmitter from neurons, hormones from endocrine cells

■ **B9.** Most sugars (such as glucose) are transported by the process of *(simple? facilitated?)* diffusion. The hormone named insulin helps to transport glucose *(into? out of?)* cells which *(raises? lowers?)* the level of glucose in blood.

Insulin does this by inserting proteins that function as _____ into cell membranes.

■ **B10.** Complete the following exercise about osmosis in blood.

a. Human red blood cells (RBCs) contain intracellular fluid that is osmotically similar to _____ NaCl.
   A. 2.0 percent   B. 0.9 percent   C. 0 percent (pure water)

b. A solution that is hypertonic to RBCs contains *(more? fewer?)* solute particles and *(more? fewer?)* water molecules than blood.

c. Which of these solutions is hypertonic to RBCs?
   A. 2.0 percent NaCl   B. 0.9 percent NaCl   C. Pure water

d. If RBCs are surrounded by hypertonic solution, water will tend to move *(into? out of?)* them, so they will *(crenate? hemolyze?)*.

e. A solution that is _____-tonic to RBCs will maintain the shape and size of the RBC.

Intravenous (IV) solutions are likely to be _____-tonic. Solutions given to persons who

are dehydrated are likely to be _____-tonic. An example of such a solution is _____ .

f. Which solution will cause RBCs to hemolyze?
   A. 2.0 percent NaCl   B.   0.9 percent NaCl   C.   Pure water
g. Which solution has the highest osmotic pressure?
   A. 2.0 percent NaCl   B.   0.9 percent NaCl   C.   Pure water

B11.   Complete this exercise on active transport.

a. Typically, the human body utilizes about _____ % of its ATP (energy) supply on active transport (pumps).

b. Name several ions commonly transported across plasma membranes by primary active

transport. _____ The most important primary active transport pump pumps Na^+ *(into? out of?)* cells. Na^+ *(does? does not?)* leak back into cells by diffusion, so the pump must perform *(only occasionally? continuously?)*. As a result of this pump, Na^+ is much more concentrated in *(intra? extra?)*-cellular fluid, whereas K^+ is much more concentrated in *(intra? extra?)*-cellular fluid.

C.   Cytoplasm   (pages 52–57)

C1.   The cell is compartmentalized by the presence of organelles. Of what advantage is this?

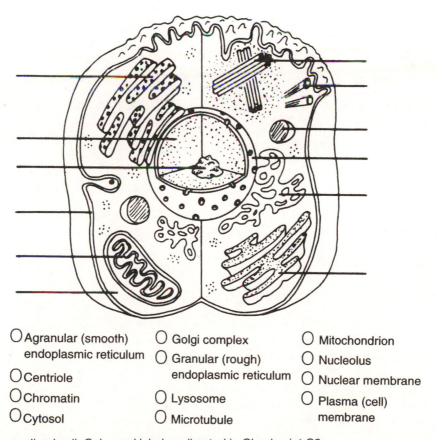

◯ Agranular (smooth)   ◯ Golgi complex   ◯ Mitochondrion
   endoplasmic reticulum ◯ Granular (rough)   ◯ Nucleolus
◯ Centriole   endoplasmic reticulum ◯ Nuclear membrane
◯ Chromatin   ◯ Lysosome   ◯ Plasma (cell)
◯ Cytosol   ◯ Microtubule   membrane

Figure LG 3.3   Generalized cell. Color and label as directed in Checkpoint C2.

■ **C2.** Color and then label all the parts of Figure LG 3.3 listed with color code ovals.

■ **C3.** Identify organelles in the box that fit descriptions below. One answer will be used twice, all others, once.

Cen. Centrioles	G. Golgi complex	Mit. Mitochondria
Cil. Cilia	L. Lysosomes	RER. Rought endoplasmic reticulum
F. Flagella	Mf. Microfilaments	SER. Smooth endoplasmic reticulum
	Mt. Microtubules	

_____ a. Site of synthesis and brief storage of proteins

_____ b. Site of synthesis of fatty acids, phospholipids, and steroids

_____ c. Site of enzymes that detoxify alcohol and other harmful chemicals

_____ d. Stacks of cisterns with vesicular ends, involved in packaging and secretion of proteins and lipids

_____ e. Release enzymes that lead to autolysis of the cell

_____ f. Cristae-containing structures, called "powerhouses of the cell" because ATP production occurs here

_____ g. Form part of cytoskeleton, involved with cell movement and contraction

_____ h. Part of cytoskeleton, provide support and give shape to cell; form flagella, cilia, centrioles, and spindle fibers; made of protein tubulin

_____ i. Help organize mitotic spindle used in cell division

_____ j. Long, hairlike structures that help move entire cell, as in sperm cells

_____ k. Short, hairlike structures that move particles over cell surfaces

■ **C4.** *For extra review.* Considering the functions of organelles listed in the above exercise, choose the answers (organelles) that fit the following descriptions. Answers in that list may be used more than once.

_____ a. Abundant in liver cells that detoxify Phenobarbital and other drugs that enter the liver.

_____ b. Present in large numbers of muscle and liver cells, which require much energy.

_____ c. Located on surface of cells of the respiratory tract; help to move mucus.

_____ d. Extensive in pancreatic cells that secrete insulin.

_____ e. The inherited condition, Tay-Sachs disease, involves absence of an enzyme within these organelles in nerve cells.

C5. Contrast functions of *peroxisomes* and *proteasomes*.

D. Nucleus; gene action: protein synthesis (pages 57–62)

■ **D1.** Identify parts of the nucleus in the box that fit descriptions below.

C. Chromosomes	NE. Nuclear envelope	NU. Nucleolus

_____ a. These rod-shaped bodies contain DNA and protein.

_____ b. Consists of RNA, DNA, and protein; site of assembly of ribosomes.

_____ c. Large pores here permit passage of ribosomes and large proteins between nucleus and cytosol.

■ **D2.** The Human Genome Project began in the year _____ and was completed by the year _____. Explain the purpose of this project.

This study of genomics determined that human chromosomes contain a total of about _____ genes.

D3. Explain the significance of protein synthesis in the cell.

■ **D4.** Review the structure of DNA and RNA in this exercise. Circle the correct answer to each question.

a. Which base is found in RNA but not in DNA?
    A. Adenine   C. Cytosine   G. Guanine   U. Uracil   T. Thymine

b. Which sugar is present in RNA but not in DNA?
    D. Dexoyribose   R. Ribose

c. DNA and RNA consist of repeating units known as:
    A. Amino acids   N. Nucleotides

d. One nucleotide consists of a base, a sugar, as well as a:
    K. Potassium   P. Phosphate

e. A section of three successive nucleotides of DNA is known as a _(base triplet? codon?)_, whereas three successive RNA nucleotides make up a _____ .

For extra review, color sugars and phosphates in Figure LG 3.4, and also see _Learning Guide,_ Chapter 2, Checkpoint D8.

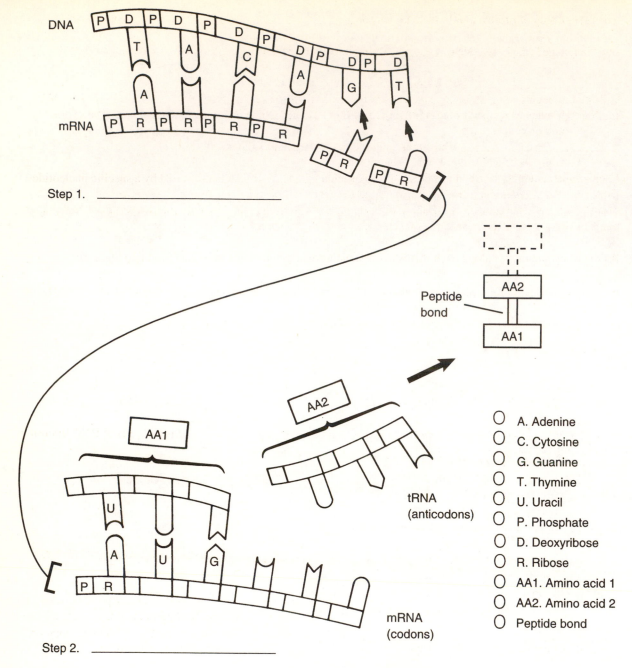

Step 1. _____

Step 2. _____

Figure LG 3.4   Protein synthesis. Label and color as directed in Checkpoint D5.

■ **D5.**   Refer to Figure LG 3.4 and describe the process of protein synthesis in this exercise.

a.   Write the names of the two major steps of protein synthesis on the two lines in Figure LG 3.4.

b.   Transcription takes place in the *(nucleus? cytoplasm?)*. In this process, a portion *(called a gene)* of one side of a double stranded DNA is copied to form a complementary strand of *(DNA? RNA?)*. Transcription requires the help

of the enzyme RNA _____ which attaches to a section of DNA known as the *(promoter? terminator?)* close to the gene to be copied. The signal to stop transcribing occurs when RNA polymerase reaches the sequence of DNA nucleotides known as the *(promoter? terminator?)*.

c.   Complete the strand of mRNA by writing letters of complementary bases. Then color the bases of both DNA and mRNA. The newly synthesized mRNA then *(stays in the nucleus? moves through a nuclear pore into the cytoplasm?)*.

32

d. The second step of protein synthesis is known as _____ because it involves translation of

   one "language" (the base code of _____ RNA) into another "language" (the correct sequence of the 20

   _____ to form a specific protein).

e. Translation begins when an mRNA attaches to the *(small? large?)* subunit of a ribosome. A *(promoter? initiator?)* tRNA attaches to a start codon (AUG) on mRNA. The ribosome becomes functional when the

   _____ ribosomal unit attaches to the small subunit.

f. Each amino acid is transported to this site by a particular _____ RNA characterized by a specific nucleotide triplet or *(codon? anticodon?)*. This portion of tRNA (for example, U A C) binds to a complementary portion of mRNA *(codon? anticodon?)*. In step 2 of Figure LG 3.4, complete the labeling of bases and color all components of tRNA and mRNA. Be sure you select colors that correspond to those used in step 1.

g. As the ribosome moves along each mRNA codon, additional amino acids are transferred into place by _____

   RNA. _____ bonds form between adjacent amino acids. Now color the amino acids and the peptide bonds that join them.

h. As the ribosome moves on to the next codon on mRNA, what happens to the "empty" tRNA?

i. When the protein is complete, synthesis is stopped by the _____ codon.

■ **D6.** Check your understanding of protein synthesis by identifying roles of DNA and three types of RNA in protein synthesis. Choose from answers in the box.

> D. DNA   M. mRNA   R. rRNA   T. tRNA

_____ a. Forms part of organelles that serve as sites of assembly of proteins.

_____ b. Formed in the transcription step of protein synthesis.

_____ c. Copied in the transcription step of protein synthesis.

_____ d. Contains promoter and terminator regions.

_____ e. Binds to an amino acid to direct it to the proper location during translation.

_____ f. Contains anticodons.

_____ g. Contains codons, including the STOP codon.

_____ h. A polyribosome refers to several ribosomes attached to the same _____ so that large quantities of a protein may be synthesized in a short time.

E.   Somatic cell division   (pages 62–65)

E1.   Once you have reached adult size, do your cells continue to divide *(Yes? No?)*. Of what significance is this fact?

■ **E2.**   Describe aspects of cell division by completing this exercise.

a. Division of any cell in the body consists of two processes: division of the _____ and

   division of the _____ .

b. In the formation of mature sperm and egg cells, nuclear division is known as _____ .
In the formation of all other body cells, that is, *somatic* cells, nuclear division is called

_____ .

c. Cytoplasmic division in both somatic and reproductive division is known as _____ .
In *(meiosis? mitosis?)* the two newly formed cells have the same hereditary material and genetic potential as the parent cell.

■ **E3.**   Carefully study the phases of the cell cycle shown in Figure 3.21, page 63, in your text. Then check your understanding of the process by identifying major events in each phase.

| A. Anaphase | M. Metaphase | T. Telophase |
| I. Interphase | P. Prophase | |

_____ a. This phase immediately follows inter-phase.

_____ b. Chromosomes condense into distinct chromosomes (chromatids) each held together by a centromere.

_____ c. Nucleoli and nuclear envelope break up, the two centrosomes move to opposite poles of the cell, and formation of the mitotic spindle begins.

_____ d. Chromatids line up with their centromeres along the metaphase plane.

_____ e. Centromeres divide; chromatids (now called chromosomes) are moved by microtubules to opposite poles of the cell. Cytokinesis begins.

_____ f. Events of this phase are essentially a reversal of prophase; cytokinesis is completed. A cleavage furrow is present.

_____ g. Organelles are reproduced during this period.

_____ h. DNA uncoils so that bases are exposed for complementary base pairing; for each DNA molecule, two are now present.

■ **E4.**   Fill in the term that fits each description of structures involved in cell division. Choose from answers in the box.

| Centromeres | Centrosomes | Chromatids | Chromatin |

a. Tangled mass containing DNA during interphase; condenses into chromosomes during prophase:

_____

b. Name given to replicated chromosomes in prophase:

_____

c. Holds chromatid pair together; these align on the metaphase plate: _____

d. Form the mitotic spindle composed of microtubules: _____

F.   Aging and cells; wellness; disorders; medical terminology   (pages 64–67)

■ **F1.**   Name two types of cells that are the least likely cells to divide after birth.

F2.   Contrast terms in each pair.

a. Geriatrics – gerontology

b. Aging genes – telomeres

c. Progeria – Werner syndrome

F3. Describe theories of aging that focus on:

a. Free radicals

b. Autoimmunity

F4. Describe mechanisms by which the following types of foods can prevent cell damage that leads to aging or cancer.

a. Green tea and tomatoes

b. Garlic, onions, broccoli, and cauliflower

■ **F5.** Complete this exercise about cancer.

a. The term _____ refers to the study of cancer. _____ and

_____ nurses are health care providers who specialize in work with clients who have cancer.

b. A term that means tumor or abnormal growth is _____ . A cancerous growth is a *(benign? malignant?)* neoplasm, whereas a noncancerous tumor is a *(benign? malignant?)* neoplasm.

c. Which type of growth is more likely to spread and to possibly cause death? *(Benign? Malignant?)* A term that

means the spread of cancer cells is _____ .

d. In the process of metastasis, cancer cells compete with normal cells for _____ . By what routes do cancers reach distant parts of the body?

e. What may cause the pain associated with cancer?

f. Explain how tissue angiogenesis factors (TAFs) enhance growth of cancer cells.

F6. Contrast terms in each pair:

a. Carcinogens/mutations

b. Oncogenes/proto-oncogenes

c. Hyperplasia/metastasis

d. Apoptosis/necrosis

■ **F7.** Match the terms in the box describing alterations in cells or tissues with the descriptions below.

D. Dysplasia	HP. Hyperplasia	HT. Hypertrophy	M. Metaplasia

_____ a. Increase in size of tissue or organ by increase in size (not number) of cells, such as growth of your biceps muscle with exercise

_____ b. Increase in size of tissue or organ due to increase in number of cells, such as a callus on your hand or breast tissue during pregnancy

_____ c. Change of one cell type to another normal cell type, such as change from single row of tall (columnar) cells lining airways to multilayers of cells as response to constant irritation of smoking

_____ d. Abnormal change in cells in a tissue as due to irritation or inflammation; may revert to normal if irritant is removed, or may progress to neoplasia

■ **F8.** Match the terms in the box with the definitions below.

An. Anaplasia	At. Atrophy	B. Biopsy

_____ a. Removal and examination of tissue from the living body for diagnosis

_____ b. Decrease in size of cells with decrease in size of tissue or organ

_____ c. Loss of tissue differentiation that occurs in most malignancies

A1. Plasma cell membrane, cytoplasm (cytosol and organelles), nucleus.

B1. (a) Lipid; phospholipids. (b) Protein; glycoproteins. (c) A, phospholipid bilayer. B, glycoprotein. C, cholesterol. D, pore. E, integral protein. (d) Region D (pore in integral protein): water, ions. Region with arrow (phospholipid layer): steroids, vitamins A and D, water, O_2, CO_2. Large proteins do not normally pass through the membrane except within vesicles.

B2. Proteins form channels with pores, transporters, receptors, enzymes, cell identity markers (when combined with carbohydrates as glycoproteins).

B3. (a) Intracellular. (b) Interstitial fluid, extracellular; homeostasis. (c) Blood, lymph. (d) Extra.

B4. A, arteries and arterioles; B, blood capillaries; C, interstitial (intercellular) fluid; D, intracellular fluid; E, interstitial (intercellular) fluid again; F, blood or lymph capillaries; G, venules and veins or lymph vessels.

B5. Kinetic; kinetic energy, ATP; passive, active.

B7. Lipid, lipid bilayer; pores in channels; K^+; may be.

B8. (a) SD. (b) FD. (c) O. (d) BFF. (e) P. (f) E.

B9. Facilitated; into, lowers; transporters.

B10. (a) B. (b) More, fewer. (c) A. (d) Out of, crenate. (e) Iso, 0.9 percent NaCl (normal saline) or 5.0 percent glucose; iso; hypo. (f) C. (g) A.

B11. (a) 40. (b) Na^+, K^+, H^+, Ca^{2+}, I^-, and Cl^-; out of; does, continuously; extra, intra.

C2.

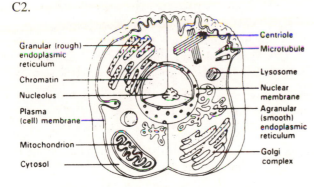

Figure LG 3.3A   Generalized cell.

C3. (a) RER. (b) SER. (c) SER. (d) G. (e) L. (f) Mit. (g) Mf. (h) Mt. (i) Cen. (j) F. (k) Cil.

C4. (a) SER. (b) Mit. (c) Cil. (d) G. (e) L.

D1. (a) C. (b) NU. (c) NE.

D2. 1990. 2003 to sequence the nucleotides in the human genome which permits early detection and treatment of genetic disorders; 30,000.

D4. (a) U. (b) R. (c) N. (d) P. (e) Base triplet, codon.

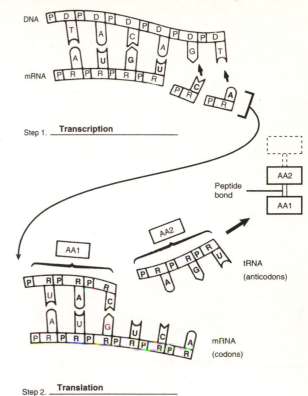

Figure LG 3.4A   Protein synthesis.

D5. (a) See Figure LG 3.4A. (b) Nucleus; RNA; polymerase, promoter; terminator. (c) See Figure LG 3.4A; moves though a nuclear pore into the cytoplasm. (d) Translation, m (messenger), amino acids. (e) Small, initiator; large. (f) t (transfer), anticodon; codon (see Figure LG 3.4A). (g) t (transfer); peptide (see Figure LG 3.4A). (h) It is released and may be recycled to transfer another amino acid. (i) Stop.

D6. (a) R. (b) M. (c) D. (d) D. (e) T. (f) T. (g) M. (h) M.

E2. (a) Nucleus, cytoplasm. (b) Meiosis; mitosis. (c) Cytokinesis; mitosis.

E3. (a) P. (b) P. (c) P. (d) M. (e) A. (f) T. (g) I. (h) I.

E4. (a) Chromatin. (b) Chromatids. (c) Centromeres. (d) Centrosomes.

F1. Skeletal muscle and neurons.

F5. (a) Oncology; oncologists, oncology. (b) Neoplasm; malignant, benign. (c) Malignant; metastasis. (d) Nutrients and space; via blood or lymph vessels or invading a body cavity. (e) Pressure on nerve, obstruction of passageway, loss of function of a vital organ. (f) These chemicals stimulate growth of blood vessels that nourish and support development of cancer cells.

F7. (a) HT. (b) HP. (c) M. (d) D.

F8. (a) B. (b) At. (c) An.

CRITICAL THINKING: CHAPTER 3

1. Explain how the structure of DNA is related to synthesis of a particular protein.
2. Discuss whether interphase is a "resting" phase or a highly active phase in the cell cycle. State your rationale.
3. Explain how replication of DNA during interphase differs from the transcription phase of protein synthesis.
4. Define phytochemicals and cruciferous vegetables, and explain how they may help to protect you against cancer.
5. Explain how DNA is the key to the uniqueness of each individual.

MASTERY TEST: ■ CHAPTER 3

Questions 1–6: Circle the letter preceding the one best answer to each question.

1. The organelle that carries out the process of autophagy in which old organelles are digested so that their components can be recycled is:
   A. Golgi
   B. Mitochondrion
   C. Centrosome
   D. Lysosome
   E. Endoplasmic reticulum

2. Choose the *false* statement about protein synthesis.
   A. Translation occurs in the cytoplasm.
   B. Messenger RNA picks up and transports an amino acid during protein synthesis.
   C. Messenger RNA travels to a ribosome in the cytoplasm.
   D. Transfer RNA is attracted to mRNA due to their complementary bases.

3. All of the following structures are part of the nucleus *except:*
   A. Nucleolus   C. Chromosomes
   B. Chromatin   D. Centrosome

4. Choose the *false* statement about genes.
   A. Genes contain DNA.
   B. Genes contain information that controls heredity.
   C. Every cell in the human body has a total of 46 genes.
   D. Genes are transcribed by messenger RNA during the first step of protein synthesis.

5. Which term refers to increase in size of a tissue related to increase in number of cells?
   A. Anaplasia   C. Hyperplasia
   B. Hypertrophy   D. Dysplasia

6. Choose the one *false* statement.
   A. Cristae are folds of membrane in mitochondria.
   B. *Crenation* is a term that means bursting of red blood cells when they are placed in hypotonic solution.
   C. Plasma and interstitial fluid are both extracellular fluids (ECFs).
   D. Microtubules are components of the structure of flagella, cilia, centrioles, and the mitotic spindle.

Questions 7–8: Circle T (true) or F (false). If the statement is false, change the underlined word or phrase so that the statement is correct.

T   F   7.   With aging, telomeres are likely to become <u>shorter</u>.

T   F   8.   A 5 percent glucose solution is <u>hypotonic</u> to a 15 percent glucose solution.

Questions 9–10: Arrange the answers in correct sequence.

_____ _____ _____   9.   Stages in mitosis from first to last after prophase:
   A. Anaphase
   B. Metaphase
   C. Telophase

_____ _____ _____   10.   Steps in protein synthesis from first to last after prophase:
   A. Formation of peptide bonds connecting amino acids
   B. Translation
   C. Transcription

_____ 11. Plasma membranes consist mainly of two types of chemicals. These are _____

and _____ .

_____ 12. Cytokinesis is another name for _____ .

_____ 13. Active processes involved in movement of substances across cell membranes are

those that use energy from the splitting of _____ .

_____ 14. White blood cells engulf large solid particles by the process of _____ .

_____ 15. The sequence of bases of mRNA that would be complementary to DNA bases in

the sequence A-T-T-C-A-C would be _____ .

ANSWERS TO MASTERY TEST: ■ CHAPTER 3

Multiple Choice
1. D
2. B
3. D
4. C
5. C
6. B

True or False
7. T
8. T

Arrange
9. B A C
10. C B A

Fill-ins
11. Phospholipids and protein
12. Division of the cytoplasm
13. ATP
14. Phagocytosis
15. U-A-A-G-U-G

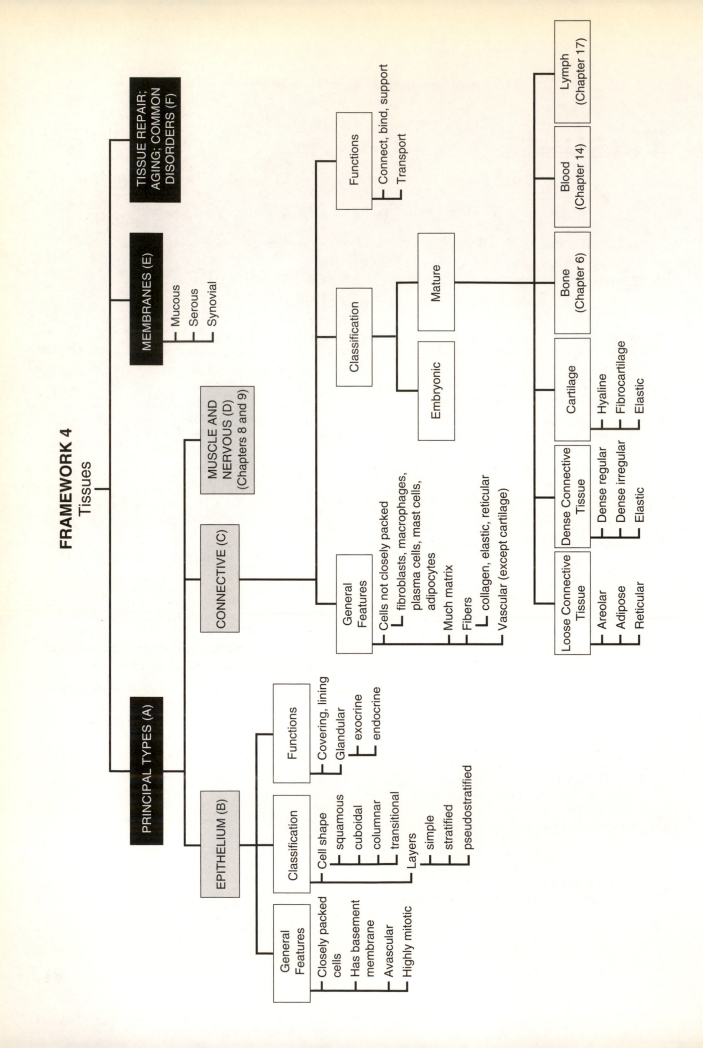

FRAMEWORK 4
Tissues

PRINCIPAL TYPES (A)

MEMBRANES (E)
- Mucous
- Serous
- Synovial

TISSUE REPAIR;
AGING; COMMON
DISORDERS (F)

EPITHELIUM (B)

General Features
- Closely packed cells
- Has basement membrane
- Avascular
- Highly mitotic

Classification
- Cell shape
  - squamous
  - cuboidal
  - columnar
  - transitional
- Layers
  - simple
  - stratified
  - pseudostratified

Functions
- Covering, lining
- Glandular
  - exocrine
  - endocrine

CONNECTIVE (C)

MUSCLE AND
NERVOUS (D)
(Chapters 8 and 9)

General Features
- Cells not closely packed
  - fibroblasts, macrophages, plasma cells, mast cells, adipocytes
- Much matrix
- Fibers
  - collagen, elastic, reticular
- Vascular (except cartilage)

Classification
- Embryonic
- Mature

Functions
- Connect, bind, support
- Transport

Loose Connective Tissue
- Areolar
- Adipose
- Reticular

Dense Connective Tissue
- Dense regular
- Dense irregular
- Elastic

Cartilage
- Hyaline
- Fibrocartilage
- Elastic

Bone
(Chapter 6)

Blood
(Chapter 14)

Lymph
(Chapter 17)

Tissues

Tissues consist of cells and the intercellular (between cells) materials secreted by these cells. All the organs and systems of the body consist of tissues that may be categorized into four classes or types: epithelium, connective, muscular, and nervous. Each tissue exhibits structural characteristics (anatomy) that determine the function (physiology) of that tissue. For example, epithelial tissues consist of tightly packed cells that form the secretory glands and also provide protective barriers, as in the membranes that cover the surface of the body and line passageways or cavities. Tissues are constantly stressed by daily wear and tear and sometimes by trauma or infection. Tissue repair is therefore an ongoing human maintenance project.

As you begin your study of tissues, carefully examine the Chapter 4 Framework and note relationships among concepts and key terms there. Also refer to the Topic Outline and Objectives; you may want to check off each objective as you complete it.

TOPIC OUTLINE AND OBJECTIVES

A. Types of tissue

☐ 1. Name the four basic types of tissue that make up the human body and state characteristics of each.

B. Epithelial tissue

☐ 2. Discuss the general features of epithelial tissue.
☐ 3. Describe the structure, location, and function of the various types of epithelia.

C. Connective tissue

☐ 4. Discuss the general features of connective tissue.
☐ 5. Describe the structure, location, and function of the various types of connective tissues.

D. Muscular tissue and nervous tissue

☐ 6. Describe the functions of muscular tissue.
☐ 7. Contrast the locations of the three types of muscular tissues.
☐ 8. Describe the functions of nervous tissue.

E. Membranes

☐ 9. Define a membrane.
☐ 10. Describe the classification of membranes.

F. Tissue repair; aging and tissues; common disorders and medical terminology

☐ 11. Describe the role of tissue repair in restoring homeostasis.
☐ 12. Describe the effects of aging on tissues.

Study each wordbyte, its meaning, and an example of its use in a term. Check your understanding by jotting meanings of wordbytes in margins. Identify other examples of terms that contain these wordbytes as you continue through the text and *Learning Guide*.

Wordbyte	Meaning	Example(s)	Wordbyte	Meaning	Example(s)
adip(o)-	fat	*adipo*se	multi-	many	*multi*cellular
-crine	to secrete	endo*crine*	path-	disease	*path*ology
endo-	within	*endo*crine	pseudo-	false	*pseudo*stratified
exo-	outside	*exo*crine	squam-	thin plate	*squam*ous
hist-	tissue	*hist*ology	strat-	layer	*strat*ified
lupus	wolf	systemic *lupus* erythematosus	uni-	one	*uni*cellular
			vasc-	blood vessel	a*vasc*ular
macro-	large	*macro*phage			

CHECKPOINTS

A. Types of tissue (page 73)

A1. Define the following terms:

a. Tissue

b. Histology

c. Pathology

A2. Complete the table about the four basic types of tissue.

Tissue	General Functions
a. Epithelial tissue	
b. Connective tissue	
c.	Movement
d.	Initiates and transmits nerve impulses that help coordinate body activities

■ **A3.** Circle the two types of tissues in Checkpoint A2 that undergo little or no mitosis after birth. In other words, those tissues grow only through *(hyperplasia? hypertrophy?)*.

B. Epithelial tissue (pages 73–82)

■ **B1.** Describe three types of cell junctions based on functions of the tissue.

■ **B2.** Name the two subtypes of epithelial tissue based on location and function.

_____   _____

   a. Now list several sites of *lining epithelium.*

   b. Name three or more types of glands composed of epithelium. _____

_____   _____

   c. In epithelium composed of many layers, apical layers are *(deepest? most superficial?).*

B3. Complete the table by describing five or more locations and functions of epithelial tissue. One is done for you.

Location	Function
a. Eyes, nose, outer layer of skin	Sensory reception
b.	
c.	
d.	
e.	

■ **B4.** Describe the structure of epithelium in this exercise.

   a. Epithelium consists mostly of *(closely packed cells? intercellular material with few cells?).*

   b. Epithelium is penetrated by *(many? no?)* blood vessels. A term meaning "lacking in blood vessels" is

   _____ . Epithelium *(has? lacks?)* a nerve supply.

   c. Epithelium *(does? does not?)* have the capacity to undergo mitosis.

   d. The structure that attaches epithelium to underlying connective tissue is known as the

   _____ membrane.

   e. Epithelium that consists of a single layer of cells is known as _____ epithelium, whereas epithelium

   that formed of two or more layers of cells is called _____ epithelium.

   f. *(Columnar? Cuboidal? Squamous?)* epithelial cells are flat. Epithelial cells that have height much greater than

   width are known as _____ cells.

■ **B5.** Study diagrams of epithelial tissue types (A–F) in Figure LG 4.1 and also Table 4.1, pages 76–81 in your text. Write the following information on lines provided on Figure LG 4.1:

   a. The name of each tissue

   b. One or more locations of each type of tissue

   c. One or more functions of each type of tissue

   d. Now color the following structures on each diagram:

      ○ Nucleus   ○ Intercellular material

      ○ Cytoplasm   ○ Basement membrane

43

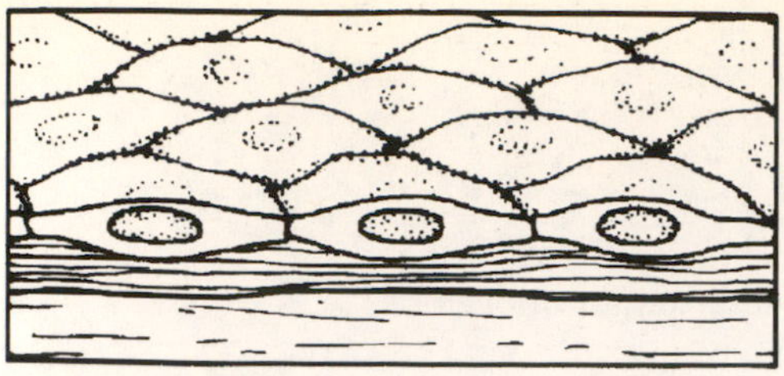

A Name _____

    Location _____

    Function _____

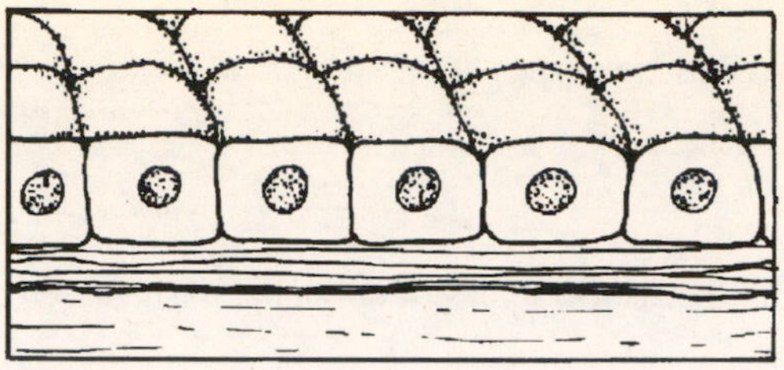

B Name _____

    Location _____

    Function _____

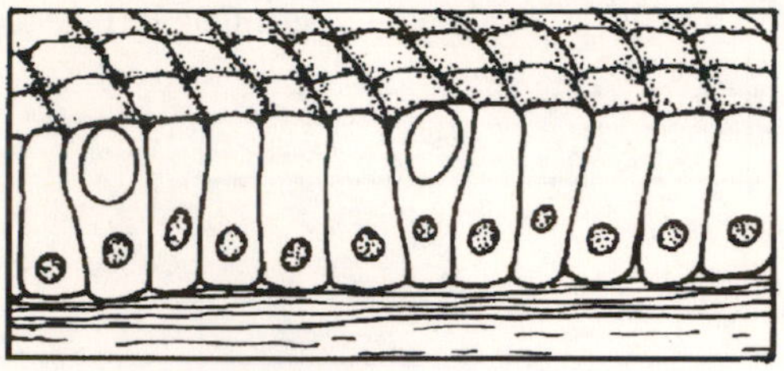

C Name _____

    Location _____

    Function _____

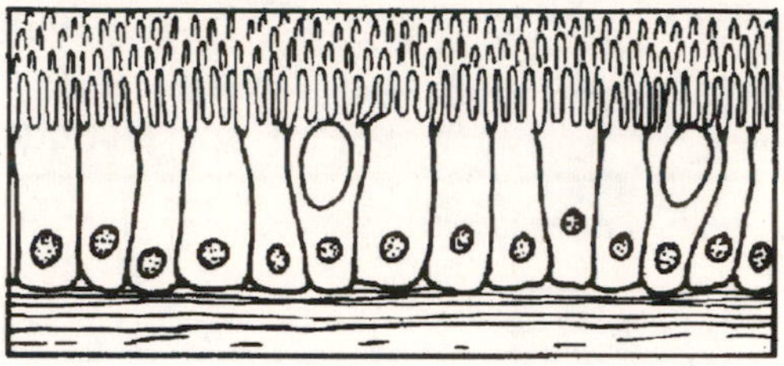

D Name _____

    Location _____

    Function _____

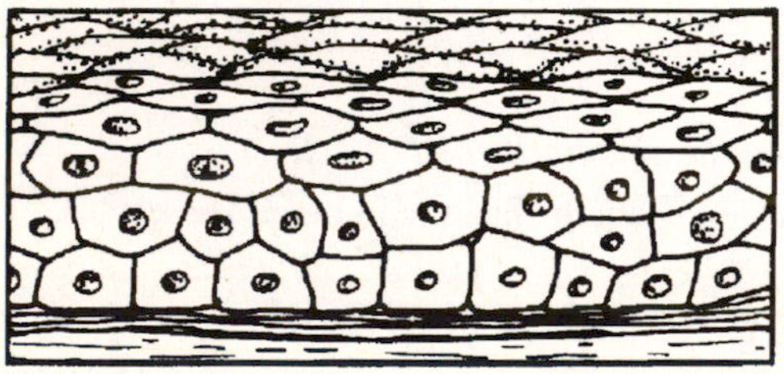

E Name _____

    Location _____

    Function _____

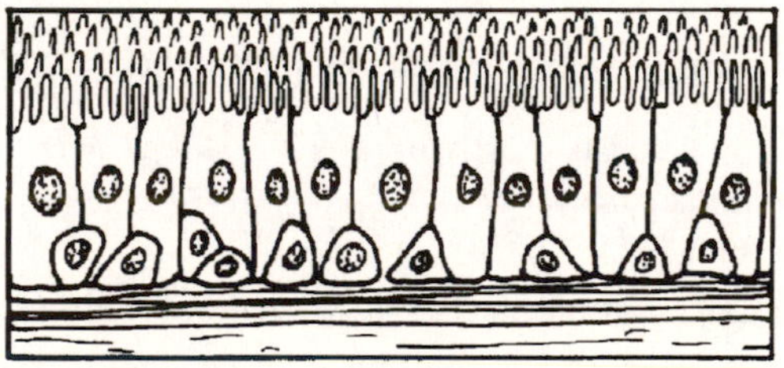

F Name _____

    Location _____

    Function _____

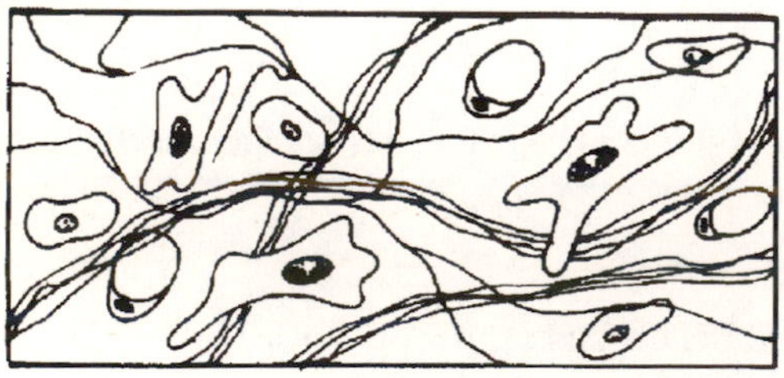

G Name _____

    Location _____

    Function _____

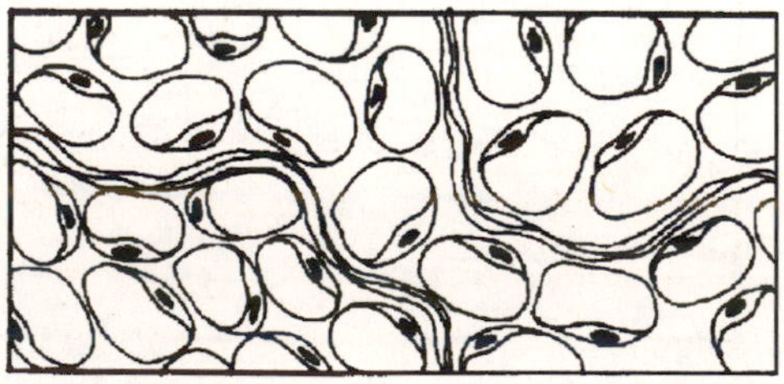

H Name _____

    Location _____

    Function _____

Figure LG 4.1   Diagrams of selected tissue types. Complete as directed.

I Name _____

   Location _____

   Function _____

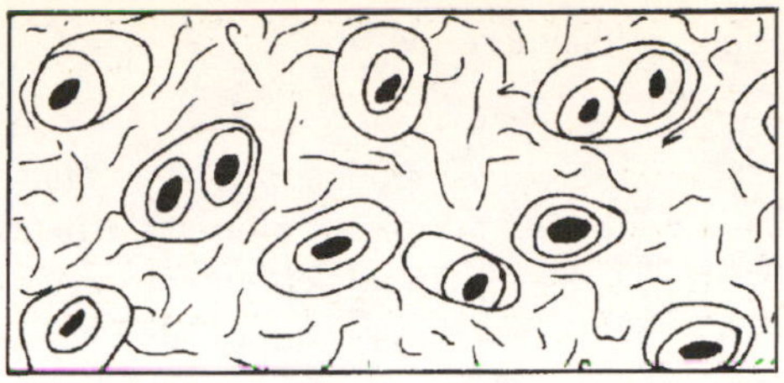

J Name _____

   Location _____

   Function _____

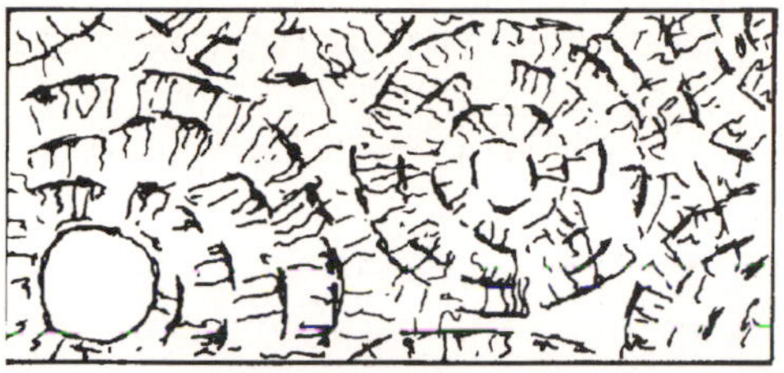

K Name _____

   Location _____

   Function _____

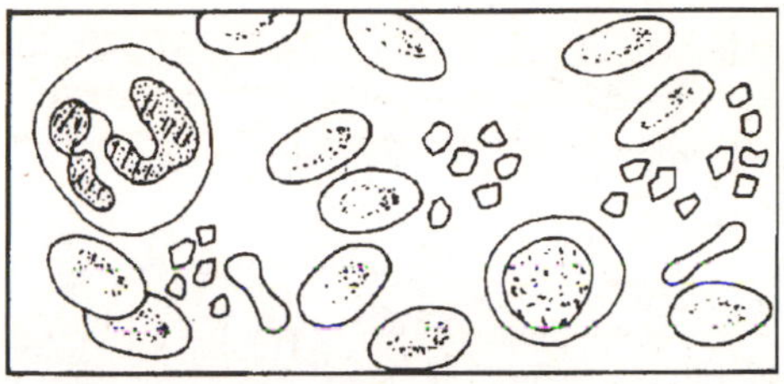

L Name _____

   Location _____

   Function _____

M Name _____

   Location _____

   Function _____

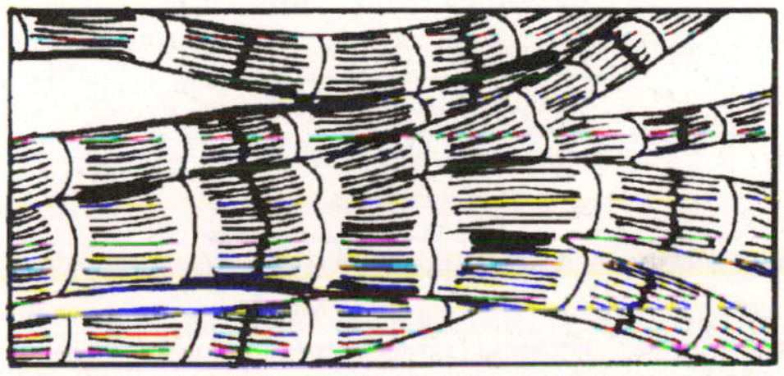

N Name _____

   Location _____

   Function _____

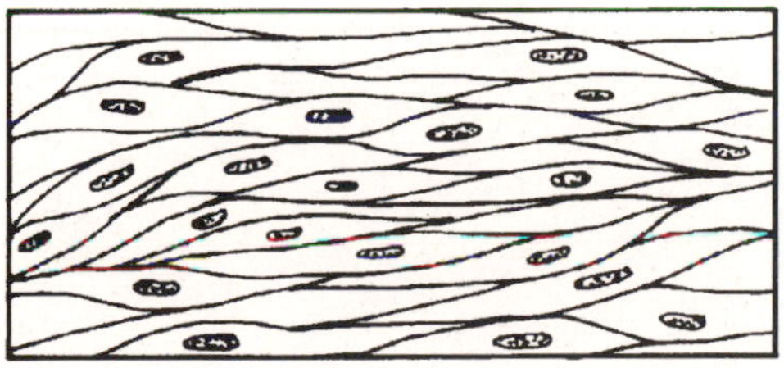

O Name _____

   Location _____

   Function _____

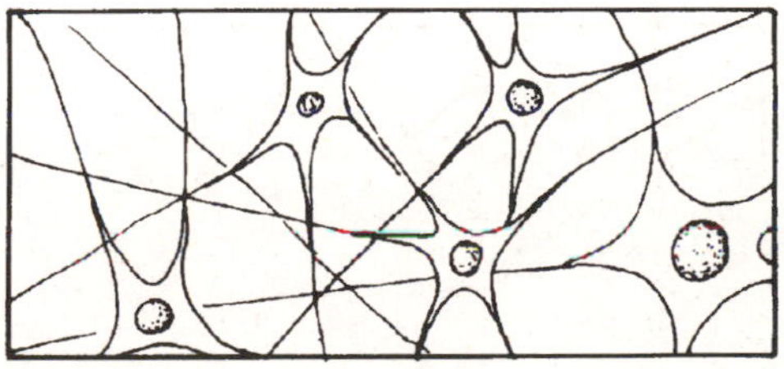

P Name _____

   Location _____

   Function _____

Figure LG 4.1 *Continued*

■ **B6.** Check your understanding of the most common types of epithelium listed in the box by writing the name of the type after the phrase that describes it.

> | | |
> |---|---|
> | Pseudostratified columnar | Simple squamous |
> | Simple columnar, ciliated | Stratified squamous |
> | Simple columnar, nonciliated | Transitional |
> | Simple cuboidal | |

a. Lines the inner surface of the stomach and intestine; may have microvilli at the apical

   surface: _____ .

b. Lines the urinary tract, permits stretching (distension)

   of the bladder: _____ .

c. Lines mouth; present on outer surface of skin:

   _____ .

d. Single layer of cube-shaped cells; found in kidney tubules and ducts of some glands:

   _____ .

e. Lines air sacs of lungs, where thin cells are required for diffusion of gases into blood:

   _____ .

f. Not a true stratified; all cells on basement membrane, but some do not reach surface of tissue:

   _____ .

g. Line upper respiratory passageways (two

   answers): _____ .

h. Endothelium that lines blood vessels and heart

   chambers: _____ .

B7. Write a sentence describing structure and function of each of these modifications of the epithelum:

a. Microvilli (found on cells lining some of the gastrointestinal tract)

b. Goblet cells (found in mucous membranes lining the gastrointestinal, respiratory, reproductive, and urinary tracts)

c. Cilia (found on epithelium lining the respiratory and parts of the female and male reproductive tracts)

■ **B8.** Keratin is a *(carbohydrate? protein?)*. What are its functions?

List three locations of nonkeratinized stratified squamous epithelium.

■ **B9.** What are *glands* and why are they studied in this section on epithelium?

■ **B10.** Write EXO before descriptions of *exocrine* glands and ENDO before descriptions of *endocrine* glands. (Endocrine glands will be studied further in Chapter 13.)

_____ a. Their products are secreted into ducts that lead either directly or indirectly to the outside of the body.

_____ b. Their products are secreted into ECF and then the blood and so stay within the body; they are ductless glands.

_____ c. Examples are glands that secrete sweat, oil, mucus, and digestive enzymes.

_____ d. Examples are glands that secrete hormones.

C. Connective tissue   (pages 82–89)

■ **C1.** Write *C* for connective tissue or *E* for epithelial tissue next to the related descriptions below.

_____ a. Consists of many cells with minimal intercellular substance (fibers and ground substance)

_____ b. Penetrated by blood vessels (vascular)

_____ c. Does not cover body surface or line passageways or cavities, but is more internally located; binds, supports, protects

■ **C2.** Identify characteristics of each connective tissue cell type by writing the abbreviation of the correct cell type by its description. (Note that several of these cell types are shown in diagrams G, H, and I of Figure LG 4.1.)

A. Adipocyte	Mac. Macrophage	P. Plasma cells
F. Fibroblast	Mas. Mast cell	

_____ a. Phagocytic cell that engulfs bacteria and cleans up debris; important during infection

_____ b. Fat cell

_____ c. Secrete fibers and ground substance of matrix

_____ d. Abundant along walls of blood vessels; produces histamine, which dilates blood vessels

_____ e. Antibody-secreting cell that is formed from a B lymphocyte; important

■ **C3.** List three categories of extracellular matrix of connective tissue.

■ **C4.** Describe the connective tissue matrix in this activity. Match the chemicals and fibers in the box with the descriptions below.

C. Collagen fibers	F. Fibrillin	R. Reticular fibers
E. Elastic fibers	H. Hyaluronic acid	

_____ a. Viscous, slippery substance that binds cells together and lubricates joints

_____ b. A protein that is defective in tissues that require elasticity in persons with Marfan syndrome

_____ c. Formed of collagen and glycoproteins; form branching networks that provide stroma for soft organs such as spleen and networks around fat, nerve, and muscle cells

_____ d. Tough fibers in bundles that provide great strength, as needed in bone and tendons; formed of the most abundant protein in the body

_____ e. Can be greatly stretched without breaking, an important quality of connective tissues forming skin, blood vessels, and lungs

C5. After you study each mature connective tissue type in Table 4.2, pages 85–89 in the text, refer to diagrams of connective tissues on Figure LG 4.1, G–L. Write the following information on lines provided on the figure. (Note: Provide brief information on Figures K and L, as tissues will be studied in greater detail in later chapters.)

■ a. The name of each tissue

  b. One or more locations of the tissue

  c. One or more functions of the tissue

  d. Now color the following structures on each diagram, using the same colors as you did for diagrams A–F:

  ○ Nucleus   ○ Matrix

  ○ Cytoplasm   ○ Fat globule (a cellular inclusion), diagram H

■ e. From this activity, you can conclude that connective tissues appear to consist mostly of *(cells? matrix?)*.

■ f. Which diagrams in Figure LG 4.1 show loose connective tissue? _____

■ g. Which two types of tissue in the figure form subcutaneous tissue? _____

■ h. Explain what accounts for the "class ring" (with a stone) appearance of adipocytes shown in H of the figure.

  i. Label fibers in Figure LG 4.1, diagrams G, I, and J. (Note that fibers *are* present in bone, but are not visible in diagram K.)

C6. Write three or more functions of fat in the body.

Read the *Focus on Wellness* on page 91 and write a paragraph on health risks of carrying extra fat tissue, especially in the abdomen.

■ **C7.** Identify one or more location of each of these categories of connective tissues:

DI-CT. Dense irregular connective tissue   DR-CT. Dense regular connective tissue
E-CT. Elastic connective tissue

_____   a. Lung tissue and elastic arteries

_____   b. Ligaments, tendons, and aponeuroses

_____   c. Periosteum, perichondrium, and heart valves

■ **C8.** Match the common types of dense regular connective tissue with the descriptions.

A. Aponeurosis	L. Ligament	T. Tendon

_____ a. Connects muscles to bones

_____ b. Holds bones together at joints

_____ c. Flat band or sheet of tissue connecting muscles to each other or to bones

■ **C9.** Do this activity about cartilage tissue.

a. Mature cartilage cells are known as _____-cytes. These are located in spaces known as

_____ ("little lakes") surrounded by a dense, rubbery matrix.

b. In general, cartilage can endure (*more? less?*) stress than other connective tissues studied so far. The type of cartilage located where strength and rigidity are especially needed, as between hipbones and in discs between vertebrae, is (*elastic? fibrous? hyaline?*). This type of cartilage contains large numbers of (*collagen? elastic?*) fibers.

c. The type of cartilage that is white and glossy and forms articular and rib cartilages is (*elastic? fibrous? hyaline?*). This type of cartilage is the (*most? least?*) abundant type of cartilage in the body.

d. Cartilage heals (*more? less?*) rapidly than bone. Explain why this is so.

■ **C10.** Complete this exercise about three other types of connective tissues.

a. Bone tissue is also known as _____ tissue. Bones serve as important storage sites for the

minerals, _____ and _____, which contribute to the rigidity of bone tissue.

b. Compare diagrams J and K in Figure LG 4.1. Which appears to have a more organized structure? (*J? K?*) which

shows _____ tissue. This organization makes possible the penetration of blood vessels (within the centers of each concentric ring system) throughout bone tissue. (See Chapter 6 for more detail.)

c. Blood consists of a liquid extracellular matrix called _____ in which cells are suspended. The major role of (*red? white?*) blood cells is to transport oxygen and carbon dioxide through the blood stream. Write

one or more function of white blood cells: _____. Platelets assist in the function of blood

_____. Blood is shown in diagram _____ of Figure LG 4.1. (See Chapter 14 for more details on blood.)

d. _____ is fluid within lymph vessels; this fluid is similar to blood plasma but contains much

(*more? less?*) protein. _____ are the main type of cells in lymph.

D. Muscular tissue and nervous tissue (page 90)

■ **D1.** Write three functions of muscle tissue.

■ **D2.** Match the muscle types listed in the box with locations below.

C. Cardiac	Sk. Skeletal	Sm. Smooth

_____ a. Found in larger blood vessels and in the walls of the stomach and intestine, urinary bladder, and uterus

_____ b. Attached by tendons to bones; examples: the biceps and triceps muscles

_____ c. Makes up most of the wall of the heart

D3. Turn to the beginning of Chapter 8 of your text and study the diagrams of the three types of muscle tissue. Label them in Figure LG 4.1, M–O, and write as much of the following information as you can on the lines provided on the figure. Fill in details once you complete Chapter 8.

■ a. The name of each muscle tissue

b. One or more locations of the tissue

c. One or more functions of the tissue

d. Color the following structures on each diagram, using the same colors as you did for diagrams A–L.

○ Nucleus   ○ Cytoplasm   ○ Intercellular material

■ e. From this activity, you can conclude that muscle tissues appear to consist mostly of (cells? intercellular material [matrix]?).

D4. Refer to diagram P in Figure LG 4.1. Write the following information on the lines provided on the figure. (More on nervous tissue in Chapter 9.)

■ a. The name of the tissue

b. One or more locations of the tissue

c. One or more functions of the tissue

E. Membranes   (pages 90–91)

E1. Complete the table about three types of membranes in the body.

Type of Membrane	Location	Example of Specific Location	Function(s)
a.	Lines body cavities leading to exterior		
b. Serous			Allows organs to glide easily over each other
c.		Lines knee and hip joints	

■ **E2.** Check your understanding of membrane types by doing this exercise.

a. The serous membrane covering the heart is known as the _____ , whereas that covering

the lungs is called the _____ . The serous membrane over abdominal organs is the

_____ .

b. The portion of serous membranes that covers organs (viscera) is called the _____ layer;

the portion lining the cavity is named the _____ layer. See Figure 18.4, page 450 in your
textbook.

c. A _____ membrane secretes a lubricating fluid known as synovial fluid, and is found lining

_____. Such a membrane (*does? does not?*) contain epithelium, so it (*is? is not?*) classified as an ep-
ithelial membrane.

d. The fourth type of membrane in the body is the skin which is also known as the _____ membrane.

F. Tissue repair; aging and tissues; common disorders and medical terminology (pages 92–93)

F1. Contrast abilities of the following tissues to heal. State your rationales.

a. Epithelium – nervous tissue

b. Cartilage – bone

c. Skeletal muscle – smooth muscle

■ **F2.** *A clinical challenge.* Circle correct answers in each clinical case.

a. Mr. Bracken's liver has been damaged after several years of alcohol abuse, effects of toxic drugs, and viral hepatitis. Ability of his liver regain function will be based more upon the capacity for the (*parenchymal? stromal?*) cells
of his liver to regenerate.

b. Rev. Carlson is hospitalized for several days following an automobile accident. The skin care nurse specialist
notes growth of *granulation tissue* at Rev. Carlson's major wound sites. This is an indicator of (*good? poor?*)
wound healing.

■ **F3.** List the three major factors that promote tissue repair.

F4. Describe aging changes involving the following components of tissues:

a. Glucose

b. Collagen fibers

c. Elastin

•

■ **F5.** Check your understanding of two tissue disorders, Sjögren's syndrome and systemic lupus erythematosus (SLE), by filling in the table in this Checkpoint.

Characteristic of Disease	Sjögren's Syndrome	SLE (Lupus)
a. Autoimmune *(Yes? No?)*		
b. Type of tissue most affected?		
c. Gender most affected *(F? M?)*		
d. Distinguishing signs/symptoms		

■ **F6.** Xenotransplantation refers to transplantation using tissues or organs from *(another human? nonhuman animal?)*.

ANSWERS TO SELECTED CHECKPOINTS: CHAPTER 4

A3. Nervous tissue and most muscular tissue; hypertrophy, which is tissue growth by enlargement of existing cells (see page 67 in your text).

B1. Tight junctions that prevent leakage between cells; junctions that provide support for an organ by preventing separation of cells in the tissue; channels that permit cell communication or impulse conduction.

B2. Covering or lining and glandular. (a) Lining airways, mouth, esophagus, stomach, intestine, urinary tract, and reproductive organs that open to the outside of the body. (b) Sweat, oil, thyroid, pancreas. (c) Most superficial.

B4. (a) Closely packed cells. (b) No; avascular; has. (c) Does. (d) Basement. (e) Simple stratified. (f) Squamous columnar

B5. (A) Simple squamous epithelium. (B) Simple cuboidal epithelium. (C) Simple columnar epithelium (nonciliated). (D) Simple columnar epithelium (ciliated). (E) Stratified squamous epithelium. (F) Pseudostratified columnar epithelium, ciliated.

B6. (a) Simple columnar, nonciliated. (b) Transitional. (c) Stratified squamous. (d) Simple cuboidal. (e) Simple squamous. (f) Pseudostratified columnar. (g) Pseudostratified columnar and simple columnar, ciliated. (h) Simple squamous epithelium.

B8. Protein; resists friction, and repels heat, chemicals, and bacteria. Examples of locations: wet surfaces such as lining of the mouth, tongue, esophagus, and vagina.

B9. One or more epithelial cells that secrete substances into ducts or blood or onto a surface. All glands, whether exocrine or endocrine, are composed of epithelium.

B10. (a) EXO. (b) ENDO. (c) EXO. (d) ENDO.

C1. (a) E. (b–c) C.

C2. (a) Mac. (b) A. (c) F. (d) Mas. (e) P.

C3. Fluid, gel, or solid.

C4. (a) H. (b) F. (c) R. (d) C. (e) E.

C5. (a) (G) Areolar connective tissue. (H) Adipose tissue. (I) Dense connective tissue, such as tendon or ligament. (J) Cartilage. (K) Bone (osseous) tissue. (L) Blood (vascular) tissue. (e) Intercellular material (matrix). (f) G, H. (g) G, H. (h) The nucleus and cytoplasm are pushed over to one side (like the stone in the ring) by the large fat droplet.

C7. (a) E–CT. (b) DR–CT. (c) DI–CT.

C8. (a) T. (b) L. (c) A.

C9. (a) Chondro; lacunae. (b) More; fibrous; collagen. (c) Hyaline; most. (d) Less; cartilage is not penetrated by blood vessels, so chemicals needed for repair must reach cartilage by diffusion from the perichondrium or other surrounding tissues.

C10. (a) Osseous; calcium (and) phosphate. (b) K, bone. (c) Plasma; red; defense (phagocytosis and immunity) and allergic reactions; clotting; L. (d) Lymph, less; lymphocytes.

D1. Produces motion, maintains posture, and generates heat.

D2. (a) Sm. (b) 5k. (c) C.

D3. (M) Skeletal muscle. (N) Cardiac muscle. (O) Smooth muscle. Cells (known as muscle fibers).

D4. Nervous tissue consisting of neurons and neuroglia.

E2. (a) Pericardium, pleura; peritoneum. (b) Visceral; parietal. (c) Synovial, freely moveable (synovial joints); does not, is not. (d) Cutaneous

F2. (a) Parenchymal. (b) Good. (c) No.

F3. Nutrition (such as protein and vitamins), good blood supply, and higher metabolic rate (as in younger persons).

F5. (a) Yes, Yes. (b) Glands (epithelium); connective tissue. (c) F, F. (d) Sjögren's: dry mouth and dry eyes; also arthritis. SLE: butterfly rash, sensitivity to light; painful joints; possibly serious damage to organs such as kidneys.

F6. Nonhuman animal.

CRITICAL THINKING: CHAPTER 4

1. The outer part (epidermis) of skin is composed of epithelium. Explain what makes epithelium suitable for this location.
2. Which tissue is likely to heal better: bone or cartilage? Explain why.
3. Serous membranes are normally relatively free of microorganisms, whereas mucous membranes and skin have many microorganisms on them. Based on this information, why may antibiotics be administered to patients undergoing stomach or intestinal surgery?
4. Explain the role of stem cells in tissue repair.

MASTERY TEST: ■ CHAPTER 4

Questions 1–7: Circle the letter preceding the one best answer to each question.

1. All of the following are secretions from exocrine glands *except*:
   A. Adrenal gland hormones
   B. Perspiration
   C. Mucus
   D. Digestive enzymes
   E. Ear wax (cerumen)
2. The type of tissue that covers body surfaces, lines body cavities, and forms glands is:
   A. Nervous
   B. Muscular
   C. Connective
   D. Epithelial
3. Neurons are cells that are part of:
   A. Nervous tissue
   B. Muscular tissue
   C. Connective tissue
   D. Epithelial tissue
4. Which statement about connective tissue is *false?*
   A. Cells are very closely packed together.
   B. Most connective tissues have an abundant blood supply.
   C. Matrix is present in large amounts.
   D. It is the most abundant tissue in the body.
5. Which tissues are avascular?
   A. Skeletal muscle and cardiac muscle
   B. Bone and dense connective tissue
   C. Cartilage and epithelium
   D. Areolar and adipose tissues
6. Which term refers to microscopic fingerlike projections that increase the surface of the plasma membrane?
   A. Microvilli
   B. Basement membrane
   C. Cilia
   D. Goblet cells
7. The serous membrane covering the stomach and liver is known as the:
   A. Pericardium
   B. Peritoneum
   C. Pleura
   D. Synovium

Questions 8–10: Circle T (true) or F (false). If the statement is false, change the underlined word or phrase so that the statement is correct.

T   F   8.   The mouth and nose are both lined with <u>mucous</u> membranes.

T   F   9.   The surface attachment between epithelium and connective tissue is called <u>basement membrane</u>.

T   F   10.   <u>Simple</u> squamous epithelium is most likely to line the areas of the body that are subject to wear and tear.

Questions 11–15: Fill-ins. Write the word or phrase that best fits the description.

_____ 11.   Ground substance and fibers together form the _____ of connective tissue.

_____ 12.   The type of epithelium that lines the inside of the urinary bladder is _____ .

_____ 13.   The kind of tissue that lines alveoli (air sacs) of lungs is _____ .

_____ 14.   The kind of tissue that contains lacunae and chondrocytes is _____ .

_____ 15.   Tissue that forms the thick surface layer of skin on hands and feet, providing extra

protection, is _____ .

ANSWERS TO MASTERY TEST: ■ CHAPTER 4

Multiple Choice

1. A
2. D
3. A
4. A
5. C
6. A
7. B

True or False

8. T
9. T
10. F. Stratified

Fill-ins

11. Matrix
12. Transitional
13. Simple squamous epithelium
14. Cartilage
15. Stratified squamous epithelium

FRAMEWORK 5
The Integumentary System

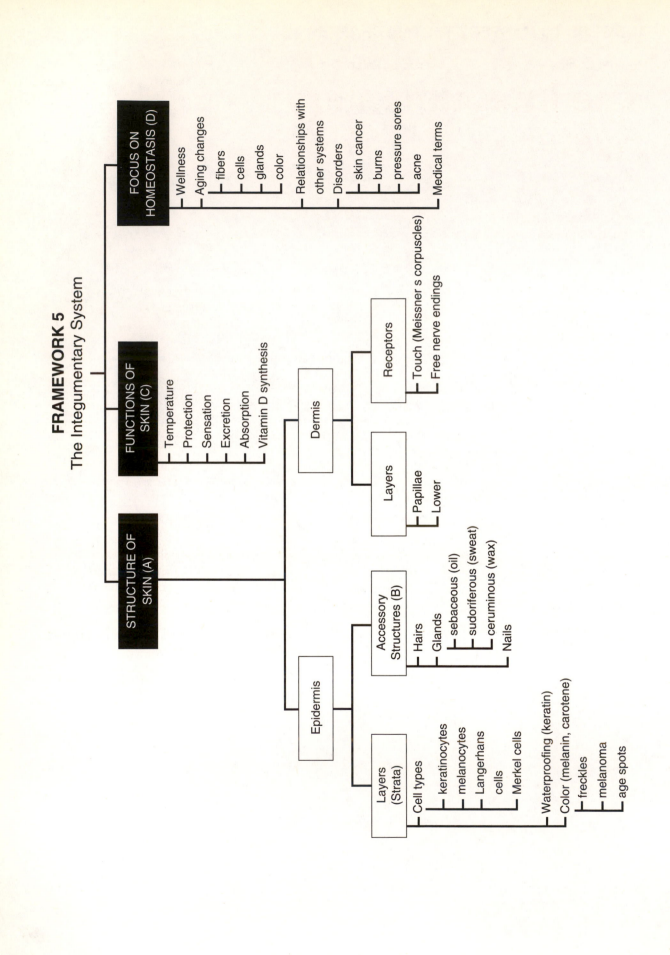

The Integumentary System

In your human body tour, you have now arrived at your first system, the integumentary system, which envelops the entire body. Composed of the skin, hair, nails, and glands, the integument serves as both a barrier and a link with the environment. The many waterproofed layers of cells, as well as the pigmentation of skin, afford protection to the body. Nails, glands, and even hairs offer additional fortification. Sense receptors and blood vessels in skin increase awareness of conditions around the body and facilitate appropriate responses. Even so, this microscopically thin surface cover may be traumatized or invaded, or simply succumb to the passage of time. The integument then demands repair and healing for continued maintenance of homeostasis.

As you begin your study of the integumentary system, carefully examine the Chapter 5 Framework and note relationships among concepts and key terms there. Also refer to the Topic Outline and Objectives; you may want to check off each objective as you complete it.

TOPIC OUTLINE AND OBJECTIVES

A. Skin

☐ 1. Describe the structure and functions of skin.
☐ 2. Explain the basis for different skin colors.

B. Accessory structures of the skin

☐ 3. Describe the structure and functions of hair, skin glands, and nails.

C. Functions of skin

☐ 4. Describe how the skin contributes to the regulation of body temperature, protection, sensation, excretion, absorption, and synthesis of vitamin D.

D. Focus on homeostasis: aging, wellness, and disorders of skin

☐ 5. Describe the effects of aging on the integumentary system.

WORDBYTES

Now become familiar with the language of this chapter by studying each wordbyte, its meaning, and an example of its use within a term. After you study the entire list, self-check your understanding by writing the meaning of each wordbyte on the line. As you continue through the *Learning Guide,* identify (and fill in) additional terms that contain the same wordbyte.

Wordbyte	Meaning	Example(s)	Wordbyte	Meaning	Example(s)
albin-	white	*albin*ism	fer-	carry	sudori*fer*ous
basale	base	stratum *basale*	lucidus	clear	stratum *lucidum*
corneum	horny	stratum *corneum*	melan-	black	*melan*oma
cut-	skin	*cut*aneous	seb-	fat	*seb*aceous
derm-	skin	*derm*atologist	sub-	under	*sub*cutaneous
epi-	over	*epi*dermis	sudor-	sweat	*sudor*iferous

CHECKPOINTS

A. Skin (pages 98–101)

■ **A1.** Name the structures included in the integumentary system.

Becky has an appointment with a doctor who specializes in skin disorders; this doctor is a specialist in _____ .

■ **A2.** Refer to Figure LG 5.1 and answer these questions about skin.

a. The outer layer of skin is named the *(dermis? epidermis?)*. It is composed of *(connective tissue? epithelium?)*.

   The inner layer of skin, called the _____ , is made of *(connective tissue? epithelium?)*.
b. Which layer is thicker *(dermis? epidermis?)*? Label both layers on Figure LG 5.1.
c. Which layer has most sensory receptors, nerves, blood vessels, and glands embedded in it? Color those structures and their related color code ovals.

d. The tissue that underlies skin is called subcutaneous, meaning _____ . It consists of two

   types of tissue, _____ and _____ . What functions does subcutaneous tissue serve?

■ **A3.** Epidermis contains four distinct cell types. Fill in the name of the cell type that fits each description.
a. Most numerous cell type, this cell produces keratin, which helps to waterproof skin:

   _____ .

b. This type of cell produces the pigments that give skin its color and that absorb UV rays to protect nuclei of skin

   cells from damaging rays: _____ .

c. These cells, function in immunity: _____ .
d. Located in the deepest layer of the epidermis, these cells contribute to the sensation of touch:

   _____ .

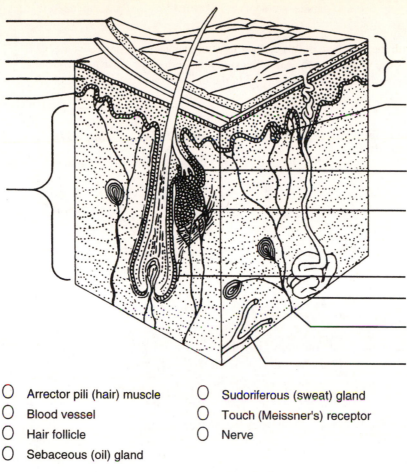

○ Arrector pili (hair) muscle   ○ Sudoriferous (sweat) gland

○ Blood vessel   ○ Touch (Meissner's) receptor

○ Hair follicle   ○ Nerve

○ Sebaceous (oil) gland

Figure LG 5.1   Structure of the skin. Label as directed in Checkpoints A2, A4, and A6.

■ **A4.**   Label each of the five layers of epidermis on the left side of Figure LG 5.1.

■ **A5.**   Match names of epidermal strata with correct descriptions. Choose from answers in the box.

B. Basale	C. Corneum	G. Granulosum	L. Lucidum	S. Spinosum

_____ a. Clear stratum present only in thick skin.

_____ b. Structum consisting of one single layer of cuboidal or columnar cells; new cells form here.

_____ c. Most superficial layer of skin; consists of many layers of flat, dead cells; excessive shedding of cells in this layer of the scalp

in the condition known as dandruff: this layer thickens during callus formation.

_____ d. Cells in this stratum secrete a lipid waterproofing sealant.

_____ e. Cells in this region exhibit spiny projections.

■ **A6.**   Describe the dermis in this exercise.

a. The dermis is located *(deep? superficial?)* to the epidermis and consists of *(epithelial? connective?)* tissue. The upper portion of the dermis projects into epidermal regions by fingerlike extensions known as dermal _____ . Meissner's corpuscles here are sensitive to *(pressure? touch?)*. Free nerve endings within papillae are sensitive to _____ .

b. Deeper regions of the dermis are composed of *(loose? dense?)* connective tissue strengthened by _____ and _____ fibers. Extreme stretching of these fibers (as in pregnancy or obesity) leads to formation of stretch marks or _____ . These fibers can also be disrupted by UVA rays that causes severe _____ .

c. Refer to Figure 5.1 in your text and name six other types of structures found within deep regions of the dermis.

■ **A7.** Explain what accounts for different skin colors by doing this exercise.

a. Dark skin is due primarily to the presence of the pigment _____ , which is in greatest

abundance in the *(dermis? epidermis?)*. In the inherited condition called _____ , this pigment is absent in hair, skin, and eyes.

b. What are freckles and age (liver) spots?

c. Carotene contributes a _____ color to skin.

d. What accounts for the pinker color of skin during blushing and acts as a cooling mechanism during exercise?

This condition is known as *(cyanosis? erythema?)*.

e. List two possible causes of the condition known as pallor.

f. Jaundiced skin has a more *(blue? red? yellow?)* hue, often due to _____ problems. Cyanotic skin appears more *(blue? red? yellow?)* due to lack of oxygen and excessive carbon dioxide in the blood vessels of the skin.

A8. *A clinical challenge.* Skin may provide clues about the health of the body; in this case at least, you *can* tell a lot about a book by its cover. Write three examples of how a physical or emotional health state may be detected by inspection of skin. (*Hint*: See question A7.)

B. Accessory structures of the skin (pages 101–104)

■ **B1.** Accessory structures of skin are all derived from *(epidermis? dermis?)*, and many extend down into *(epidermis? dermis?)*. Name the three types of accessory structures of skin.

■ **B2.** What is the main function of hair? _____ Complete this activity about hair structure.

a. A hair is composed of *(living cells? dead cells? secretions from cells?)*.

b. Arrange the parts of a hair from deepest to most superficial: _____ _____ _____

A. Shaft   B. Bulb   C. Root

c. Surrounding the root of a hair is the hair _____ . The _____ is the part of a hair follicle where cells undergo mitosis permitting growth of a new hair. What is the function of the papilla of the hair?

60

d. *A clinical challenge.* When chemotherapeutic agents are used to treat cancer, hair loss is a common side effect, because anticancer drugs target (*rapidly? slowly?*)-growing cells—whether they are cancer cells or normal human hair follicle cells.

e. What causes "goose bumps"?

f. Hair color is due mostly to the pigment (*carotene? hemoglobin? melanin?*). Gray hair is due to decrease in _____ ,

   and white hair results from accumulation of _____ in the hair.

■ **B3.** Check your understanding of skin glands by stating whether the following descriptions refer to *sebaceous, sudoriferous,* or *ceruminous* glands.

a. Sweat glands: _____

b. Glands leading directly to hair follicle; secrete sebum, which keeps hair and skin from drying out:

   _____

c. Line the outer ear canal; secrete earwax: _____

d. When enlarged, they form blackheads, pimples, or boils: _____

e. Their principal functions are to regulate body temperature and to eliminate wastes: _____

f. Eccrine versions of these glands are most prominent on forehead, palms, and soles, whereas apocrine types are

   found in armpits, groin, breasts, and in bearded areas of males. _____

B4. Look at one of your own nails and Figure 5.4, page 104 in your text. Identify these parts of your nail: *free edge, nail body, lunula,* and *cuticle.*

■ **B5.** Answer these questions about nails and hair.

a. Are nails formed of cells (*Yes? No?*) Do nails contain keratin? (*Yes? No?*) What type of tissue forms nails? (*Dermis? Epidermis?*)

b. How is the function of the *nail matrix* similar to that of the *matrix of a hair?*

c. Why does the nail body appear pink, yet the lunula and free edge appear white?

C. Functions of skin (pages 104–106)

C1. Skin may be one of the most underestimated organs in the body. What functions does your skin perform while it is "just lying there" covering your body? List five functions on the lines provided.

_____   _____

_____   _____

_____

■ **C2.** When the body temperature is too hot, as during vigorous exercise, nerve messages from the brain inform sweat glands to *(in? de?)*-crease sweat production. In addition, blood vessels in the skin are directed to *(dilate? constrict?)*. As a result, body temperature _____ increases.

C3. Describe how each of the following components of the integumentary system provide protection.
a. Keratin

b. Melanin

c. Lipids from lamellar granules

d. Acidic pH of sweat

e. Hair and nails

■ **C4.** *A clinical challenge.* Fill in each blank with an example of a medication that can be applied transdermally for each purpose listed below.
a. To minimize motion sickness: _____

b. For chest pain (angina) related to heart disease: _____

c. To relieve severe pain associated with cancer: _____

d. To provide contraception: _____

e. For smoking cessation: _____

D. Focus on homeostasis: aging, wellness, and disorders of skin   (pages 105–108)

■ **D1.** Complete the table relating observable changes in aging of the integument to their causes.

Changes	Causes
a. Wrinkles; skin springs back less when gently pinched	
b.	Macrophages become less efficient; decrease in number of Langerhans cells
c.	Loss of subcutaneous fat
d. Dry, easily broken skin	
e.	Decrease in number and size of melanocytes

■ **D2.** Briefly describe how each of the following treatments is designed to reduce or reverse effects of aging on skin.

a. Microdermabrasion

b. Chemical peel

c. Dermal fillers

d. Botulinum toxin (Botox©)

■ **D3.** Check your understanding of effects of sun on skin by completing this Checkpoint.
  a. UVA rays makes up about (*5% 95%*) of the ultraviolet radiation that reaches the earth. These rays (*are? are not?*) absorbed by the ozone layer; they do penetrate skin layers and (*are? are not?*) by melanocytes, UVA rays are known as (*burning? tanning?*) rays.

63

b. UVB rays are partially absorbed by the ozone layer. These rays (*do? do not?*) penetrate skin layers as well as UVA rays. UVB rays (*do? do not?*) cause sunburn and skin damage associated with aging.

c. Which type of rays are thought to cause skin cancer? (*UVA? UVB?*)

d. Heightened reactions to UV light (natural are from tanning salons) following consumption of certain medications

or contact with certain chemicals is a phenomenon known as photo-_____. List four or more signs

or symptoms of this condition: _____ _____ _____ _____.
List several categories of triggers that can lead to photosensitivity.

■ **D4.** Identify characteristics of the three different classes of burns by completing this exercise.

a. Which type of burn is more serious? (*First-degree? Third-degree?*) Which type of burn involves damage to only surface layers of the epidermis? (*First-degree? Third-degree?*)

b. In which type of burn is skin more likely to form blisters? (*First-degree? Second-degree?*)

c. Which type of burn is most likely to cause hair loss and possibly loss of sensation? (*First-degree? Third-degree?*)

D5. Mrs. Chad, age 49, has second degree burns to the posterior surfaces of both arms and the entire posterior surface of her trunk and buttocks. Estimate the percent of her body surface that is burned. _____ % Does she qualify as having a *major burn*? (*Yes? No?*)

■ **D6.** Match the name of the disorder with the description given.

Ac. Acne	D. Decubitus ulcers	N. Nevus
Al. Albinism	H. Hives	P. Pruritus
B. Basal cell carcinoma	I. Impetigo	W. Wart
C. Cold sore	M. Malignant melanoma	

_____ a. Absence of skin pigmentation

_____ b. Mole

_____ e. Rapidly metastasizing form of cancer

_____ f. The most common form of skin cancer

_____ g. Itching

_____ h. Reddened, elevated patches that may itch; may result from allergic reactions to food or medications

_____ c. Staphylococcal or streptococcal infection, which may become epidemic in nurseries

_____ d. Inflammation of sebaceous glands especially in chin area; occurs under hormonal influence

_____ i. Caused by type I herpes simplex virus (HSV)

_____ j. Growth caused by papilloma virus; usually not cancerous

_____ k. Pressure sore or bed sore caused by decreased circulation through skin overlying a bony projection such as a heel or elbow

D7. Write the meaning of each letter in the mnemonic, "ABCD," that is helpful in assessing skin lesions for malignant melanoma:

A_____   B_____   C_____   D_____

D8. Contrast the following pairs of terms.

a. Topical/intradermal

b. Corn/abrasion

64

A1. Skin, hair, nails, glands, and specialized receptors for sensation; dermatology.

A2. (a) Epidermis; epithelium; dermis, connective tissue. (b) Dermis. See Figure LG 5.1A. (c) Dermis. See Figure LG 5.1A. (d) Under skin; areolar, adipose; anchors skin to underlying tissues and organs.

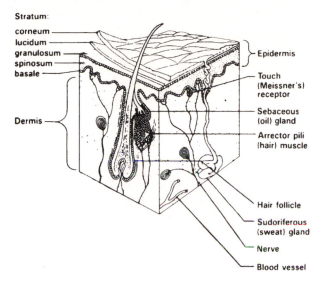

Figure LG 5.1A   Structure of the skin.

A3. (a) Keratinocyte. (b) Melanocyte. (c) Langerhans cell. (d) Merkel cell.

A4. See Figure LG 5.1A; quickly; keratin; dandruff.

A5. (a) L. (b) B. (c) C. (d) G. (e) S.

A6. (a) Deep, connective; papillae; touch; hot, cold, pain, tickle, and itch. (b) Dense, collagen (and) elastic; striae; wrinkling. (c) Blood vessels, nerve endings, pressure receptors, hair follicles, oil glands and sweat glands.

A7. (a) Melanin, epidermis; albinism; (b) Patches of melanin. (c) Yellow. (d) Widening (dilation) of blood vessels; erythema. (e) shock or anemia (f) Yellow, liver; blue.

B1. Epidermis, dermis; hair, nails, and glands.

B2. Protection, as well as contributing to appearance. (a) Dead cells. (b) B C A. (c) Follicle; matrix; contains blood vessels to nourish the growing hair. (d) Rapidly. (e) Pulling on hairs by smooth muscle (arrector pili muscles). (f) Melanin; melanin, air bubbles.

B3. (a) Sudoriferous. (b) Sebaceous. (c) Ceruminous. (d) Sebaceous. (e, f) Sudoriferous.

B5. (a) Yes; yes; epidermis. (b) They bring about growth of nails and hairs, respectively. (c) The pink color is related to visibility of blood vessels deep to the nail body; the free edge has no tissue (so no blood vessels) deep to it, whereas the thickened stratum basale obscures blood vessels deep to the lunula.

C2. In; dilate; de.

C4. (a) Scopolamine. (b) Nitroglycerin. (c) Fentanyl. (d) Ethinyl estradiol and norelgestromin. (e) Nicotine.

D1.

Changes	Causes
a. Wrinkles; skin springs back less when gently pinched	**Elastic fibers thicken into clumps and fray**
b. **increased susceptibility to skin infections and skin breakdown**	Macrophages become less efficient; decrease in number of Langerhans cells
c. **Loss of body heat; increased likelihood of skin breakdown, as in decubitus uclers**	Loss of subcutaneous fat
d. Dry, easily broken skin	**Decreased secretion of sebum by sebaceous glands, decreased fluids**
e. **Gray or white hair; atypical skin pigmentation**	Decrease in number and size of melanocytes

D3. (a) 95%; are not, are; tanning. (b) Do not; do. (c) Both UVA and UVB. (d) Photosensitivity; erythema, pruritis, blistering, peeling, hives, shock; see text.

D4. (a) Third-degree; first-degree. (b) Second-degree. (c) Third-degree.

D6. (a) Al. (b) N. (c) I. (d) Ac. (e) M. (f) B. (g) P. (h) H. (i) C. (j) W. (k) D.

CRITICAL THINKING: CHAPTER 5

1. Nathan, age 21, has a receding hairline associated with male-pattern baldness. What causes this type of baldness? Nathan is considering using a minoxidil (Rogaine), a product that he has seen advertised to treat hair loss. How does this product increase hair growth? Is it ingested or applied topically? Would it be more effective if used now or after he has lost most of the hair on his head?

2. Emily, age 78, notices that the skin on her arms, hands and feet is increasingly dry and cracking. What factors are likely to cause this condition?

What steps are advisable to improve the health of her skin?

3. Three friends, all age 19, have different hair color: Miji's is black, Sarah's is red, and Stacy has white hair and white eyebrows. What factors may account for these hair colors? (*Hint:* None of the young women uses hair color.)

4. Kelsey is having open-heart surgery this week. The nurse educator explains that Kelsey will need to remove her nail polish. Explain.

Questions 1–5: Circle the letter preceding the one best answer to each question.

1. "Goose bumps" occur as a result of:
   A. Contraction of arrector pili muscles
   B. Secretion of sebum
   C. Contraction of elastic fibers in the bulb of the hair follicle
   D. Contraction of papillae

2. Select the one *false* statement about the stratum basale.
   A. It is the one layer of cells that can undergo cell division.
   B. It consists of a single layer of squamous epithelial cells.
   C. It does normally contain keratinocytes.
   D. It is the deepest layer of the epidermis.

3. Select the one *false* statement.
   A. Epidermis is composed of epithelium.
   B. Dermis is composed of connective tissue.
   C. White hair can be due to presence of air bubbles in the hair.
   D. Carotene is a pigment that gives red color to skin.

4. The medical specialty that deals with diagnosis and treatment of skin disorders is:
   A. Oncology   C. Cytology
   B. Myology   D. Dermatology

5. Select the cell that functions in immunity:
   A. Merkel cell   C. Melanocyte
   B. Langerhans cell   D. Keratinocyte

Questions 6–7: Arrange the answers in correct sequence.

_____ _____ _____ 6. From most superficial to deepest:
   A. Dermis
   B. Epidermis
   C. Subcutaneous tissue

_____ _____ _____ 7. From most superficial to deepest:
   A. Stratum lucidum
   B. Stratum corneum
   C. Stratum basale

Questions 8–10: Circle T (true) or F (false). If the statement is false, change the underlined word or phrase so that the statement is correct.

T   F   8. Hairs are <u>noncellular structures composed entirely of nonliving substances secreted by follicle cells</u>.

T   F   9. <u>Both epidermis and dermis contain</u> blood vessels (are vascular).

T   F   10. <u>Hairs, sweat glands, and oil glands</u> are normally found on palms and soles.

Questions 11–15: Fill-ins. Complete each sentence with the word or phrase that best fits.

_____ 11. Skin contains a chemical that, under the influence of ultraviolet radiation, leads to formation of vitamin _____ .

_____ 12. The cells that are sloughed off as skin cells and undergo keratinization are those of the stratum _____ .

_____ 13. Constant exposure to friction or pressure stimulates formation of a _____ , which is an abnormal thickening of the epidermis.

_____ 14. The black, brown, or tan color of skin is due primarily to a pigment named _____ .

_____ 15. When the temperature of the body increases, nerve messages from brain to skin decrease body temperature by _____ .

ANSWERS TO MASTERY TEST: ■ CHAPTER 5

Multiple Choice

1. A
2. B
3. D
4. D
5. B

Arrange

6. B A C
7. B A C

True or False

8. F. Composed of fused, dead, keratinized cells
9. F. Only the dermis contains
10. F. Sweat glands

Fill-ins

11. D
12. Corneum
13. Callus
14. Melanin
15. Stimulating sweat glands to secrete and blood vessels to dilate

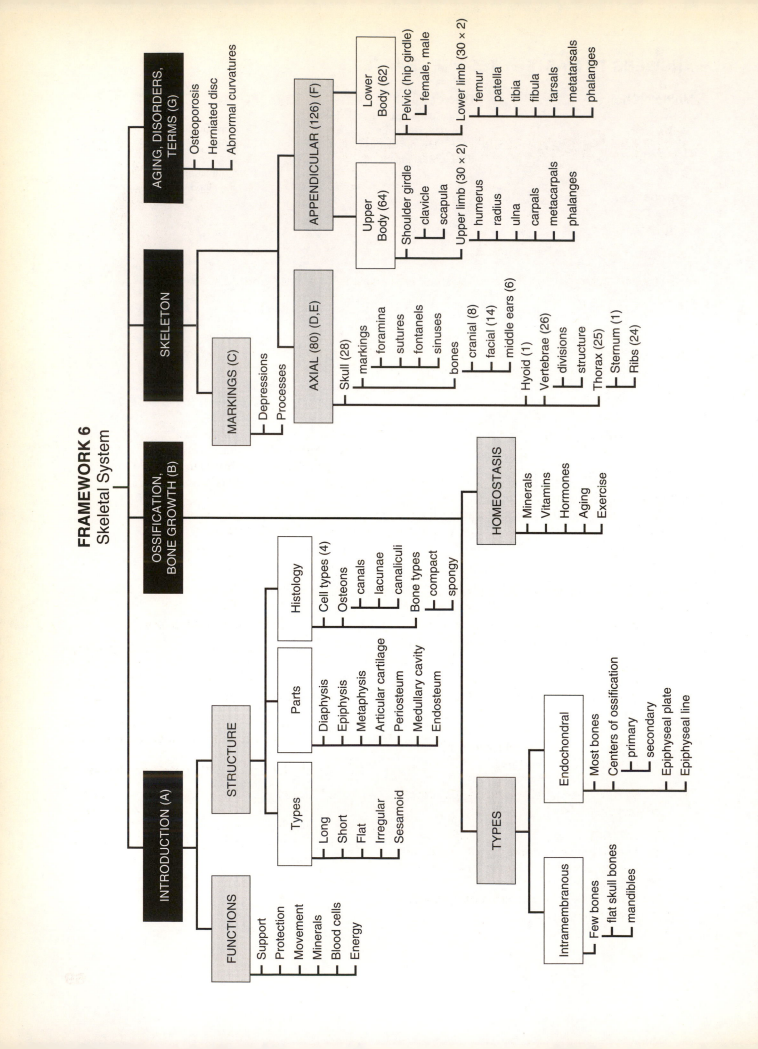

FRAMEWORK 6
Skeletal System

The Skeletal System

CHAPTER

6

The skeletal system provides the framework for the entire body, affording strength, support, and firm anchorage for the muscles that move the body. In this chapter you will first explore the tissue that forms the skeleton. Microscopically, bones present a variety of cell types intricately arranged in an osseous sea of calcified intercellular matrix. Bone growth, or ossification, continues through life. Essential nutrients, hormones, and exercise regulate the growth and maintenance of the skeleton.

The 206 bones of the human skeleton are classified into two divisions—axial and appendicular—based on their locations. The axial skeleton is composed of 80 bones that immediately surround the axis of the body, primarily the skull bones, back bone, and bones that surround the thorax. The appendicular skeleton includes 126 bones that form the framework of the limbs, or extremities, together with bones of the shoulders and hips.

As you begin your study of bone tissue, carefully examine the Chapter 6 Framework and note relationships among concepts and key terms there. Also refer to the Topic Outline and Objectives; you may want to check off each objective as you complete it.

TOPIC OUTLINE AND OBJECTIVES

A. Introduction: functions, types of bone, gross structure of bone tissue, histology

- ☐ 1. Discuss the functions of bone and the skeletal system.
- ☐ 2. Classify bones on the basis of their shape and location.
- ☐ 3. Describe the parts of a long bone.
- ☐ 4. Describe the histological features of bone tissue.

B. Bone formation and homeostasis: effects of exercise

- ☐ 5. Explain the steps involved in ossification.
- ☐ 6. Describe the factors involved in bone growth and maintenance and how hormones regulate calcium homeostasis.
- ☐ 7. Describe how exercise and mechanical stress affect bone tissue.

C. Divisions of the skeletal system

- ☐ 8. Group the bones of the body into axial and appendicular divisions.

D. Axial skeleton: skull

- ☐ 9. Name the cranial and facial bones and indicate their locations and major structural features.

E. Axial skeleton: hyoid, vertebrae, thorax

- ☐ 10. Identify the regions and normal curves of the vertebral column and describe its structural and functional features.
- ☐ 11. Identify the bones of the thorax and their principal markings.

F. Appendicular skeleton; female and male skeletons

☐ 12. Identify the bones of the pectoral (shoulder) girdle and their principal markings.

☐ 13. Identify the bones of the upper limb and their principal markings.

☐ 14. Identify the bones of the pelvic (hip) girdle and their principal markings.

☐ 15. List the skeletal components of the lower limb and their principal markings.

☐ 16. Identify the principal structural differences between female and male skeletons.

G. Focus on homeostasis; aging and the skeleton; common disorders and medical terminology

☐ 17. Describe the effects of aging on the skeletal system.

WORDBYTES

Study each wordbyte, its meaning, and an example of its use in a term. Check your understanding by jotting meanings of wordbytes in margins. Identify other examples of terms that contain these wordbytes as you continue through the text and *Learning Guide*.

Wordbyte	Meaning	Example(s)	Wordbyte	Meaning	Example(s)
acro-	tip	*acro*mion process	lambd-	L-shaped	*lambd*oid suture
cap-	head	*cap*itulum	lumb-	loin	*lumb*osacral
-blast	germ, to form	osteo*blast*	meta-	beyond	*meta*tarsal
cervic-	neck	*cervic*al collar	optic	eye	*optic* foramen
-clast	to break	osteo*clast*	os-, osteo-	bone	*os*sification, *osteo*cyte
chondr-	cartilage	peri*chondr*ium	peri-	around	*peri*osteum
costa-	rib	*costa*l cartilage	-physis	to grow	pubic sym*physis*
crist-	crest	*crist*a galli	semi-	half	*semi*lunar notch
ethm-	sieve	*ethm*oid bone	sym-	together	pubic *sym*physis

CHECKPOINTS

A. Introduction: functions, types of bone, gross structure of bone tissue, histology (pages 114–118)

■ **A1.** List six functions of the skeletal system.

_____   _____

_____   _____

_____   _____

■ **A2.** Refer to Figure LG 6.1. Use the KEY to the figure to locate each bone in the list below. Classify each of the following bones by writing *long, short, flat,* or *irregular* next to the name of each bone.

a. Femur: _____   d. Scapula: _____

b. Fibula: _____   e. Hipbone: _____

c. Carpals: _____

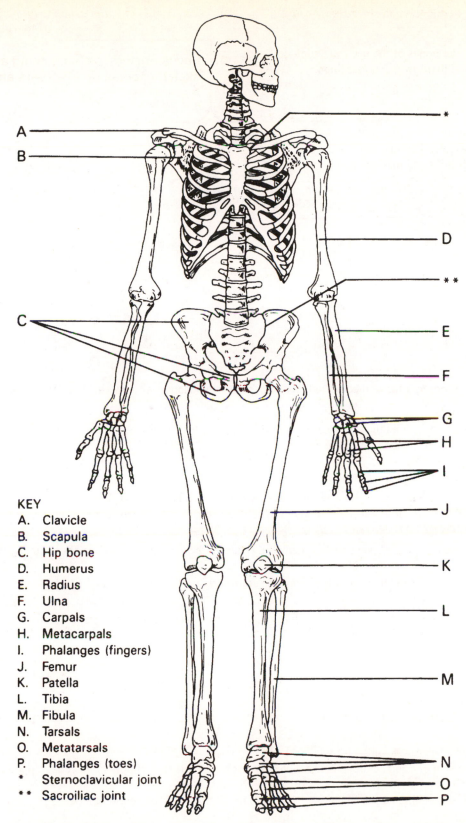

KEY
A. Clavicle
B. Scapula
C. Hip bone
D. Humerus
E. Radius
F. Ulna
G. Carpals
H. Metacarpals
I. Phalanges (fingers)
J. Femur
K. Patella
L. Tibia
M. Fibula
N. Tarsals
O. Metatarsals
P. Phalanges (toes)
* Sternoclavicular joint
** Sacroiliac joint

Figure LG 6.1 Anterior view of the skeleton. Color, label, and answer questions as directed in Checkpoints A2, C3, C4, E7, E9, F1, F3, F6, F8, and F9.

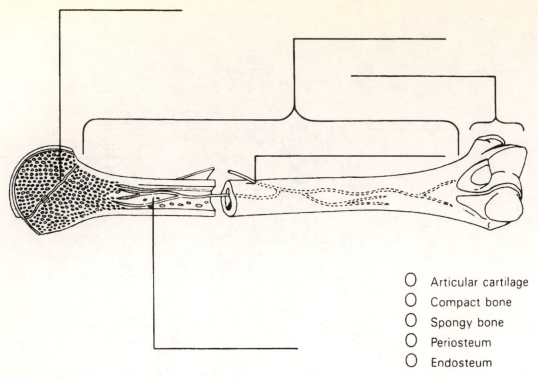

○ Articular cartilage
○ Compact bone
○ Spongy bone
○ Periosteum
○ Endosteum

Figure LG 6.2 Diagram of a long bone that has been partially sectioned lengthwise. Color and label as directed in Checkpoints A3 and F4.

■ **A3.** On Figure LG 6.2, label the *diaphysis, epiphysis,* and *medullary (marrow) cavity.* Then use color code ovals to color the parts of a long bone listed on the figure.

■ **A4.** Describe the components of bone in this exercise.

a. Which bone cells are mature? (*Osteoblasts? Osteocytes?*) Which bones break down bone? (*Osteoblasts? Osteoclasts?*) Which are the only bone cells to undergo cell division, and these cells develop into osteoblasts? (*Osteoblasts? Osteocytes? Osteogenic cells?*)

b. Like all connective tissues, bone consists mainly of *(cells? intercellular material?).* About 50% of bone consists

of crystalline mineral salts that provide *(flexibility? hardness?).* Collagen fibers make up about _____ % of bone and provide *(flexibility?).*

c. Bone *(is completely solid? contains some spaces?).* Write two advantages of that structural feature of bone.

■ **A5.** Refer to Figure LG 6.3 and complete this exercise about bone structure.

a. Compact bone is arranged in concentric circle patterns known as _____ . Each individual ring of bones is

known as a concentric _____ , labeled with letter _____ in that figure.

b. The osteon pattern of compact bone permits blood vessels and nerves to supply bone cells trapped in hard bone

tissue. Blood vessels and nerves penetrate bone from the periosteum, labeled with letter _____ in the figure.

These structures then pass through horizontal canals, labeled _____ , and known as perforating (volkmann's) canals. These vessels and nerves finally pass into microscopic channels known as central (haversian) canals,

labeled _____ , in the center of each osteon. Color blood vessels red and blue in the figure.

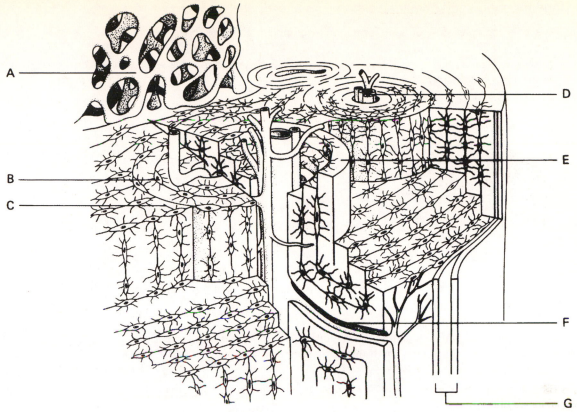

A ————

B ————

C ————

———— D

———— E

———— F

———— G

Figure LG 6.3 Osteons (haversian systems) of compact bone. Identify lettered structures and color as directed in Checkpoint A5.

c. Mature bone cells, known as _____ , are located relatively far apart in bone tissue. These

are present in "little lakes," or _____ , labeled _____ in the figure. Color ten lacunae green.

d. Minute canals, known as canaliculi, are labeled with letter _____ . What is the function of these channels?

■ **A6.**   Do this exercise about spongy bone.

a. Spongy bone is arranged in *(osteons? trabeculae?),* which may be defined as:

b. Spongy bone makes up most of the *(diaphyses? epiphyses?)* of long bones. Spongy bone *(is not? is also?)* located within bones that are short, flat, and irregular in shape.

c. *A clinical challenge.* Red blood cell formation (hemopoiesis) normally takes place in *(all? only certain areas of?)* spongy bone tissue. Name four or more bones in which this process takes place.

_____   _____

_____   _____

■ **A7.**   Identify which diagnostic information is likely to be gained by regions of bone scans that show:

a. "Hot spots"

b. "Cold spots"

B. Bone formation and homeostasis: effects of exercise (pages 118–124)

■ **B1.**   Check your understanding of bone formation by completing this exercise.

a. Bone formation is also called _____ . Circle periods of time when this process normally occurs:
   A. In embryonic and fetal life
   B. In childhood and adolescence
   C. In adulthood
   D. During fracture repair

b. Name the two methods of bone formation:

   _____ and _____ .

c. Which bones of the body form by the intramembranous process? _____ and

   _____ . In this process, osteo-*(blasts? clasts? cytes? genic cells?)* in the fibrous
   membrane secrete calcium salts. Flat bones formed by this method have a structure much like a flattened jelly
   sandwich: two outer layers *(compact? trabecular?)* bone (much like bread) enclosing a layer of *(compact? trabecular?)* bone (much like the jelly).

d. In endochondral ossification, bone replaces an initial "model" made of _____ . Identify
   the four major steps in this process by filling in the four long lines (I–IV) below with the correct terms selected
   from those in the box below.

Development	Growth
Secondary ossification centers	Primary ossification center

   I.   _____ of the cartilage model   _____ _____

   II. _____ of the cartilage model   **A**_____ _____

   III. Development of the _____   _____ _____ _____

   IV. Development of the _____   _____

e. Now fill in details of the process by placing the following statements of events in the correct sequence. Write the
   letters on the lines to the right of the correct phase (I–IV) above. One is done for you.
   A. Cartilage cells enlarge within the center of the diaphysis. Surrounding matrix calcifies and deprives chondro-
       cytes of nutrients.
   B. Many cartilage cells die; in this way spaces are formed within the cartilage model.
   C. Blood vessels penetrate the perichondrium, stimulating cells there to form osteoblasts. A collar of bone forms
       and gradually thickens around the diaphysis. The membrane covering the developing bone is now called the
       periosteum.
   D. A hyaline cartilage model of future bone is laid down by differentiation of mesenchyme into chondroblasts.
   E. A perichondrium develops around the cartilage model.
   F. Secondary ossification centers develop in epiphyses, forming spongy bone there about the time of birth. Hya-
       line cartilage remains as the epiphyseal plate for as long as the bone grows. Hyaline cartilage also remains as
       articular cartilage.
   G. Capillaries grow into spaces within the diaphysis and osteoblasts deposit bone matrix over disintegrating calci-
       fied cartilage. In this way spongy bone forms within the diaphysis at the primary ossification center.
   H. Osteoclasts break down the newly formed spongy bone in the very center of the bone, thereby leaving the
       medullary (marrow) cavity.

■ **B2.**   Complete this activity about maturation of the skeleton.

a. The epiphyseal *(plate? line?)* is the bony region in bones of adults that marks the original cartilaginous epiphyseal
   (plate? line?). Once the epiphyseal *(plate? line?)* forms, a bone can no longer increase its *(length? thickness?)*

b. As bones lengthen, they (*do? do not?*) increase in thickness. Which cells add layers of bones to the outside of the bone? (*Osteoblasts? Osteoclasts?*) Which cells present in the endosteum destroy bone to hollow out additional bone marrow? (*Osteoblasts? Osteoclasts?*)

c. Bone remodeling involves two processes, namely, _____ and _____. The role of osteo-clasts is to (*add? remove?*) mineral salts and collagen fibers. Osteo- _____ present in periosteum lay down new bone. Excessive action of these cells can lead to thick bumps (called _____) on bone.

■ **B3.** List the type of fracture that fits each description below. Choose from these answers:

    Closed   Complete   Open   Partial

a. The broken ends of the fractured bone do break through the skin; also known as a compound fracture: _____

b. A cracked bone that does not break entirely across the bone: _____

■ **B4.** Match the hormones listed in the box with their function in both growth and remodeling, and in calcium homeostasis.

hGH. Human growth hormone	PTH. Parathyroid hormone
IGF. Insulin-like growth factor	Calc. Calcitriol (vitamin D)

_____ a. Produced by the parathyroid gland, it increases osteoclast activity, causing bone destruction and increase in blood level of calcium.

_____ b. Promotes absorption of calcium from foods in the intestine into blood stream.

_____ c. Stimulates kidneys to retain Ca^{++} in the blood stream by decreasing urinary output of Ca^{++}.

_____ d. Secreted by the anterior pituitary; promotes growth of tissues such as bone.

_____ e. Produced by bone and by liver in response to hGH stimulation.

■ **B5.** Circle the correct answers about the effects of exercise on bones.

a. Exercise (*strengthens? weakens?*) bones. A healing bone that is not exercised is likely to become (*stronger? weaker?*) during the period that it is in a cast.

b. The pull of muscles on bones, as well as the tension on bones as they support body weight during exercise, causes (*increased? decreased?*) production of the protein collagen.

C. Divisions of the skeletal system   (pages 124–125)

■ **C1.**   Describe the bones in the two principal divisions of the skeletal system by completing this exercise. It may help to refer to Figure LG 6.1, page 71.

a. Bones that lie along the axis of the body are included in the *(axial? appendicular?)* skeleton.

b. The axial skeleton includes the following groups of bones. Indicate how many bones are in each category.

_____ Skull (cranium, face)   _____ Vertebrae (backbone)

_____ Earbones (ossicles)   _____ Sternum

_____ Hyoid   _____ Ribs

c. The total number of bones in the axial skeleton is _____ .

d. The appendicular skeleton consists of bones in which parts of the body?

e. Write the number of bones in each category. Note that you are counting bones on one side of the body only.

_____ Left shoulder girdle   _____ Left hipbone

_____ Left upper extremity (arm, forearm, wrist,   _____ Left lower extremity (thigh, kneecap, leg,
         hand)   foot)

f. There are _____ bones in the appendicular skeleton.

g. In the entire human body there are _____ bones.

C2.   *For extra review.* Color light blue all bones in the axial skeleton in Figure LG 6.1. (Check your accuracy by referring to Figure 6.6, page 125 in your text.)

D. Axial skeleton: skull (pages 125–132)

■ **D1.** Answer these questions about the cranium.

a. What is the main function of the cranium? _____ .

b. Which bones make up the cranium (rather than the face)?

○ Ethmoid bone
○ Frontal bone
○ Lacrimal bone
○ Mandible bone
○ Maxilla bone
○ Nasal bone
○ Occipital bone
○ Parietal bone
○ Sphenoid bone
○ Temporal bone
○ Zygomatic bone

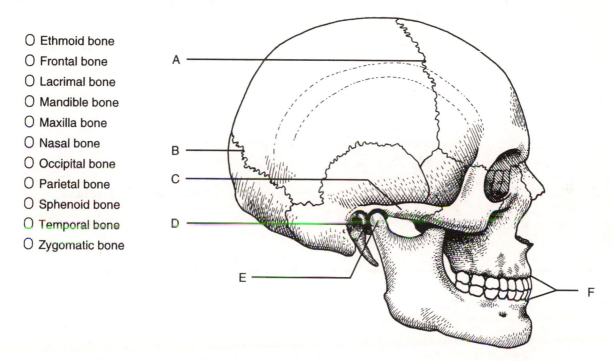

Figure LG 6.4   Skull bones. Skull viewed from right side. Color and label as directed in Checkpoints D2 and D5.

D2.   After you study Figure 6.7, pages 126–127 in your text, color the skull bones in Figure LG 6.4. Be sure to color the corresponding color code oval for each bone listed in the figure.

■ **D3.**   Check your understanding of locations and functions of skull bones by writing the name of each skull bone next to its description.

_____ a. This bone forms the lower jaw, including the chin, together with the temporal bones, forms the right and left "TMJ joints."

_____ b. These are the cheek bones; they also form lateral walls of the orbit of the eye.

_____ c. Tears pass through tiny foramina in these bones; they are the smallest bones in the face.

_____ d. The bridge of the nose is formed by these bones.

_____ e. Organs of hearing (internal part of ears) and mastoid air cells are located in and protected by these bones.

_____ f. This bone sits directly over the spinal column; it contains the foramen through which the spinal cord connects to the brain.

_____ g. The name means "wall." The bones form most of the roof and much of the side walls of the skull.

_____ h. These bones form most of the roof of the mouth (hard palate) and contain the sockets into which upper teeth are set.

_____ i. L-shaped bones form the posterior parts of the hard palate and nose.

_____ j. Commonly called the forehead, it provides protection for the anterior portion of the brain.

_____ k. A fragile bone, it forms much of the roof and internal structure of the nose.

_____ l. It serves as a "keystone," binding together many of the other bones of the skull. It is shaped like a bat, with the wings forming part of the sides of the skull and the legs at the back of the nose.

_____ m. This bone forms the inferior part of the septum, dividing the nose into two nostrils.

_____ n. Two delicate bones form the lower parts of the side walls of the nose.

■ **D4.** Complete the table describing major markings of the skull.

Marking	Bone	Function
a. Condylar process	Mandible	
b.		Forms superior portion of septum of nose
c.		Site of pituitary gland
d.		Largest hole in skull; passageway for spinal cord
e.	(2)	Bony sockets for teeth
f.		Passageway for sound waves to enter ear
g. Olfactory foramina		

■ **D5.** Now label markings A–F on Figure LG 6.4.

■ **D6.** Do this exercise about bony structures related to the nose.

a. Name four bones that contain paranasal sinuses.

_____ _____ _____

_____

b. List three functions of paranasal sinuses.

c. Most internal portions of the nose are formed by a bone that includes these markings: paranasal sinuses, conchae,

and the superior portion of the nasal septum. Name the bone: _____ .

d. What functions do nasal conchae perform?

e. Name the bone that forms the inferior portion of the nasal septum. _____ . The anterior

portion of the nasal septum consists of flexible _____ .

■ **D7.** Describe these special features of the skull.

a. Define *sutures*.

Between which two bones is the sagittal suture located? _____

b. "Soft spots" of a newborn baby's head are known as _____ . What is the location of the

largest one? _____

E. Axial skeleton: hyoid, vertebrae, thorax   (pages 132–138)

■ **E1.** Identify the location of the hyoid bone on yourself. Place your hand on your throat and swallow. Feel your larynx move upward? The hyoid sits just *(superior? inferior?)* to the larynx at the level of the mandible.

■ **E2.** In what way is the hyoid unique among all the bones of the body?

■ **E3.** Do this exercise on the vertebral column.

   a. First write the names of the regions of the vertebral column on the lines below. Arrange from superior to inferior. One is done for you.

<u>Cervical</u> _____ **(7)**   _____ ( )

_____ ( )   _____ ( )

_____ ( )

   b. Now write in parentheses the number of vertebrae in each region. One is done for you.

   c. Which two regions normally develop anteriorly concave curvatures?

      _____ _____

   d. What special names are given to the first two cervical vertebrae? _____

      _____ Which of these vertebrae contains the marking known as the dens? _____ What is the function of the dens?

   e. What are functions of the cartilage discs between vertebrae?

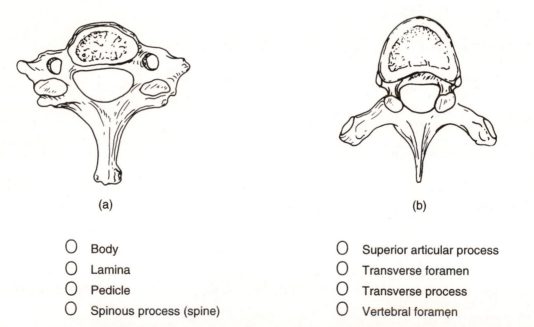

          (a)   (b)

◯ Body   ◯ Superior articular process
◯ Lamina   ◯ Transverse foramen
◯ Pedicle   ◯ Transverse process
◯ Spinous process (spine)   ◯ Vertebral foramen

Figure LG 6.5 Typical vertebrae. (a) Cervical vertebra, superior view. (b) Thoracic vertebra, superior view. Color and label as directed in Checkpoint E4.

■ **E4.** On Figure LG 6.5, color the parts of vertebrae and corresponding color code ovals. As you do this, notice differences in size and shape between the two vertebral types in the figure.

■ **E5.** Identify distinctive features of vertebrae in each region.

C. Cervical	L. Lumbar	S. Sacral	T. Thoracic

_____ a. Small body, foramina for vertebral blood vessels that pass vertically through transverse processes

_____ b. Only vertebrae that articulate with ribs

_____ c. Massive body, blunt spinous process and articular processes directed medially or laterally

_____ d. Long spinous processes that point inferiorly

_____ e. Articulate with the two hipbones

E6. Name the structures that compose the thoracic cage. Why is it called a "cage"?

E7. On Figure LG 6.1, color the three parts of the sternum.

■ **E8.** Name one clinical procedure in which pressure is applied to the sternum.

■ **E9.** Complete this exercise about ribs. Again refer to Figure LG 6.1.

a. There are a total of _____ ribs (_____ pairs) in the human skeleton.

b. Posteriorly, all ribs articulate (form joints) with _____ .

c. Anteriorly, ribs numbered _____ to _____ attach to the sternum directly by means of strips of hyaline

cartilage, called _____ cartilage. These ribs are called *(true? false?)* ribs. Color these ribs on Figure LG 6.1, leaving the costal cartilages white.

d. Ribs 8 to 12 are called _____ . Color these ribs a different color, again leaving the costal cartilages white. Do these ribs attach directly to the sternum? *(Yes? No?)* if so, in what manner?

e. Ribs _____ and _____ are called floating ribs. Add a third color for these ribs. Why are they so named?

f. What occupies intercostal spaces? _____

F. Appendicular skeleton; female and male skeletons (pages 139–146)

■ **F1.** Do this exercise about the pectoral girdle.

a. Which bones form the pectoral girdle? _____ _____

b. Do these bones articulate (form a joint) with vertebrae or ribs? _____

c. The pectoral girdle is part of the *(axial? appendicular?)* skeleton. Identify the point (marked by *) on Figure LG 6.1 at which the shoulder girdle articulates with the axial skeleton. Name the two bones forming that joint. Palpate (press and feel) the bones at this joint on yourself.

_____ _____

■ **F2.** Write the name of the bone that articulates with each of these markings on the scapula.

a. Acromion process: _____

b. Glenoid cavity: _____

■ **F3.** List the bone (or groups of bones) in the upper limb from proximal to distal. Indicate how many of each bone there are. Two are done for you. Refer to Figure LG 6.1 to check your answers.

a. **Humerus** _____ **(1)**   d. _____ **( )**

b. _____ **( )**   e. **Metacarpals** _____ **(5)**

c. _____ **( )**   f. _____ **( )**

■ **F4.** Name the marking that fits each description. Write *H, U,* or *R* to indicate whether the marking is part of the humerus, ulna, or radius. One has been done for you.

a. Articulates with glenoid cavity: **head** _____ (__**(H)**__).

b. Rounded head that articulates with radius: _____ (_____)

c. Posterior depression that receives the olecranon process: _____ (_____)

d. Half-moon-shaped curved area that articulates with trochlea: _____ (_____)

e. Slight depression in which the head of the radius pivots: _____ (_____)

f. Biceps brachii muscle attaches here: _____ (_____)

g. The deltoid muscle attaches here: _____ (_____)

h. Also known as the elbow: _____ (_____)

i. *For extra review,* label these markings on Figure LG 6.2: *head, trochlea, capitulum, coronoid fossa,* and *radial fossa.*

■ **F5.** Answer these questions about bones of the wrist and hand.

a. Wrist bones are called _____ . There are *(5? 7? 8? 14?)* of them in each wrist.

b. Which bones do you "hold in the palm of your hand"? _____ There are *(5? 7? 8? 14?)* of them in each hand.

c. How many phalanges are in your thumb? _____ In your entire hand? _____

d. Arrange the parts of one of your phalanges from most proximal to most distal: *base, head, shaft.*

_____ → _____ → _____

e. *For extra review.* Trace an outline of your wrist and hand on separate paper. Draw in bones and label each one.

■ **F6.** Describe the pelvic bones in this exercise.

a. Name the bones that form the pelvic girdle. _____

Which bones form the pelvis? _____

b. Which of these bones is/are part of the axial skeleton? _____

c. Locate the point (**) on Figure LG 6.1 at which the pelvic girdle portion of the appendicular skeleton articulates

with the axial skeleton. Name the bones involved in that joint. _____ and

_____

■ **F7.** Answer the following questions about hipbones.

a. Each coxal (hip) bone originates as three bones that fuse early in life. These bones are the

_____ , _____ , and _____ . At what

location do the bones fuse? _____

b. The largest of the three bones is the _____ . A ridge along the superior border is called the

iliac _____ . Locate this on yourself.

c. What are the functions of these markings?

Acetabulum: _____

Pubic symphysis: _____

Lesser (true) pelvis: _____

d. Which is more superior in location: pelvic *(inlet? outlet?).*

■ **F8.** Refer to Figure LG 6.1 and list the bones (or groups of bones) in the lower extremity from proximal to distal. Indicate how many of each bone there are. One is done for you.

a. **Femur** **(1)**   e. _____ **( )**

b. _____ **( )**   f. _____ **( )**

c. _____ **( )**   g. _____ **( )**

d. _____ **( )**

F9. Contrast the size, location, and names of the bones of the upper and lower extremities by coloring the bones on Figure LG 6.1 as follows. Color on one side of the figure only.

    Humerus and femur (red)   Carpals and tarsals (blue)

    Patella (brown)   Metacarpals and metatarsals (orange)

    Ulna and tibia (green)   Phalanges (purple)

    Radius and fibula (yellow)

■ **F10.** Circle the term that correctly indicates the location of these parts of the lower extremity.

a. The head is the *(proximal? distal?)* epiphysis of the femur.

b. The greater trochanter is located on the *(lateral? medial?)* aspect of the femur.

c. The outer portion of the ankle is the *(lateral? medial?)* malleolus, which is part of the *(tibia? fibula?)*.

d. The tibial condyles are more *(concave? convex?)* than the femoral condyles.

e. The lateral condyle of the femur articulates with the *(fibula? lateral condyle of the tibia?)*.

f. The tibial tuberosity is *(proximal? distal?)* to the patella.

g. The tibia is *(medial? lateral?)* to the fibula.

■ **F11.** The tarsal bone that is most superior in location (and that articulates with the tibia and fibula) is the

_____ . The largest and strongest of the tarsals is the _____ .

F12. Answer these questions about the arch of the foot.

a. How is the foot maintained in an arched position?

b. Locate each of these arches on your own foot. *For extra review:* List the bones that form each arch.
Longitudinal

Transverse

c. What causes flatfoot?

■ **F13.** Now that you have seen all the bones of the appendicular skeleton, complete this table relating common and anatomical names of bones.

Common Name	Anatomical Name
a. Shoulder blade	
b.	Pollex
c. Collarbone	
d. Heel bone	
e.	Olecranon
f. Kneecap	
g.	Tibial crest
h. Toes	
i. Palm of hand	
j. Wrist bones	

■ **F14.** Identify specific differences in pelvic structure in the two sexes by placing F before characteristics of the female pelvis and M before structural descriptions of the male pelvis.

_____ a. Shallow greater pelvis   _____ d. Acetabulum small

_____ b. Heart-shaped pelvic inlet   _____ e. Pelvic outlet comparatively small

_____ c. Pubic arch greater than 90° angle

G. Focus on homeostasis; aging and the skeleton; common disorders and medical terminology (pages 147–151)

G1. Identify which body system is assisted by the functions of the skeletal system listed in the table below. The first one is done for you.

Skeletal System Function	Other Body System Assisted
a. Bones of the skull and vertebral column protect the brain and spinal cord.	**Nervous system**
b. Pelvic bones protect ovaries and uterus of female, and prostate of male.	
c. Ribs and vertebrae protect kidneys; pelvis protects urinary bladder.	
d. Red marrow of bones provides sites for hemopoiesis. Long bones offer protection to adjacent blood vessels.	
e. Skeleton provides attachment sites for tendons of skeletal muscles.	
f. Ribs protect lungs and allow for expansion of thorax.	
g. Middle ear bones provide lever system that amplifies sound.	

■ **G2.** Complete this exercise about skeletal changes that occur in the normal aging process.

a. The amount of calcium in bone *(de?, in?)*-creases with age. As a result, bones of the elderly are likely to be *(stronger? weaker?)* than bones of younger people. This change occurs at a younger age in *(men? women?)*.

b. Another component of bones that decreases with age is _____ . What is the significance of this change?

G3. Describe osteoporosis in this exercise.

a. Osteoporotic bones exhibit _____-creased bone mass and _____-creased susceptibility to fractures. This

disorder is associated with a loss of _____ hormones. These hormones stimulate osteo-

_____ activity.

b. Circle the category of people at higher risk for osteoporosis in each pair:

younger people/older people   short, thin people/tall, large-build people
men/women   blacks/whites

c. List three factors in daily life that will help to prevent osteoporosis. (*Hint:* Focus on diet and activities.)

■ **G4.** Check your understanding of bone disorders by matching the terms in the box with related descriptions.

| Osteogenic sarcoma | Osteomalacia | Osteomyelitis | Osteopenia |

a. Decreased bone mass: c. Soft bones due to inadequate calcification:

_____   _____

b. Malignant bone tumor: d. Bone infection:

_____   _____

G5. Contrast the terms related to bones.

a. Scoliosis/kyphosis/lordosis

b. Osteoarthritis/Anthroplasy

A1. Support, protection, movement, homeostasis of minerals such as calcium, storage of energy as in fat in bone marrow, production of blood cells.

A2. (a, b) Long. (c) Short. (d) Flat. (e) Irregular.

A3.

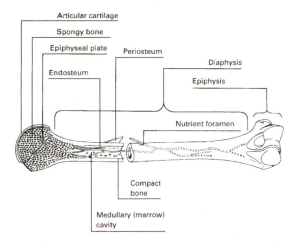

Figure LG 6.2A Diagram of a developing long bone that has been partially sectioned lengthwise.

A4. (a) Osteocytes; osteoclasts; osteogenic cells. (b) Intercellular material; hardness; 25, flexibility. (c) Contains some spaces; provides channels for blood vessels and makes bones lighter in weight.

A5. (a) Osteons (haversian systems); lamella, E. (b) G; F; D. (c) Osteocytes; lacunae, B. (d) C; contain extensions of osteocytes bathed in extracellular fluid (ECF); they are sites of extensions of osteocytes and provide routes for nutrients and oxygen to reach osteocytes and for wastes to diffuse away.

A6. (a) Trabeculae, irregular latticework of plates of bone tissue containing osteocytes surrounded by red marrow. (b) Epiphyses; is also. (c) Only certain areas of; hipbones, ribs, sternum, vertebrae, skull bones, and epiphyses of long bones such as femurs.

A7. (a) Bone cancer, abnormal bone growth, or abnormal bone healing. (b) Degenerative bone disease, fracture, infection, rheumatoid arthritis.

B1. (a) Ossification; A–D. (b). intramembranous, endochondral. (c) Flat bones of the skull and the mandible (lower jaw bone); blasts; compact trabecular. (d, e) Cartilage; (I) Development; D E. (II) Growth: A B. (III) Primary ossification center: C G H. (IV) Secondary ossification centers: F.

B2. (a) Line, plate; line, length. (b) Do; osteoblasts; osteoclasts. (c) Resorption and deposition; remove; blasts; spurs.

B3. (a) Open. (b) Partial.

B4. (a) PTH. (b) Vit D. (c) PTH. (d) Ins. (e) hGH.

B5. (a) Strengthens; weaker. (b) Increased.

C1. (a) Axial. (b) 22 skull, 6 earbones (studied in Chapter 12), 1 hyoid, 26 vertebrae, 1 sternum, 24 ribs. (c) 80. (d) Shoulder girdles, upper extremities, hipbones, lower extremities. (e) 2 left shoulder girdle, 30 left upper extremity, 1 left hipbone, 30 left lower extremity. (f) 126. (g) 206.

D1. (a) Protects the brain. (b) Frontal, parietals, occipital, temporals, ethmoid, sphenoid.

D3. (a) Mandible. (b) Zygomatic (malar). (c) Lacrimal. (d) Nasal. (e) Temporal. (f) Occipital. (g) Parietal. (h) Maxilla. (i) Palatine. (j) Frontal. (k) Ethmoid. (l) Sphenoid. (m) Vomer. (n) Inferior concha.

D4.

Marking	Bone	Function
a. Condylar process	Mandible	**Articulates with temporal bone (in TMJ)**
b. **Perpendicular plate**	Ethmoid	Forms superior portion of septum of nose
c. **Sella turcica**	Sphenoid	Site of pituitary gland
d. **Foramen magnum**	Occipital	Largest hole in skull; passageway for spinal cord
e. **Alveolar processes**	(2) **Maxillae and mandible**	Bony sockets for teeth
f. **External auditory meatus**	Temporal	Passageway for sound waves to enter ear
g. Olfactory foramina	Ethmoid	**Openings in cribriform plate for olfactory (smell) nerves**

D5. A, coronal suture. B, lambdoid suture. C, zygomatic arch. D, external auditory meatus. E, condylar process. F, alveolar processes.

D6. (a) Frontal, ethmoid, sphenoid, maxilla. (b) Warm and humidify air since air sinuses are lined with mucous membrane; serve as resonant chambers for speech and other sounds; make the skull lighter in weight. (c) Ethmoid. (d) Like paranasal sinuses, they are covered with mucous membrane, which warms, humidifies, and filters air entering the nose. (e) Vomer; cartilage.

D7. (a) Immovable, fibrous joints between skull bones; between parietal bones. (d) Fontanels; between frontal and parietal bones.

E1. Superior.

E2. It articulates (forms a joint) with no other bones.

E3. (a, b) Cervical (7), thoracic (12), lumbar (5), sacrum (1), coccyx (1). (c) Thoracic and sacral. (d) Atlas (C1) and axis (C2); axis; atlas pivots around the dens when the head is rotated, as if to say, "no." (e) Discs support, absorb shock, and permit movement and flexibility.

E4.　See Figure LG 6.5A.

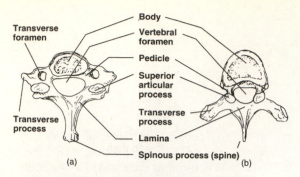

Figure LG 6.5A Typical vertebrae. (a) Cervical vertebra, superior view. (b) Thoracic vertebra, superior view.

E5.　(a) C. (b) T. (c) L. (d) T. (e) S.
E8.　Cardiopulmonary resuscitation (CPR).
E9.　(a) 24, 12. (b) Bodies of thoracic vertebrae. (c) 1, 7, costal; true. (d) False; yes; their costal cartilages attach to the sternum indirectly or not at all. (e) 11, 12; they have no anterior attachment to the sternum. (f) Intercostal muscles, nerves, arteries, and veins.
F1.　(a) Two clavicles and two scapulas. (b) No. (c) Appendicular; clavicles and manubrium of sternum.
F2.　(a) Clavicle. (b) Humerus.
F3.　(b) Ulna, 1. (c) Radius, 1. (d) Carpals, 8. (f) Phalanges, 14.
F4.　(b) Capitulum (H). (c) Olecranon fossa (H). (d) Trochlear notch (U). (e) Radial notch (U). (f) Radial tuberosity (R). (g) Deltoid tuberosity (H). (h) Olecranon process (U). See Figures 6.19–20, page 140–141 in the text.

F5.　(a) Carpals; 8. (b) Metacarpals; 5. (c) 2; 14. (d) Base, shaft, head. (e) Refer to Figure 6.21 page 141 in your text.
F6.　(a) Two hipbones; hipbones plus sacrum and coccyx. (b) Sacrum and coccyx. (c) Sacrum and iliac portion of hipbone (sacroiliac joint).
F7.　(a) Ilium, ischium, pubis; acetabulum. (b) Ilium; crest. (c) Socket for the head of the femur; point of union anteriorly of the two pubic bones; contains and protects pelvic organs, including urinary bladder, rectum, and internal reproductive organs, such as uterus or prostate. (d) Inlet.
F8.　(b) Patella, 1. (c) Tibia, 1. (d) Fibula, 1. (e) Tarsals, 7. (f) Metatarsals, 5. (g) Phalanges, 14.
F10.　(a) Proximal. (b) Lateral. (c) Lateral, fibula. (d) Concave. (e) Lateral condyle of the tibia. (f) Distal. (g) Medial.
F11.　Talus; calcaneus.
F13.　(a) Scapula. (b) Thumb. (c) Clavicle. (d) Calcaneus. (e) Elbow. (f) Patella. (g) Shinbone. (h) Phalanges. (i) Metacarpals. (j) Carpals.
F14.　(a) F. (b) M. (c) F. (d) F. (e) M.
G1.　(b) Reproductive system. (c) Urinary system. (d) Cardiovascular system. (e) Muscular system. (f) Respiratory system. (g) Sensory system: hearing.
G2.　(a) De; weaker; women. (b) Protein; bones are more brittle and vulnerable to fracture.
G3.　(a) De, in; estrogen; blast. (b) Older people; women; short, thin people; whites. (c) Adequate intake of calcium, vitamin D, weight-bearing exercise, and not smoking; estrogen replacement therapy (ERT) may be advised for some women.
G4.　(a) Osteopenia. (b) Osteogenic sarcoma. (c) Osteomalacia. (d) Osteomyelitis.

CRITICAL THINKING: CHAPTER 6

1. One process of bone formation involves development of bone from fibrous connective tissue membranes rather than from cartilage. Tell which bones form by this process, and explain how this might be advantageous to the developing human body.

2. Contrast compact bone with spongy bone according to structure and locations.

3. Discuss probable effects of these hormonal changes upon bone development and health:
   a. removal of both ovaries in a 26-year-old woman
   b. excessive production of parathyroid hormone (PTH) in a 55-year-old-woman

4. Consider the structure and functions of the axis and atlas. How do these bones help you to move your head and neck? What parts of these bones (and related joints) are most vulnerable to injury?

5. A fractured elbow can involve disruption of any/all of three joints. Describe these.

6. Identify functional differences that can be attributed to differences in location and structure of the carpal and tarsal bones.

7. What characteristics of the female pelvis make it more suitable for childbirth than the male pelvis?

8. Which joint is the point (bilaterally) at which the entire upper appendage attaches to the axial skeleton? At which joint (bilaterally) does the entire lower extremity attach to the axial skeleton? Speculate about what additional structures help to prevent the appendages from falling away from the axial skeleton.

MASTERY TEST: ■ CHAPTER 6

Questions 1–5: Circle the letter preceding the one best answer to each question.

1. Choose the one *false* statement.
   A. Osteoclasts are bone-destroying cells.
   B. In a long bone, the primary ossification center is located in the diaphysis, whereas the secondary center of ossification is in the epiphysis.
   C. Canaliculi are tiny canals containing blood that nourishes bone cells in lacunae.
   D. Another name for the greater pelvis is the false pelvis.

2. Choose the one *false* statement.
   A. The scapula bones do not articulate (form joints with) the vertebrae.
   B. The olecranon is a marking on the ulna, and the olecranon fossa is a marking on the humerus.
   C. The fibula articulates with the femur, tibia, talus, and calcaneus.
   D. In anatomical position, the radius is lateral to the ulna.

3. All of the following are bones in the lower extremity *except:*
   A. Talus   D. Ulna
   B. Tibia   E. Fibula
   C. Calcaneus

4. The humerus articulates with all of these bones *except:*
   A. Ulna   C. Clavicle
   B. Radius   D. Scapula

5. All of these bones are included in the axial skeleton *except:*
   A. Rib   D. Hyoid
   B. Sternum   E. Ethmoid
   C. Clavicle

Questions 6–10: Arrange the answers in correct sequence.

_____ _____ _____   6. From superior to inferior:
- A. Atlas
- B. Axis
- C. Occipital bone

_____ _____ _____   7. According to size of the bones, from largest to smallest:
- A. Femur
- B. Ulna
- C. Humerus

_____ _____ _____   8. From proximal to distal:
- A. Phalanges
- B. Metacarpals
- C. Carpals

_____ _____ _____   9. From most superficial to deepest:
- A. Endosteum
- B. Periosteum
- C. Compact bone

_____ _____ _____ 10. From anterior to posterior:
- A. Ethmoid bone
- B. Sphenoid bone
- C. Occipital bone

Questions 11–15: Fill-ins. Write the word or phrase that best completes the statement.

_____ 11. _____ is a term that refers to the shaft of the bone.

_____ 12. _____ is a disorder primarily of older women associated with decreased estrogen level; it is characterized by weakened bones.

_____ 13. _____ is the connective tissue covering over cartilage in adults and also over the embryonic cartilaginous skeleton.

_____ 14. The majority of bones formed by intramembranous ossification are located in the _____ .

_____ 15. The sternum is used for a marrow biopsy because _____ .

Multiple Choice

1. C
2. C
3. D
4. C
5. C

Arrange

6. C A B
7. A C B
8. C B A
9. B C A
10. A B C

Fill-ins

11. Diaphysis
12. Osteoporosis
13. Perichondrium
14. Skull
15. It contains red marrow and it is readily accessible

FRAMEWORK 7
Joints

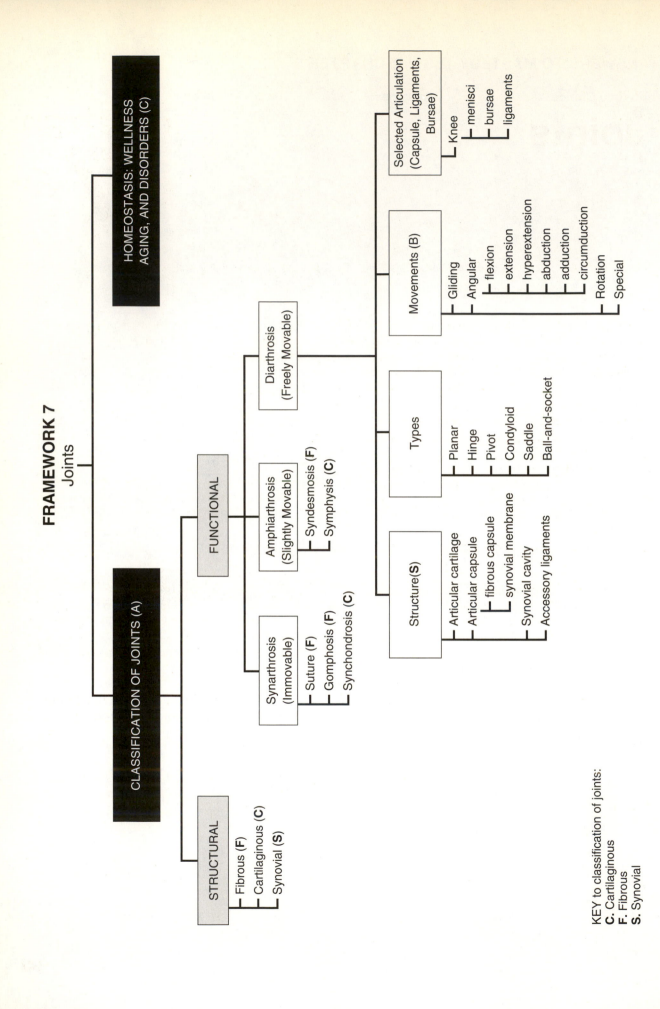

HOMEOSTASIS: WELLNESS AGING, AND DISORDERS (C)

CLASSIFICATION OF JOINTS (A)

STRUCTURAL
- Fibrous (**F**)
- Cartilaginous (**C**)
- Synovial (**S**)

FUNCTIONAL

Synarthrosis (Immovable)
- Suture (**F**)
- Gomphosis (**F**)
- Synchondrosis (**C**)

Amphiarthrosis (Slightly Movable)
- Syndesmosis (**F**)
- Symphysis (**C**)

Diarthrosis (Freely Movable)

Structure(**S**)
- Articular cartilage
- Articular capsule
  - fibrous capsule
  - synovial membrane
- Synovial cavity
- Accessory ligaments

Types
- Planar
- Hinge
- Pivot
- Condyloid
- Saddle
- Ball-and-socket

Movements (B)
- Gliding
- Angular
  - flexion
  - extension
  - hyperextension
  - abduction
  - adduction
  - circumduction
- Rotation
- Special

Selected Articulation (Capsule, Ligaments, Bursae)
- Knee
  - menisci
  - bursae
  - ligaments

KEY to classification of joints:
C. Cartilaginous
F. Fibrous
S. Synovial

Joints

In the last chapter you learned a great deal about the 206 bones in the body. Separated, or disarticulated, these bones would constitute a pile as disorganized as the rubble of a ravaged city deprived of its structural integrity. Fortunately, bones are arranged in precise order and held together in specific conformations—articulations or joints—that permit bones to function effectively. Joints may be classified by function (how movable) or by structure (the type of tissue forming the joint). Synovial joints are most common; their structure, movements, and types will be discussed. A detailed examination of one important synovial joint (the knee) is also included. Joint disorders such as arthritis are usually not life-threatening but plague much of the population, particularly the elderly. An introduction to joint disorders completes this chapter.

As you begin this chapter, carefully examine the Chapter 7 Topic Outline and check off each objective after you meet it. Also note relationships among concepts and key terms in the Framework.

TOPIC OUTLINE AND OBJECTIVES

A. Classifications of joints

☐ 1. Describe how the structure of a joint determines its function.

☐ 2. Describe the structural and functional classes of joints.

☐ 3. Describe the structure and functions of the three types of fibrous joints.

☐ 4. Describe the structure and functions of the two types of cartilaginous joints.

☐ 5. Describe the structure and the six subtypes of synovial joints.

B. Types of movement at synovial (diarthrotic) joints

☐ 6. Describe the types of movements that can occur at synovial joints.

☐ 7. Describe the principal structures and functions of the knee joint.

C. Homeostasis: wellness, aging, common disorders, and medical terminology

☐ 8. Explain the effects of aging on joints.

WORDBYTES

Study each wordbyte, its meaning, and an example of its use in a term. Check your understanding by jotting meanings of wordbytes in margins. Identify other examples of terms that contain these wordbytes as you continue through the text and *Learning Guide*.

Wordbyte	Meaning	Example(s)	Wordbyte	Meaning	Example(s)
amphi-	both	*amphi*arthrotic	-osis	condition of	syndesm*osis*
arthr-	joint	osteo*arthr*itis	ov-	egg	syn*ov*ial fluid
articulat-	joint	*articulat*ion	rheumat-	a watery flow	*rheumat*ologist
cruci-	cross	*cruci*ate	sutur-	seam	coronal *suture*
gompho-	bolt, nail	*gompho*sis	syn-	together	*syn*arthrotic
-itis	inflammation	osteo*arthritis*	syndesmo-	band or ligament	*syndesmo*sis

CHECKPOINTS

A. Classification of joints (pages 157–160)

■ **A1.** Define the term *articulation (joint)*.

List three structural features that affect movement at a joint.

■ **A2.** Name three classes of joints based on structure.

■ **A3.** Fill in the blanks below to name three classes of joints according to the amount of movement they permit.

a. Synarthrosis: _____

b. _____ : slightly movable

c. _____ : freely movable

■ **A4.** Describe synarthrotic (immovable) joints by completing this exercise.

a. Fibrous joints *(have? lack?)* a joint cavity. They are held together by _____ connective tissue.

b. One type of fibrous joint is a _____ found between skull bones. Such joints are *(freely?*

slightly? im-?) movable, or _____-arthrotic.

c. Which of the following is a site of a *gomphosis?*
   A. Joint at distal ends of tibia and fibula
   B. Epiphyseal plate
   C. Attachment of tooth by periodontal ligament to tooth socket in maxilla or mandible

94

d. Synchondroses involve *(hyaline? fibrous?)* cartilage between regions of bone. An example is the

_____ cartilage between diaphysis and epiphysis of a growing bone. This cartilage *(persists through life? is replaced by bone during adult life?)*. Synchondroses are *(somewhat movable? immovable?)*.

■ **A5.** Describe amphiarthrotic joints by completing this exercise.

a. The distal end of the tibia/fibula joint is a fibrous joint. It is *(more? less?)* mobile than a suture and is therefore

_____-arthrotic. It is called a _____ .

b. Fibrocartilage is present in the type of joint known as a _____ . These joints permit some

movement, so are called _____-arthrotic. Two locations of symphyses are

_____ and _____ .

■ **A6.** What structural features of synovial joints make them more freely movable than fibrous or cartilaginous joints?

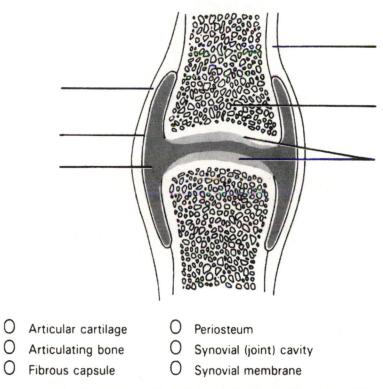

O Articular cartilage	O Periosteum
O Articulating bone	O Synovial (joint) cavity
O Fibrous capsule	O Synovial membrane

Figure LG 7.1   Structure of a generalized synovial joint. Color as directed in Checkpoint A7.

■ **A7.** On Figure LG 7.1, color the indicated structures.

■ **A8.** Select the parts of a synovial joint (listed in the box) that fit the descriptions below.

A. Articular cartilage	F. Fibrous capsule	SF. Synovial Fluid
B. Bursa	L. Ligaments	SM. Synovial membrane

_____ a. Hyaline cartilage that covers ends of articulating bones but does not bind them together.

_____ b. With the consistency of uncooked egg white or oil, it lubricates the joint and nourishes the avascular articular cartilage.

_____ c. Connective tissue membrane that lines synovial cavity and secretes synovial fluid.

_____ d. Parallel fibers in some fibrous capsules; bind bones together.

_____ e. Together these form the articular capsule (two answers).

_____ f. Fluid-filled sac that cushions movements.

■ **A9.** *A clinical challenge. Address these cases involving joint problems.*

a. While playing soccer, Susanna "tore a knee cartilage and damaged her ACL." Dr. Meyer will perform an arthroscopy and possibly a meniscectomy and ACL repair. Explain how these procedures can help Susanna.

b. David and Mark who have both been in excellent health describe intense pain localized in the right shoulder and elbow after spending a weekend on a carpentry project in their new house. Rest and use anti-inflammatory medications help to relieve the pain. Suggest a possible cause of their pain.

B. Types of movements at synovial (diarthrotic) joints   (pages 160–166)

■ **B1.** From the terms listed in the box, choose the one that fits the type of movement in each case. Not all answers will be used.

Abd. Abduction	Ex. Extension	PF. Plantar flexion
Add. Adduction	F. Flexion	Pron. Pronation
C. Circumduction	G. Gliding	Prot. Protraction
D. Dorsiflexion	HE. Hyperextension	RL. Rotation: lateral
El. Elevation	I. Inversion	RM. Rotation: medial

_____ a. Decrease in angle between anterior surfaces of bones (or between posterior surfaces at knee and toe joints)

_____ b. Simplest kind of movement that can occur at a joint; no angular or rotary motion involved

_____ c. State of entire body when it is in anatomical position

_____ d. Movement away from the midline of the body

_____ e. Movement of a bone around its own axis

■ **B2.** Complete the table below to correctly classify types of movements in the box above within the four main categories of movements at synovial joints listed below. Fill in all lines provided. One is done for you.

Main Category	Types of Movements					
a. Angular	<u>Abd</u>	_____	_____	_____	_____	_____
b. Special	_____	_____	_____	_____	_____	_____
c. _____	(Omit)					
d. Rotation(al)	_____	_____				

■ **B3.** Perform the action described. Then identify the type of movement. Select answers from those in the box for Checkpoint B1.

_____ a. Describe a cone with your arm, as if you are winding up to pitch a ball. The movement at your shoulder is called:

_____ b. Stand in anatomical position (palms forward). Now turn your palms backward. This action is called:

_____ c. Move your fingers from fingers-together to fingers-apart position. This action is known as _____ of fingers.

_____ d. Raise your shoulders, as if to shrug them. This movement is called _____ of shoulders.

_____ e. Stand on your toes. This action at the ankle joint is:

_____ f. Grasp a ball in your hand. Your fingers are performing the type of movement called:

_____ g. Sit with the soles of your feet pressed together. In this position, your feet are performing the action called:

_____ h. Thrust your jaw outward (gently!). This action is known as _____ of the mandible.

_____ i. Turn your head from right to left to say "No."

_____ j. Nod your head as if to say "Yes."

■ **B4.** Identify the kinds of movements shown in Figure LG 7.2 (next two pages). Write the name of the movement below each figure. Use the following answers: abduction (ABD), adduction (ADD), extension (E), flexion (F), and hyperextension (HE). (*Note*: In Chapter 8, you will add names of muscles that contract to cause each of these movements.)

■ **B5.** Refer to Figure 7.9 (page 164 of the text) and check your understanding of types of synovial joints by choosing the type of joint that fits the description. (Answers may be used more than once.)

B. Ball-and-socket	H. Hinge	Pla. Planar
C. Condyloid	Piv. Pivot	S. Saddle

_____ a. Sternoclavicular and acromioclavicular joints; permit gliding movements.

_____ b. Examples include atlas–axis joint (movement of which allows you to say "no"), and joint between head of radius and radial notch at proximal end of ulna (movement of which permits pronation and supination.

_____ c. Allows movement in all three planes as a rounded head moves within a cup-like depression.

_____ d. Hip and shoulder joints.

_____ e. Elbow, ankle, and joints between phalanges.

_____ f. Thumb joint located between metacarpal of thumb and carpal bone (trapezium); bones of this joint mimic a rider sitting in a saddle.

■ **B6.** Complete this exercise describing the knee joint. Consult Figure 7.10, page 166 in the text.

a. The tendon of the quadriceps femoris muscle inserts into the tibia by means of the _____ ligament.

b. The _____ ligament extends from the femur to the fibula.

c. The (*anterior? posterior?*) cruciate ligament is torn in about 70 percent of serious knee injuries.

d. Two cartilages named the medial and lateral _____ are shaped much like two C-shaped stadiums that face each other. What is their function?

e. _____ are saclike structures filled with fluid that reduce friction at joints.

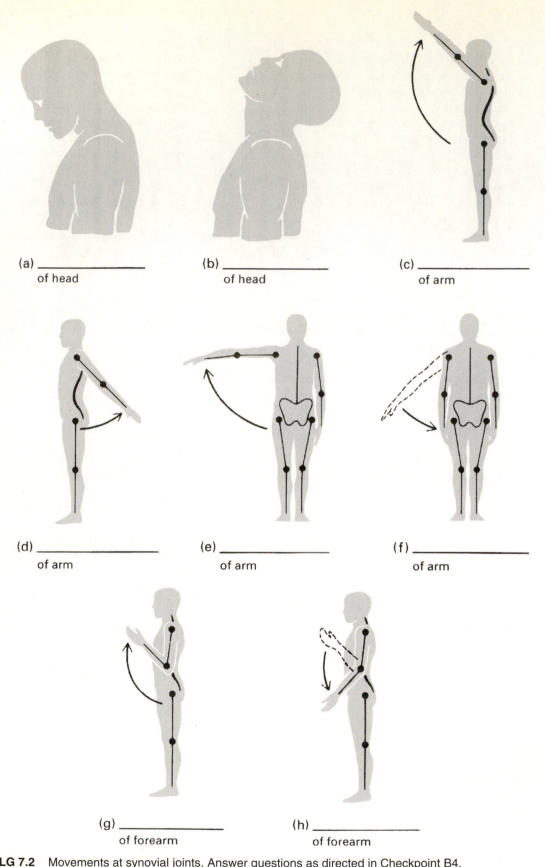

(a) _____
   of head

(b) _____
   of head

(c) _____
   of arm

(d) _____
   of arm

(e) _____
   of arm

(f) _____
   of arm

(g) _____
   of forearm

(h) _____
   of forearm

Figure LG 7.2   Movements at synovial joints. Answer questions as directed in Checkpoint B4.

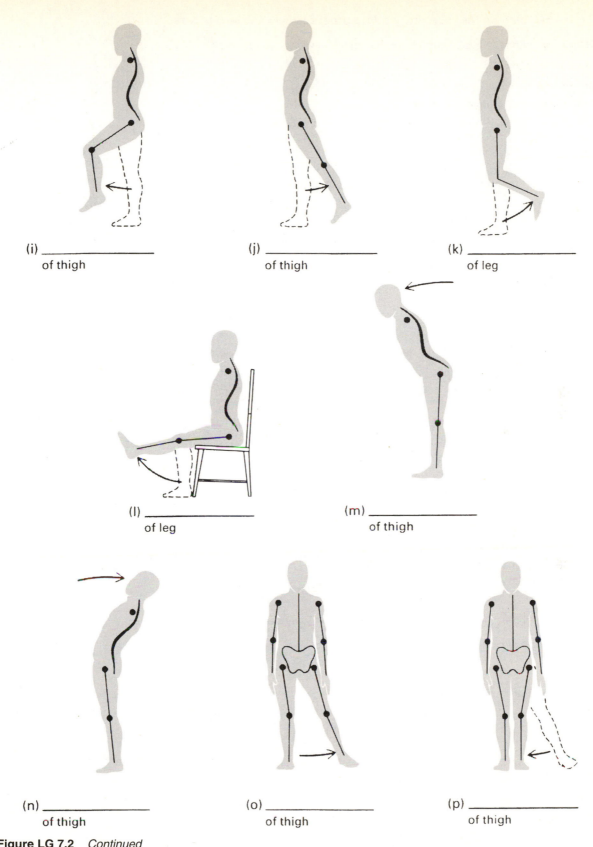

(i) _____
of thigh

(j) _____
of thigh

(k) _____
of leg

(l) _____
of leg

(m) _____
of thigh

(n) _____
of thigh

(o) _____
of thigh

(p) _____
of thigh

Figure LG 7.2 *Continued*

C. Homeostasis: wellness, aging and disorders (pages 165 and 167–168)

■ **C1.** Describe common aging changes of joints in this activity.
  a. *(Increase? Decrease?)* of synovial fluid
  b. *(Thicker? Thinner?)* articular cartilage
  c. *(Osteoarthritis? Rheumatoid arthritis?)*

■ **C2.** *A clinical challenge.* Mrs. Snyderman, age 71, has a diagnosis of osteoarthritis. Explain how this condition may have led to her:
  a. Hunched-over position (kyphosis) and loss of two inches in height

  b. Need for knee replacement surgery on her right knee

■ **C3.** How are *arthritis* and *rheumatism* related? Choose the correct answer.
  a. Arthritis is a form of rheumatism.
  b. Rheumatism is a form of arthritis.

■ **C4.** List several forms of arthritis.

Which is the most common form of arthritis? _____

Name one symptom common to all forms of this ailment. _____

C5. Carpal tunnel syndrome (CTS) is a condition that affects the (wrist? foot?). Write several symptoms of CTS.

This condition is an example of an RMI, an acronym that means _____

_____ . RMI's develop *(quickly? slowly over a period of time?)*.

■ **C6.** State common leisure or occupational activities that can lead to these joint conditions:
  a. Epicondylitis or "mouse elbow"

  b. Rotator cuff injury

  c. Dislocated knee

  d. Rupture of the tibial collateral ligament

  e. Carpal tunnel syndrome

■ **C7.** Complete this exercise describing disorders involving articulations.

   a. *Luxation*, or _____ , is displacement of a bone from its joint with tearing of ligaments. _____ refers to partial dislocation.

   b. _____ is a term that means *joint pain*.

   c. A *(sprain? strain?)* is an overstretching of a muscle. A *(sprain? strain?)* involves more serious injury to joint structures.

ANSWERS TO SELECTED CHECKPOINTS: CHAPTER 7

A1. Point of contact between bones, cartilage and bone, or teeth and bone. Shape of articulating bones, flexibility of connective tissues binding bone, and position of ligaments, tendons, and muscles associated with the joint.

A2. Fibrous, cartilaginous, synovial.

A3. (a) Immovable. (b) Amphiarthrosis. (c) Diarthrosis.

A4. (a) Lack; fibrous. (b) Suture; im-, syn. (c) C. (d) Hyaline; epiphyseal; is replaced by bone during adult life; immovable.

A5. (a) More, amphi; syndesmosis. (b) Symphysis; amphi-; discs between vertebrae, symphysis pubis between hipbones.

A6. The space (synovial cavity) between the articulating bones and the absence of tissue between those bones (which might restrict movement) make the joints more freely movable.

A7.

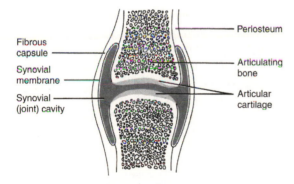

Fibrous capsule

Synovial membrane

Synovial (joint) cavity

Perlosteum

Articulating bone

Articular cartilage

Figure LG 7.1A   Structure of a generalized synovial joint.

A8. (a) A. (b) SF. (c) SM. (d) L. (e) SM and F. (f) B.

A9. (a) Dr. Meyer can view the interior knee joint structure via the pencil-thin lighted arthroscope. The damaged menicus (knee cartilage: Figure 7-10) can be removed and the anterior cruciate ligament (ACL) repaired with this same minimally invasive surgical procedure. (b) Bursitis or tendonitis.

B1. (a) F. (b) G. (c) Ex. (d) Abd. (e) RL and RM.

B2.

Main Category	Types of Movements					
a. Angular	**Abd**	Add	C	Ex	F	
b. Special	D	El	I	PF	Pron	Prot
c. Gliding	(Omit)					
d. Rotation(al)	RL	RM				

B3. (a) C. (b) Pron. (c) Abd. (d) El. (e) PF. (f) F. (g) I. (h) Prot. (i) RM and RL. (j) F (followed by Ex if the head is then lifted back into anatomical position).

B4. (a) F. (b) HE. (c) F. (d) HE. (e) ABD. (f) ADD. (g) F. (h) E. (i) F, while leg also slightly flexed. (j) HE, with leg extended. (k) F, with thigh extended. (l) E, while thigh flexed. (m) F. (n) HE. (o) ABD. (p) ADD.

B5. (a) Pla. (b) Piv. (c) B. (d) B. (e) H. (f) S.

B6. (a) Patellar. (b) Fibular collateral. (c) Anterior. (d) Menisci; increase stability of an otherwise quite unstable joint. (e) Bursae.

C1. (a) Decrease. (b) Thinner. (c) Osteoarthritis

C2. (a) Degenerative changes in bodies of her thoracic vertebrae (and possibly intervertebral discs). (b) Thinning of cartilage in the knee joint.

C3. A.

C4. Rheumatoid arthritis (RA), osteoarthritis (OA), and gout; osteoarthritis; pain.

C5. Wrist; numbness, tingling, and pain in some or all fingers; repetitive motion injury; slowly over a period of time.

C6. (a) Tennis, backhand or computer (keyboard) work. (b) Baseball (pitching), volleyball, tennis or racketball, swimming, violin-playing. (c) Contact sports such as football. (d) Football or knee-twisting falls in skiing. (e) Activities requiring repetitive wrist actions such as assembly line or keyboard work, or long term playing of music instruments (percussion, piano, violin).

C7. (a) Dislocation. (b) Arthralgia. (c) Strain, sprain.

CRITICAL THINKING: CHAPTER 7

1. Contrast structure of fibrous and synovial joints.
2. Contrast structure, mobility, and locations of these types of joints: syndesmosis, synchondrosis, and symphysis.
3. Based on the shape of the articulating bones, the knee joint is quite unstable, Describe structures that provide some stability for this joint.

4. Which joint is usually more stable? Shoulder? Hip? Describe several anatomical factors that account for this difference.

MASTERY TEST: ■ CHAPTER 7

Questions 1–8: Circle T (true) or F (false). If the statement is false, change the underlined word or phrase so that the statement is correct.

T F   1. A fibrous joint is one in which there is <u>no joint cavity and bones are held together by fibrous connective tissue.</u>

T F   2. <u>Sutures, syndesmoses, and symphyses</u> are kinds of fibrous joints.

T F   3. <u>Ball-and-socket, gliding, pivot, and condyloid joints</u> are all diarthrotic joints.

T F   4. Abduction is movement <u>away from</u> the midline of the body.

T F   5. When you touch your toes, the major action you perform at your hip joint is called <u>hyperextension.</u>

T F   6. In synovial joints synovial membranes <u>cover the surfaces of articular cartilages.</u>

T F   7. Bursae are <u>saclike structures that reduce friction</u> at joints.

T F   8. <u>Rheumatoid arthritis</u> is also known as "wear and tear" arthritis since it is likely to occur after decades of pressure on joints.

Questions 9 and 10: Circle the letter preceding the one best answer to each question.

9. All of these structures are associated with the knee joint *except:*
   A. Glenoid labrum
   B. Patellar ligament
   C. Infrapatellar bursa
   D. Medial meniscus
   E. Fibular collateral ligament

10. A suture is found between:
   A. The two pubic bones
   B. The two parietal bones
   C. Radius and ulna
   D. Diaphysis and epiphysis
   E. Tibia and fibula (distal ends)

Questions 11–15: Fill-ins. Write the word or phrase that best fits the description.

_____ 11. Name a repetitive motion injury at the wrist.

_____ 12. _____ is the forcible wrenching or twisting of a joint with partial rupture of it, but without dislocation.

_____ 13. The action of pulling the jaw back from a thrust-out position so that it becomes in line with the maxilla is the movement called _____ .

_____ 14. Another name for a freely movable joint is _____ .

_____ 15. The type of joint between the atlas and axis and also between proximal ends of the radius and ulna is a _____ joint.

True or False

1. T
2. F. Sutures, syndesmoses, and gomphoses
3. T
4. T
5. F. Flexion
6. F. Do not cover surfaces of articular cartilages, which may be visualized as the floor and ceiling of a room; but synovial membranes do line the rest of the inside of the joint cavity (much like wallpaper covering the four walls of the room).
7. T
8. F. Osteoarthritis

Multiple Choice

9. A
10. B

Fill-ins

11. CTS
12. Sprain
13. Retraction
14. Diarthrotic
15. Pivot (or synovial or diarthrotic)

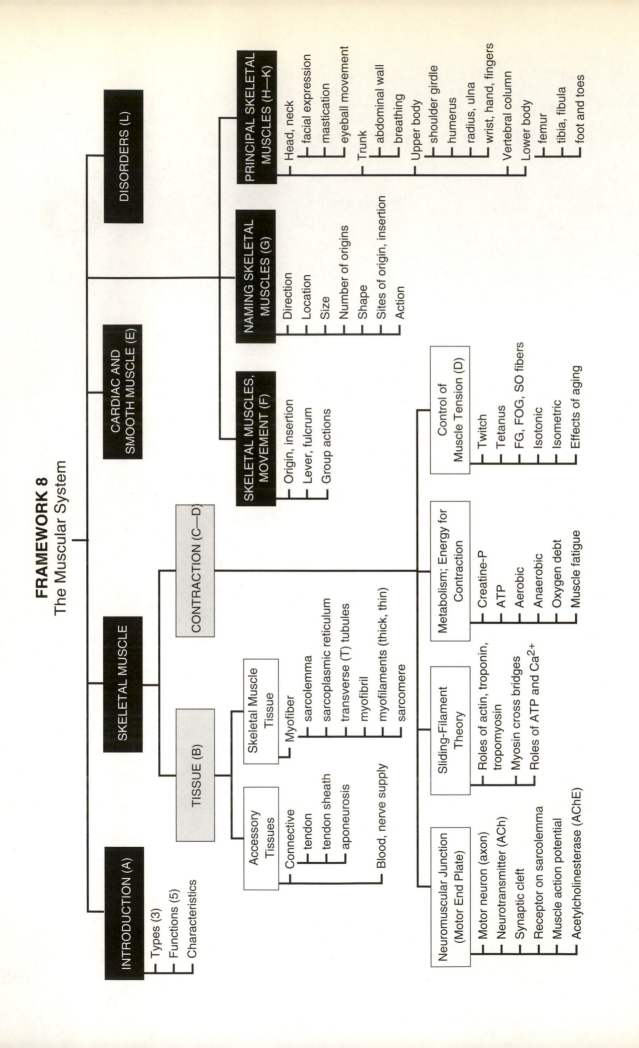

FRAMEWORK 8
The Muscular System

The Muscular System

CHAPTER
8

Muscles are making it possible for you to read this paragraph: facilitating movement of your eyes and head, maintaining your posture, and producing heat to keep you comfortable. Muscles do not work in isolation, however. Connective tissue binds muscle cells into bundles and attaches muscle to bones. Blood delivers the oxygen and nutrients for muscle work. And nerves initiate a cascade of events that culminate in muscle contraction.

A tour of the mechanisms of muscle action finds the visitor caught in the web of muscle protein filaments sliding back and forth as the muscle contracts and relaxes. Neurotransmitters, as well as calcium and power-packing ATP, serve as chief regulators. Muscles are observed contracting in a multitude of modes, from twitch to tetanus and isotonic to isometric.

More than 600 different muscles attach to the bones of the skeleton. Muscles are arranged in groups that work much as a symphony, with certain muscles quiet while others are performing. The results are smooth, harmonious movements rather that erratic, haphazard discord.

Much information can be gained by a careful initial examination of two aspects of each muscle: its name and location. A muscle name may offer such clues as size or shape, direction of fibers, or points of attachment of the muscle. A look at the precise location of the muscle and the joint it crosses—combined with logic—can usually lead to a correct understanding of action(s) of that muscle.

To organize your study of muscle tissue, glance over the Chapter 8 Topic Outline and check off each objective after you meet it. Also glance over the Chapter 8 Framework now. Be sure to refer to the Framework frequently and note relationships among key terms in each section.

TOPIC OUTLINE AND OBJECTIVES

A. Overview of muscular tissue

☐ 1. Describe the types, functions, and characteristics of muscular tissue.

B. Skeletal muscle tissue

☐ 2. Explain the relation of connective tissue components, blood vessels, and nerves to skeletal muscles.

☐ 3. Describe the histology of a skeletal muscle cell.

C. Contraction and relaxation of skeletal muscle; muscle metabolism

☐ 4. Explain how skeletal muscle fibers contract and relax.

☐ 5. Describe the sources of ATP and oxygen for muscle contraction.

☐ 6. Define muscle fatigue and list its possible causes.

D. Control of muscle tension; effects of exercise and aging on muscle

☐ 7. Explain the three phases of a twitch contraction.
☐ 8. Describe how frequency of stimulation and motor unit recruitment affect muscle tension.
☐ 9. Compare the three types of skeletal muscle fibers.
☐ 10. Distinguish between isotonic and isometric contractions.
☐ 11. Describe the effects of exercise and aging on skeletal muscle tissue.

E. Cardiac and smooth muscle tissue

☐ 12. Describe the structure and function of cardiac muscle tissue.
☐ 13. Describe the structure and function of smooth muscle tissue.

F. How skeletal muscles produce movement

☐ 14. Describe how skeletal muscles cooperate to produce movement.

G. Naming skeletal muscles

☐ 15. List and describe the ways that skeletal muscles are named.

H. Principal skeletal muscles that move head and trunk

I. Principal skeletal muscles that move the upper extremity

J. Principal skeletal muscles of the neck and deep back

K. Principal skeletal muscles that move the lower extremity

☐ 16. Describe the location of skeletal muscles in various regions of the body and identify their functions.

L. Focus on homeostasis and common disorders

WORDBYTES

Study each wordbyte, its meaning, and an example of its use in a term. Check your understanding by jotting meanings of wordbytes in margins. Identify other examples of terms that contain these wordbytes as you continue through the text and *Learning Guide*.

Wordbyte	Meaning	Example(s)	Wordbyte	Meaning	Example(s)
a-	not	*a*trophy	-lemma	rind, skin	sarco*lemma*
-algia	pain	fibromy*algia*	maxi-	large	gluteus *maxi*mus
apo-	from	*apo*neurosis	metric	length	iso*metric*
bi-	two	*bi*ceps	mini-	small	gluteus *mini*mus
brachi-	arm	*brachi*alis	myo-	muscle	*myo*fiber, *myo*sin
bucc-	mouth, cheek	*bucc*inator	mys-	muscle	epi*mys*ium
-ceps	head	tri*ceps*	or-	mouth	orbicularis *or*is
-cnem-	leg	gastro*cnem*ius	rect-	straight	*rect*us femoris
delt-	Greek D (Δ)	*delt*oid	sarco-	flesh, muscle	*sarco*lemma
dys-	bad, difficult	muscular *dys*trophy	serra-	toothed, notched	*serra*tus anterior
erg-	work	syn*erg*istic	syn-	together	*syn*ergistic
gastro-	stomach, belly	*gastro*cnemius	teres-	round	*teres* major
glute-	buttock	*glute*us medius	-tonic	tension	iso*tonic*
grac-	slender	*grac*ilis	tri-	three	*tri*ceps femoris
iso-	same	*iso*metric	troph-	nourishment	muscular dys*troph*y
-issimus	the most	lat*issimus* dorsi	vastus-	great	*vastus* lateralis
lat-	broad	*lat*issimus dorsi			

CHECKPOINTS

A. Overview of muscular tissue (page 173)

■ **A1.** Complete the table about the three types of muscle tissue. Return to Figure 4.1.

Muscle Type	Skeletal	Cardiac	Smooth
a. Striated (Yes? No?)			
b. Voluntary (Yes? No?)			No
c. Can regenerate (Yes? No?)			
d. Locations		Wall of the Heart	

■ **A2.** List five functions of muscle tissue that are important for maintenance of homeostasis.

B. Skeletal muscle tissue (pages 173–177)

■ **B1.** Arrange the following terms (connective tissue) in correct sequence according to the amount of muscle surrounded: *endomysium, epimysium, perimysium.*

_____ → _____ → _____

　　(entire muscle)　　　　　(bundle of muscle fibers)　　　(individual muscle fiber)

For extra review. Color these three connective tissues on Figure LG 8.1(a).

■ **B2.** A tendon is a continuation of deep fascia; a tendon connects:

A. Bone to bone　　　　　　B. Muscle to bone

Name the tendon that inserts your calf muscle into your heel bone: _____

■ **B3.** Skeletal muscle tissue *(is? is not?)* vascular tissue. State two or more functions of blood vessels that supply muscles.

■ **B4.** Refer to Figure LG 8.1, and do this exercise about muscle structure.

a. Arrange the following terms in correct order from largest to smallest in size: *filaments (thick or thin), myofibrils, muscle fiber (cell).*

_____ → _____ → _____

　(largest)　　　　　　　　　　　　　　　　　　　　　　　(smallest)

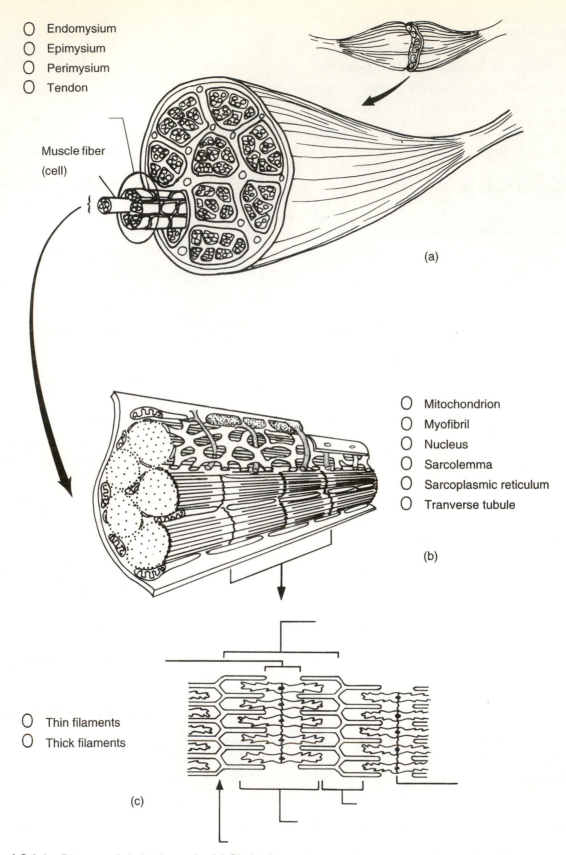

- ◯ Endomysium
- ◯ Epimysium
- ◯ Perimysium
- ◯ Tendon

Muscle fiber
(cell)

(a)

- ◯ Mitochondrion
- ◯ Myofibril
- ◯ Nucleus
- ◯ Sarcolemma
- ◯ Sarcoplasmic reticulum
- ◯ Tranverse tubule

(b)

- ◯ Thin filaments
- ◯ Thick filaments

(c)

Figure LG 8.1   Diagram of skeletal muscle. (a) Skeletal muscle cut to show cross section and longitudinal section with connective tissues. (b) Section of one muscle cell (myofiber). (c) Detail of sarcomere of muscle cell. Color and label as indicated in Checkpoints B1 and B4.

b. Using the leader lines provided, label the following structures:
Figure LG 8.1(b): *sarcomere*
Figure LG 8.1(c): *sarcomere, A band, I band, H zone, Z disc*

c. Sarcoplasmic reticulum (SR) is comparable to *(endoplasmic reticulum? ribosomes? mitochondria?)* in nonmuscle cells. *(Ca^{2+}? K^+?)* stored in SR is released to sarcoplasm as the trigger for muscle contraction. Transverse (T) tubules *(are? are not?)* continuous with the sarcolemma and lie *(parallel? perpendicular?)* to SR.

d. Color all structures indicated by color code ovals in Figure LG 8.1.

■ **B5.** After you study Figure 8.3, page 177 in your text, contrast types of filaments in skeletal muscle fibers.

a. Thin filaments are attached to _____ . These filaments are composed mostly of *(actin? myosin?)* molecules that are twisted into a helix. Thin filaments also contain two other proteins that cover

myosin-binding sites in relaxed muscle. These proteins are named _____ and _____ .

b. Each *(thick? thin?)* filament is composed of many myosin molecules. Each molecule is shaped like two inter-

twined *(footballs? golf clubs?)*. The rounded head of the "club" is called a _____ , and it attaches to *(actin? troponin?)* as muscles begin contraction.

C. Contraction and relaxation of skeletal muscle; muscle metabolism   (pages 177–183)

■ **C1.** Summarize one theory of muscle contraction in this Checkpoint.

a. In order to effect muscle shortening (or contraction), heads (crossbridges) of _____

filaments act like oars pulling on _____ molecules of thin filaments.

b. As a result, *(thick? thin?)* myofilaments are moved toward the center of the sarcomere. Because the thick and thin myofilaments *(shorten? slide?)* to decrease the length of the sarcomere (and move Z discs closer to each other),

this process of muscle contraction is known as the _____ mechanism.

■ **C2.** Do this activity about the nerve supply of skeletal muscles.

a. The combination of a motor neuron plus the muscle fibers (cells) it innervates is called a

_____ . An example of a motor unit that is likely to consist of just two or three muscle fibers (cells) precisely innervated by one neuron is *(external eye muscles? calf [gastrocnemius] muscles controlling walking?)*.

b. In other words, motor neurons form at least two, and possibly thousands, of branches called

_____ , each of which supplies an individual skeletal muscle fiber (cell). When the motor neuron "fires," *(just one muscle supplied by one axon? all muscle fibers within that motor unit?)* will be stimulated to contract.

c. Refer to Figure LG 8.2, in which we zoom in on the portion of a motor unit at which a branch of one neuron stim-

ulates a single muscle fiber (cell). A nerve impulse travels along an axon (at letter _____) toward one muscle

fiber (letter _____). The axon terminal is enlarged at its end into a bulb-shaped synaptic _____ (at letter B).

d. The nerve impulse causes synaptic vesicles (letter _____) to release the neurotransmitter (letter _____)

named _____ into the synaptic cleft (letter _____).

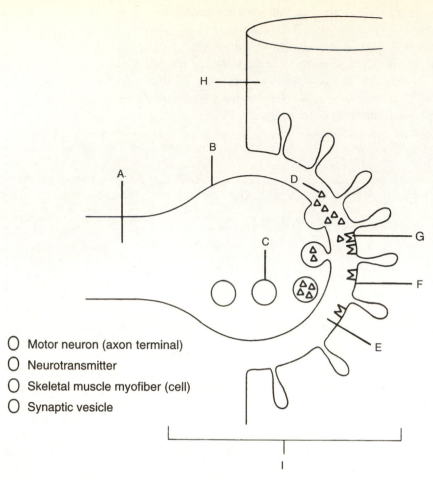

○ Motor neuron (axon terminal)
○ Neurotransmitter
○ Skeletal muscle myofiber (cell)
○ Synaptic vesicle

Figure LG 8.2   Diagram of a neuromuscular junction. Refer to Checkpoints C2 and C3.

e. The region of the muscle fiber membrane (sarcolemma) close to the axon terminal is called a *(motor end plate?*

   neuromuscular junction [NMJ]?) (letter F). This site contains specific _____ (letter G) that recognize and bind to ACh.

f. The effect of ACh is to cause Na$^+$ channels in the sarcolemma to *(open? close?)*, so that Na$^+$ enters the muscle fiber. As a result, a muscle action potential is initiated, leading to *(contraction? relaxation?)* of the muscle fiber.

g. The combination of the axon terminals and the motor end plate is known as _____ (letter I).

C3.   *For extra review.* Color the parts of Figure LG 8.2 indicated by color code ovals.

■ **C4.**   Write IN or DE to indicate whether each of the following chemicals is likely to increase nerve excitation across the neuromuscular junction leading to muscle contraction, or decrease such activity at the NMJ, leading to paralysis.

_____ a. Acetylcholine (ACh)

_____ b. Acetylcholinesterase (AChase)

_____ c. Anticholinesterase (anti-AChase)

_____ d. Curare that blocks ACh receptors on muscle cells

_____ e. The food poisoning toxin of *Clostridium botulinum* (as from improperly canned green beans) or Botox which blocks release of ACh

■ **C5.** Complete this exercise describing the principal events that occur during muscle contraction and relaxation.

a. In Checkpoint C2, we discussed stimulation of a muscle fiber at a neuromuscular junction. The muscle action potential (impulse) spreads from the sarcolemma via *(myosin? T tubules?)* to SR.

b. In a relaxed muscle the concentration of calcium ions (Ca^{2+}) in sarcoplasm is *(high? low?)*. The effect of the

nerve impulse and neurotransmitter is to cause Ca^{2+} to pass from storage areas in the _____
out to the sarcoplasm surrounding filaments.

c. In relaxed muscle, myosin crossbridges are not attached to the actin in thin filaments but

_____ is bound to myosin crossbridges, while the

tropomyosin-_____ complex blocks binding sites on actin.

d. The released calcium ions attach to *(myosin? troponin?)*, causing a structural change that leads to exposure of binding sites on *(myosin? actin?)*.

e. Breakdown of ATP (on myosin crossbridges) occurs via enzyme action. Energy derived from ATP activates

myosin crossbridges to bind to and move _____ . The oarlike action of myosin cross-

bridges (heads of golf clubs) on actin is called a _____ stroke.

f. Repeated power strokes slide actin filaments *(toward? away from?)* the H zone, and so shorten the sarcomere (and entire muscle).

g. Relaxation of a muscle occurs when a synaptic cleft enzyme named _____ destroys ACh. This terminates impulse conduction over the muscle. Calcium ions are then brought back from sarcoplasm into

_____ by *(active transport pumps? diffusion?)*.

h. With a low level of Ca^{2+} now in the sarcoplasm surrounding filaments, _____ reforms from ADP on myosin cross bridges, while tropomyosin–troponin complex once again blocks binding sites on

_____ . As a result, thick and thin filaments detach, slip back into normal position, and the

muscle is said to _____ .

i. After death, a supply of ATP *(is? is not?)* available. Explain why the condition of rigor mortis results.

C6. Explain the following related to muscles.

a. The meaning of *ROM*

b. Why *warmup* should precede stretching

c. How *muscle tone* is maintained

d. The meaning of *flaccid muscles*

e. Factors that may lead to atrophy.

f. Factors that may lead to hypertrophy

111

■ **C7.** Complete this exercise about energy sources for muscle contraction.

a. Breakdown of *(ADP? ATP?)* provides the energy that muscles use for contraction. Recall from Checkpoint C5e that ATP is attached to *(actin? myosin?)* crossbridges and so is available to energize the power stroke. Complete the chemical reaction showing ATP breakdown.

ATP

b. Name one other chemical that can provide energy for more prolonged muscle contraction. _____

c. Complete the reaction in Figure LG 8.3 to show how the creatine phosphate and ADP transport "vehicles" can meet and transfer the high-energy phosphate "trailer" so that more ATP is formed for muscle work.

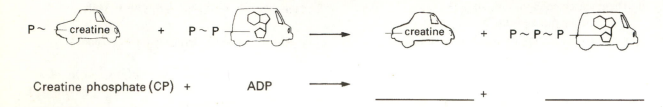

Creatine phosphate (CP) +   ADP   ⟶   _____   +   _____

Figure LG 8.3   High-energy molecules for muscle contraction: the (~P) "trailer" trade-off. Complete figure as indicated in Checkpoint C7c.

d. ADP and creatine phosphate provide only enough energy to power muscle activity for about *(an hour? 15 minutes? 15 seconds?)*. After that, muscles turn first to a series of anaerobic pathways known as

_____ that break glucose into _____ acid which converts to

_____ acid if oxygen is not available. Glycolysis provides enough energy for

about _____ seconds of muscle activity.

e. Exercising for more than several minutes depends largely on *(aerobic? anaerobic?)* pathways.

One substance in muscle that stores oxygen until oxygen is needed by mitochondria

is _____ . This protein is structurally somewhat like the

_____-globoin molecule in blood that also binds to and stores oxygen.

Both of these molecules have a _____ color that accounts for the color of blood and also of red

muscle. The additional oxygen that must be breathed in after active exercise is called _____ .

■ **C8.**   *For extra review.* Complete this Checkpoint to identify metabolic processes used to supply energy for maximal exercise of varying durations. Select answers from those in the box.

| AER. Aerobic cellular respiration | CP. Creatine phosphate |
| ANAER. Anaerobic cellular respiration | |

_____ a. 15–30 sec (100-meter dash)

_____ b. 30–40 sec (300-meter race); does not require O_2 and will lead to lactic acid formation if O_2 is limited

_____ c. Longer than 10 minutes (2 miles to a 26-mile marathon); a series of reactions within mitochondria that do require O_2

C9. List three or more factors that can lead to muscle fatigue.

D. Control of muscle tension; effects of exercise and aging on muscle (pages 183–186; 189)

■ **D1.** Match the terms in the box with definitions below.

C. Contraction period	M. Myogram	T. Twitch
L. Latent period	R. Relaxation period	

_____ a. Rapid, jerky response to a single stimulus

_____ b. Recording of a muscle contraction

_____ c. Period between application of a stimulus and start of a contraction; Ca^{2+} is released from the SR during this time

_____ d. Period when powerstrokes cease

_____ e. Upward tracing as muscle shortens

■ **D2.** Match the types of contraction listed in the box with the related descriptions below.

MT. Muscle tone	FT. Fused tetanus	T. Twitch	UT. Unfused tetanus

_____ a. Sustained partial contraction of some portions of skeletal muscle; some fibers contracted, others not

_____ b. A sustained but wavering contraction due to increased frequency of stimuli

_____ c. Smooth contraction due to rapid stimulation (greater than 90 stimuli/second)

■ **D3.** Contrast three different types of skeletal muscle fibers (or cells) present in your own muscles by completing this exercise. Circle correct answers and fill in answers using muscle fiber types listed in the box.

FG. Fast glycolytic	FOG. Fast oxidative-glycolytic	SO. Slow oxidative

a. Recall (from Checkpoints C7-C8 above) that muscles involved in fairly short bursts of activity, such as 30-second sprints use *(oxidative? glycolytic?)* pathways to generate ATP. It is logical that such muscles

used for short duration, but that fatigue rapidly, will contain many cells known as _____ fibers.

b. For endurance activities, such as maintenance of posture, muscles use *(oxidative? glycolytic?)* pathways, and are

known as _____ fibers. As a result, these cells contain *(few? many?)* mitochondria, which are sites of oxidative metabolism. And because these muscle fibers require much access to oxygen, these cells *(do? do not?)* contain myoglobin, and therefore appear *(red? white?)*.

c. Muscle fibers known as _____ fibers utilize both aerobic and anaerobic metabolism. These are intermediate in resistance to fatigue.

d. Most human skeletal muscles are *(a mixture of all three types? of only one type?)*. However, based on their functions, they are likely to have a preponderance of only one or two types.

■ **D4.** Contrast isometric and isotonic contractions by doing this exercise.

a. A contraction in which a muscle shortens while tension (tone) of the muscle remains constant is known as an *(isometric? isotonic?)* contraction.

113

b. In an isometric contraction the muscle length *(shortens? stays about the same?)* and tension of the muscle *(increases? stays the same?)*.

D5. Describe the effects of the following factors on muscles:

a. The value of anabolic steroids for athletes

b. Normal aging in a person who is a "couch potato"

c. Normal aging in a person who exercises several times a week

E. Cardiac and smooth muscle tissue   (pages 186–187)

E1. Compare the structure and function of skeletal and cardiac muscle by completing this table. State the significance of characteristics that have asterisks (*).

Characteristic	Skeletal Muscle	Cardiac Muscle
a. Number of nuclei per myofiber	Several (multinucleate)	
b. Number of mitochondria per myofiber (More or Fewer)*		More, because heart muscle requires constant generation of energy
c. Striated appearance due to alternated actin and myosin filaments (Yes or No)	Yes	
d. Arrangement of muscle fibers (Parallel or Branching)		
e. Nerve stimulation required for contraction (Yes or No)*		No, so heart can contract without nerve stimulation, but nerves or chemicals can alter heart rate

■ **E2.** Define *autorhythmicity* of cardiac muscle.

At rest the heart normally beats at a steady _____ beats/min, provided the heart has a constant supply of _____

and _____ . The heart *(can? cannot?)* use lactic acid as a source of ATP.

■ **E3.** Contrast smooth muscle with the muscle tissue you have already studied by circling or filling in the correct answers in this paragraph.

a. Smooth muscle fibers are *(cylinder-shaped? tapered at each end?)* with *(several nuclei? one nucleus?)* per cell.

114

b. They do contain actin and myosin as well as _____ filaments. However, due to the irregular arrangement of these filaments, smooth muscle tissue appears *(striated? nonstriated or "smooth"?)*.

Smooth muscle *(has? lacks?)* sarcomere structure. Intermediate fibers stretch between _____ bodies which are similar to *(H zones? Z disks?)* of skeletal muscle.

c. In general, smooth muscle holds a contraction for a *(shorter? longer?)* period of time than skeletal muscle does.

E4. Compare the two types of smooth muscle.

Muscle Type	Structure	Spread of Stimulus	Locations
a. Visceral		Impulse spreads and causes contraction of adjacent fibers	
b.			Blood vessels, iris of eye

■ **E5.** Smooth muscle *(is? is not?)* normally under voluntary control. List three chemicals released in the body or other factors that can also lead to smooth muscle contraction or relaxation.

F. How skeletal muscles produce movement (pages 188–189)

■ **F1.** Refer to Figure 8.12, page 189 in your text and consider flexion of your own forearm as you do this learning activity.

a. In Figure 8.12 of your text identify the exact point at which the muscle causing flexion attaches to the forearm. It is the *(proximal? distal?)* end of the *(humerus? radius? ulna?)*. This indicates that this is the site where the muscle exerts its effort, and it is the insertion end of the muscle. (More about insertions in a minute.)

b. Each skeletal muscle is attached to at least two bones. As the muscle shortens, one bone stays in place and so is called the *(origin? insertion?)* end of the muscle. What bone in the figure appears to serve as the origin bone?

_____ .

■ **F2.** Refer again to Figure 8.12, page 189 in the text and do this exercise about how muscles of the body work in groups.

a. The muscle that contracts to cause flexion of the forearm is called a _____ . An example

of a prime mover in this action would be the _____ muscle.

b. The triceps brachii must relax as the biceps brachii flexes the forearm. The triceps is an extensor. Because its action is opposite to that of the biceps, the triceps is called *(a synergist? an agonist? an antagonist?)* of the biceps.

c. What would happen if the flexors of your forearm were functional, but not the antagonistic extensors?

d. What action would occur if both the flexors and extensors contracted simultaneously?

e. Muscles that assist or cooperate with the prime mover to cause a given action are known as

_____ , whereas muscles that stabilize a bone (such as the scapula) so that prime movers

and synergists can move another bone (such as the humerus) are called _____ .

G. Naming skeletal muscles   (pages 189–191)

■ **G1.**   Review the Wordbyte section above. Match the names of the following muscles with their meanings.

A. Large muscle of the buttock region	D. The broadest muscle of the back
B. Belly-shaped muscle in leg	E. Large muscle in medial thigh area
C. Thigh muscle with four origins	F. Muscle that raises the upper eyelid

_____ a. Latissimus dorsi   _____ d. Quadriceps femoris

_____ b. Vastus medialis   _____ e. Gastrocnemius

_____ c. Gluteus maximus   _____ f. Levator palpebrae superioris

■ **G2.**   As you study the names of muscles, you will find that most of them provide a good description of the muscle as indicated in Table 8.2, page 191 of the text. For each of the following, indicate the type of clue that each part of the name gives. The first one is done for you.

A. Action	N. Number of origins
D. Direction of fibers	P. Points of attachment of origin and insertion
L. Location	S. size or shape

__DL__ a. Rectus abdominus   _____ d. Sternocleidomastoid

_____ b. Flexor carpi ulnaris   _____ e. Adductor longus

_____ c. Biceps brachii

H.   Principal skeletal muscles that move head and trunk   (pages 194–197)

■ **H1.**   After studying Exhibit 8.1 (page 194) in your text, check your understanding of the muscles of facial expression. Write the name of the muscle that answers each description. Locate muscles in Figure LG 8.4(a) O–R. Cover the key and write the name of each facial muscle next to its lettered leader line.

a. Allows you to show surprise by raising your eyebrows and forming horizontal forehead wrinkle:

_____

b. Muscle surrounding opening of your mouth; allows you to use your lips in kissing and in speech:

_____

c. Muscle for smiling and laughing since it draws the outer portion of the mouth upward and outward:

_____

d. Circular muscle around eye; closes eye: _____

■ **H2.**   Place your index finger and thumb on the origin and insertion of each of the muscles that move your lower jaw. (Refer to Exhibit 8.2 and Figure 8.14, page 195 in the text, for help.) Then do this learning activity.

a. Two large muscles help you to close your mouth forcefully, as in chewing. Both of these act by (*lowering the maxilla? elevating the mandible?*). The _____ covers your temple and the

_____ covers the ramus of the mandible.

b. Refer to Figure LG 8.4(a). Muscles A and B are used primarily for (*facial expressions? chewing?*), whereas muscles O, P, Q, and R are used mainly for (*facial expressions? chewing?*).

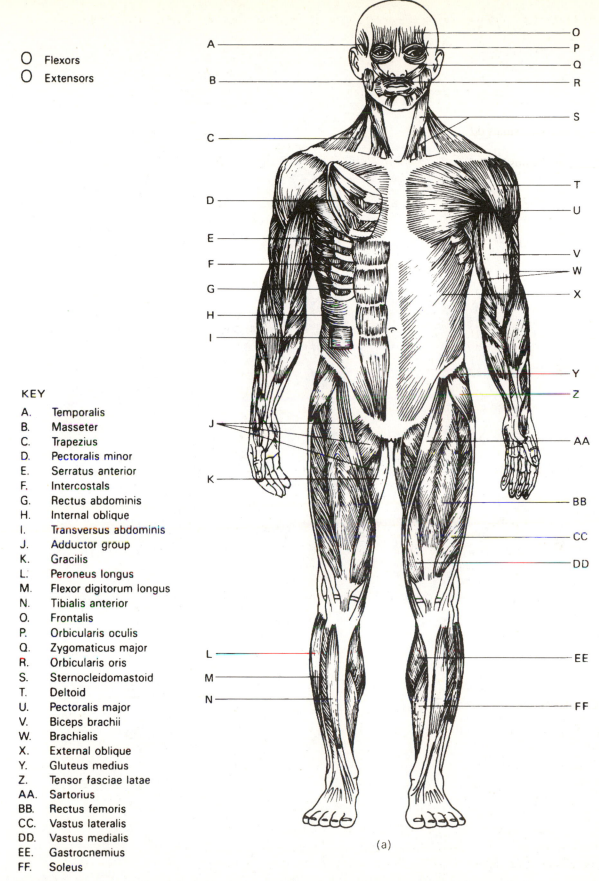

O Flexors
O Extensors

KEY

A. Temporalis
B. Masseter
C. Trapezius
D. Pectoralis minor
E. Serratus anterior
F. Intercostals
G. Rectus abdominis
H. Internal oblique
I. Transversus abdominis
J. Adductor group
K. Gracilis
L. Peroneus longus
M. Flexor digitorum longus
N. Tibialis anterior
O. Frontalis
P. Orbicularis oculis
Q. Zygomaticus major
R. Orbicularis oris
S. Sternocleidomastoid
T. Deltoid
U. Pectoralis major
V. Biceps brachii
W. Brachialis
X. External oblique
Y. Gluteus medius
Z. Tensor fasciae latae
AA. Sartorius
BB. Rectus femoris
CC. Vastus lateralis
DD. Vastus medialis
EE. Gastrocnemius
FF. Soleus

(a)

Figure LG 8.4 Major muscles of the body. Some deep muscles are shown on the left side of the figure. All muscles on the right side are superficial. Label and color as directed. (a) Anterior view. (b) Posterior view.

O Flexors
O Extensors

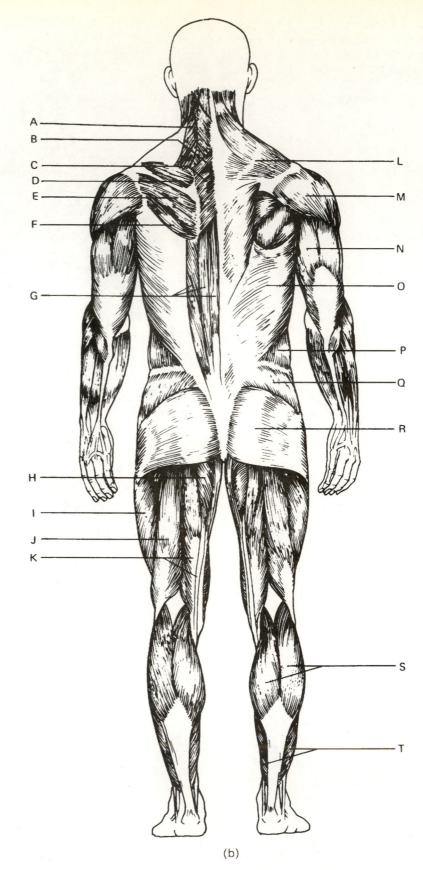

(b)

KEY

A. Levator scapulae
B. Rhomboideus minor
C. Rhomboideus major
D. Supraspinatus
E. Infraspinatus
F. Teres major
G. Erector spinae
H. Adductors
I. Vastus lateralis
J. Biceps femoris
K. Semimembranosus and
   semitendinosus
L. Trapezius
M. Deltoid
N. Triceps brachii
O. Latissimus dorsi
P. External oblique
Q. Gluteus medius
R. Gluteus maximus
S. Gastrocnemius
T. Soleus

Figure LG 8.4   *Continued*

■ **H3.** To review actions of eye muscles (Exhibit 8.3, page 197), work with a study partner. One person moves the eyes in a particular direction; the partner then names the eye muscles used for that action. *Note:* Will both eyes use muscles of the same name? For example, as you look to your right, will the lateral rectus muscles attached to both eyes contract?

■ **H4.** Each half of the abdominal wall is composed of *(two? three? four?)* muscles. Describe these in the exercise below.

a. Just lateral to the midline is the rectus abdominis muscle. Its fibers are *(vertical? horizontal?),* attached inferiorly

to the _____ and superiorly to the _____ . Contraction of this muscle permits *(flexion? extension?)* of the vertebral column.

b. List the remaining abdominal muscles that form the sides of the abdominal wall from most superficial to deepest.

_____ → _____ → _____
       most superficial   deepest

c. Do all three of these muscles have fibers running in the same direction? *(Yes? No?)* Of what advantage is this?

■ **H5.** Answer these questions about muscles used for breathing.

a. The diaphragm is _____-shaped. Its oval origin is located _____ . Its insertion is not into bone, but rather into dense connective tissue forming the roof of the diaphragm; this tissue

is called the _____ .

b. Contraction of the diaphragm flattens the dome, causing the size of the thorax to *(increase? decrease?),* as occurs during *(inspiration? expiration?).*

c. The name *intercostals* indicates that these muscles are located _____ . Which set is used during expiration? *(Internal? External?)*

H6. *For extra review.* Cover the key to Figure LG 8.4(a) and write labels for muscles F, G, H, I, and X.

I. Principal skeletal muscles that move the upper extremity   (pages 198–209)

■ **I1.** From the list in the box, select the names of muscles that move the shoulder girdle as indicated.

LS. Levator scapulae	SA. Serratus anterior
PM. Pectoralis minor	T. Trapezius
RM. Rhomboideus major	

_____ a. Superiorly (elevation)   _____ c. Toward vertebrae (adduction)

_____ b. Inferiorly (depression)   _____ d. Away from vertebrae (abduction)

■ **I2.** Note on Figure LG 8.4(a) and (b) that the only two muscles listed in Checkpoint I1 that are superficial are the

_____ and a small portion of the _____ . The others are all deep muscles.

■ **I3.** On Figure LG 8.4(a) and (b) identify pectoralis major, deltoid, and latissimus dorsi muscles. All three of these muscles are *(superficial? deep?).* They are all directly involved with movement of the *(shoulder girdle? humerus?*

radius/ulna?). Identify two muscles on Figure LG 8.4(b) that are parts of the rotator cuff: _____ _____

119

■ **14.** Write next to each muscle name listed below the letters of *all points of origin and insertion* and *all actions* that apply. Place one answer on each line provided.

Points of Origin or Insertion	Actions (of Humerus)
C. Clavicle	Ab. Abducts
H. Humerus	Ad. Adducts
I. Ilium	EH. Extension, hyperextension
RC. Ribs or costal cartilages	F. Flexion
Sc. Scapula	
St. Sternum	
VS. Vertebrae and sacrum	

Muscles   **Origins and Insertions**   **Actions (of Humerus)**

a. Pectoralis major   _____ _____ _____ _____   _____ _____

b. Deltoid   _____ _____ _____   _____ _____ _____

c. Latissimus dorsi   _____ _____ _____ _____   _____ _____

■ **I5.** Combine your knowledge of muscle actions with your knowledge of movements at joints from Chapter 7. Return to Figure LG 7.2, page 98 of the guide. Write the name(s) of one or two muscles that produce each of the actions (a–h). Write muscle names next to each figure.

■ **I6.** Complete the table describing three muscles that move the forearm.

Muscle Name	Origin	Insertion	Action on Forearm
a.		Radius (anterior)	
b. Brachialis			
c.			Extension

■ **I7.** Complete this exercise about muscles that move the wrist and fingers.

a. Examine your own forearm, palm, and fingers. There is more muscle mass on the *(anterior? posterior?)* surface. You therefore have more muscles that can *(flex? extend?)* your wrist and fingers.

b. Locate the flexor carpi ulnaris muscle on Figure 8.21, page 209 in your text. What action does it have other than

flexion of the wrist? _____ What muscles would you expect to abduct the wrist?

_____

■ **I8.** *For extra review* of muscles that move the upper extremities, write the name of one or more muscles that fit these descriptions.

a. Controls action at the elbow for a movement such as the downstroke in hammering a nail:

_____

b. Turns your hand from palm down to palm up position: _____

c. Originates from upper eight or nine ribs; inserts on scapula; moves scapula laterally:

_____

d. Used when a baseball is grasped: _____

e. Antagonist to serratus anterior: _____

120

f. Largest muscle of the chest region; used to throw a ball in the air (flex humerus) and to adduct arm:

_____

g. Raises or lowers scapula, depending on which portion of the muscle contracts:

_____

J. Principal skeletal muscles of the neck and deep back (pages 210–211; 192-193)

■ **J1.** Using a mirror, find the origin and insertion of your left sternocleidomastoid muscle. (See Figure 8.13(a), page 192 in your text.) The muscle contracts when you pull your chin down and to the right; this diagonal muscle of your neck will then be readily located. Note that the left sternocleidomastoid pulls your face toward the *(same? opposite?)* side. It also *(flexes? extends?)* the head.

■ **J2.** Describe the muscles that make up the sacrospinalis (erector spinae). Locate and label on Figure LG 8.4(b).

a. These muscles have attachments between _____ .

b. They are *(flexors? extensors?)* of the vertebral column, and so are *(synergists? antagonists?)* of the rectus abdominis muscles.

K. Principal skeletal muscles that move the lower extremity (pages 212–217)

K1. Cover the keys in Figure LG 8.4(a) and (b) and identify by size, shape, and location the major muscles that move the lower extremity.

■ **K2.** Now match muscle names in the box with their descriptions below.

Ad. Adductor group	GMax. Gluteus maximus	QF. Quadriceps femoris
Gas. Gastrocnemius	Ham. Hamstrings	Sar. Sartorius

_____ a. Consists of four heads: rectus femoris and three vastus muscles (lateralis, medialis, and intermedius).

_____ b. This muscle mass lies in the posterior (flexor) compartment of the thigh; antagonist to quadriceps femoris.

_____ c. Large muscle mass of the buttocks; antagonist to the iliopsoas.

_____ d. The only one of these muscles located in the leg (between knee and ankle), it forms the "calf." Attaches to calcaneus by "Achilles tendon."

_____ e. Forms the medial compartment of the thigh; moves femur medially.

_____ f. Crossing the femur obliquely, it moves lower extremity into "tailor position."

■ **K3.** Complete the table by marking an X below each action produced by contraction of the muscles listed. (Some muscles will have two answers.)
Key to actions in table: Ab, Abduct; Ad, adduct; EH, extend or hyperextend; F, flex.

	Movements of Thigh (Hip Joint)				Movement of Leg (Knee)	
	Ab	Ad	EH	F	EH	F
a. Iliopsoas						
b. Gluteus maximus and medius						
c. Adductor mass						
d. Tensor fasciae latae						
e. Quadriceps femoris						
f. Hamstrings						

■ **K4.** What muscles cause the actions (i)–(p) shown in Figure LG 7.2 (page 99)? Write the muscle names next to each diagram.

■ **K5.** Perform these actions of your feet and toes. Feel which muscles are contracting. Then match names of actions with descriptions.

DF. Dorsiflex	Ex. Extend toes	In. Invert foot
Ev. Evert foot	F. Flex toes	PF. Plantar flex

_____ a. Jump, as if to touch ceiling

_____ b. Walk around on your heels

_____ c. Curl toes down

_____ d. Press soles of feet medially against each other

■ **K6.** To review details of leg muscles that move the foot, complete this table. Note that muscles within a compartment tend to have similar functions. Use the same key for foot and toe actions as in the box for Checkpoint K5.

	DF PF	In Ev	F Ex
Posterior compartment: a. Gastrocnemius and soleus b. Tibialis posterior c. Flexor digitorum longus			
Lateral compartment: d. Peroneus (longus and brevis)			
Anterior compartment: e. Tibialis anterior f. Extensor digitorum longus			

■ **K7.** *For extra review* of all muscles, color flexors and extensors using color code ovals on Figure LG 8.4(a) and (b). This activity will allow you to see on which sides of the body most muscles with those actions are located. Note that some muscles are flexors *and* extensors—at different joints. (See Mastery Test question 13 also.)

L. Focus on homeostasis and common disorders (pages 218–220)

L1. Describe the relationship between the muscular system and the following body systems:

a. Digestive

b. Urinary

c. Reproductive

d. Nervous

L2. Describe the following disorders by naming involved muscles and identifying the problem. (Hint: refer to exhibits in the text.)

a. Bell's palsy (Exhibit 8.1, page 194).

b. Impingement syndrome (Exhibit 8.7, page 204).

c. Pulled groin (Exhibit 8.11, page 212).

d. Shinsplints (Exhibit 8.13, page 216).

■ **L3.** Describe myasthenia gravis in this exercise.

a. In order for skeletal muscle to contract, a nerve must release the chemical _____ at the

neuromuscular junction. Normally, ACh binds to _____ on the muscle fiber membrane.

b. It is believed that a person with myasthenia gravis produces _____ that bind to these receptors, making them unavailable for ACh binding. Therefore ACh *(can? cannot?)* stimulate the muscle, which is weakened.

L4. Define each of these types of abnormal muscle contractions.

a. Spasm

b. Fibrillation

c. Tic

ANSWERS TO SELECTED CHECKPOINTS: CHAPTER 8

A1.

Muscle Type	Skeletal	Cardiac	Smooth
a. Striated *(Yes? No?)*	Yes	Yes	No
b. Voluntary *(Yes? No?)*	Yes	No	**No**
c. Can regenerate *(Yes? No?)*	Very limited	No	Yes
d. Locations	Attached to skeleton	**Heart**	Blood vessels Airways Digestive organs Urine pathways Uterus

A2. Motion or movement; stabilizing body positions, as in maintaining posture and regulating organ volume, exemplified by contraction of the heart to pump out blood; generation of heat to alter body temperature; movement of substances within the body.

B1. Epimysium $\rightarrow$ perimysium $\rightarrow$ endomysium. See Figure LG 8.1A.

B2. B; Achilles (or calcaneal) tendon.

B3. Is; (1) provide nutrients and oxygen for generation of ATP, (2) remove wastes.

B4. (a) Muscle fiber (cell or myofiber) $\rightarrow$ myofibrils $\rightarrow$ filaments (thick or thin). (b) See Figure LG 8.1A. (c) Endoplasmic reticulum; Ca^{2+}; are, perpendicular. (d) See Figure LG 8.1A.

B5. (a) Z discs; actin; tropomyosin and troponin. (b) Thick; golf clubs; crossbridge, actin.

C1. (a) Myosin, actin. (b) Thin; slide, sliding filament.

123

C2. (a) Motor unit; external eye muscles. (b) Axons; all muscle fibers within that motor unit. (c) A, H; terminal. (d) C, D, acetylcholine (ACh), E. (e) Motor end plate; receptors. (f) Open; contraction. (g) Neuromuscular junction (NMJ).

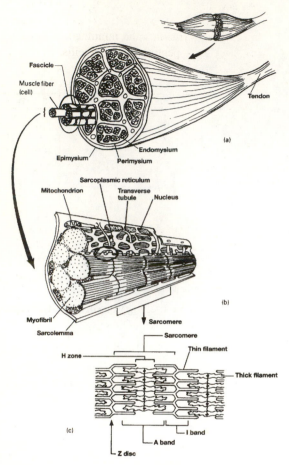

Fascicle
Muscle fiber (cell)
Tendon
Epimysium
Endomysium
Perimysium
(a)

Sarcoplasmic reticulum
Mitochondrion
Transverse tubule
Nucleus
(b)
Myofibril
Sarcolemma
Sarcomere

Sarcomere
Thin filament
H zone
Thick filament
(c)
I band
A band
Z disc

Figure LG 8.1A Diagram of skeletal muscle. (a) Skeletal muscle cut to show cross section and longitudinal section with connective tissues. (b) Section of one muscle cell (myofiber). (c) Detail of sarcomere of muscle cell.

C4. (a) IN. (b) DE. (c) IN. (d) DE. (e) DE.
C5. (a) T tubules. (b) Low; sarcoplasmic reticulum. (c) ATP, troponin. (d) Troponin, actin. (e) Actin; power. (f) Toward or even across; (g) Acetylcholinesterase (AChE); SR, active transport pumps. (h) ATP, actin (or thin filament); relax (or lengthen). (i) Is not; Myosin crossbridges stay attached to actin and the muscles remain in a state of partial contraction (rigor mortis) for 1–2 days.
C7. (a) ATP; myosin; ATP → ADP + P + energy. (b) Creatine phosphate (CP). (c) CP + ADP → creatine + ATP. (d) 15 seconds; glycolysis, pyruvic, lactic; 30–40. (e) Aerobic; myoglobin; hemo; red; oxygen debt or recovery oxygen uptake.
C8. (a) CP. (b) ANAER. (c) AER.

C9. Decrease in any of the following: Ca^{2+} in sarcoplasm, creatine phosphate, O_2, or glycogen; increase in either lactic acid or ADP.
D1. (a) T. (b) M. (c) L. (d) R. (e) C.
D2. (a) MT. (b) UT. (c) FT.
D3. (a) Glycolytic; FG. (b) Oxidative, SO; many; do, red. (c) FOG. (d) A mixture of all three types.
D4. (a) Isotonic. (b) Stays about the same, increases.
E2. Ability to contract without external stimulation (because of the heart's own intrinsic pacemaker); 60–100, oxygen and nutrients; can.
E3. (a) Tapered at each end, one nucleus. (b) Intermediate; nonstriated or "smooth"; lacks; dense, Z disks. (c) Longer.
E5. Is not; hormones, pH or temperature changes, O_2 and CO_2 levels, certain ions.
F1. (a) Proximal, radius. (b) Origin; scapula.
F2. (a) Prime mover (agonist); biceps brachii. (b) An antagonist. (c) Your forearm would stay in the flexed position. (d) None: each opposing muscle would negate the action of the other. (e) Synergists, fixators.
G1. (a) D. (b) E. (c) A. (d) C. (e) B. (f) F.
G2. (b) A, P, L. (c) N, L. (d) P. (e) A, S.
H1. (a) Frontalis. (b) Orbicularis oris. (c) Zygomaticus major. (d) Orbicularis oculi.
H2. (a) Elevating the mandible; temporalis, masseter. (b) Chewing, facial expressions.
H3. No. The left eye contracts its medial rectus, while the right eye uses its lateral rectus and exerts some tension on both oblique muscles.
H4. Four. (a) Vertical; pubis and symphysis pubis; ribs 5 to 7 and sternum; flexion. (b) External oblique → internal oblique → transversus abdominis. (c) No; strength is provided by the three different directions.
H5. (a) Dome; around the bottom of the rib cage and on lumbar vertebrae; central tendon. (b) Increase, inspiration. (c) Between ribs (costa); internal.
I1. (a) LS, RM, upper fibers of T. (b) PM, lower fibers of T. (c) RM, T. (d) SA.
I2. Trapezius, serratus anterior.
I3. Superficial; humerus; D E.
I4. (a) C, St, RC, H; Ad, F. (b) C, H, Sc; Ab, EH (posterior fibers), F (anterior fibers). (c) I, RC, VS, H; Ad, EH.
I5. (a) Sternocleidomastoid. (b) Erector spinae. (c) Pectoralis major, deltoid (anterior fibers), biceps brachii. (d) Latissimus dorsi, teres major, deltoid (posterior fibers), triceps brachii. (e) Deltoid and supraspinatus. (f) Pectoralis major, latissimus dorsi, teres major, infraspinatus. (g) Biceps brachii, brachialis, brachioradialis. (h) Triceps brachii.

I6.

Muscle Name	Origin	Insertion	Action on Forearm
a. Biceps brachii	Scapula (2 sites)	Radius (anterior)	Flexion, supination (*Note*: also flexes humerus)
b. Brachialis	Anterior humerus	Ulna	Flexion
c. Triceps brachii	Scapula and 2 sites on posterior of humerus	Posterior of ulna (olecranon)	Extension (also extension of humerus)

I7. (a) Anterior; flex. (b) Adducts wrist; those lying over radius, such as flexor and extensor carpi radialis.

I8. (a) Triceps brachii. (b) Supinator and biceps brachii. (c) Serratus anterior. (d) Flexor digitorum superficialis and profundus. (e) Trapezius and rhomboids. (f) Pectoralis major. (g) Trapezius.

J1. Opposite; flexes.

J2. (a) Hipbone (ilium), ribs, and vertebrae. (b) Extensors, antagonists.

K2. (a) QF. (b) Ham. (c) GMax. (d) Gas. (e) Ad. (f) Sar.

K3.

	Movements of Thigh (Hip Joint)				Movement of Leg (Knee)	
	Ab	Ad	EH	F	EH	F
a. Iliopsoas				X		
b. Gluteus maximus and medius	X		X			
c. Adductor mass		X		X		
d. Tensor fasciae latae	X			X		
e. Quadriceps femoris				X	X	
f. Hamstrings			X			X

K4. (i) Iliacus + psoas (iliopsoas), rectus femoris, adductors, sartorius; (j) Gluteus maximus, hamstrings, adductor magnus (posterior portion); (k) Hamstrings, gracilis, sartorius, gastrocnemius; (l) Quadriceps femoris; (m) Same as (i), but bilateral and all four heads of quadriceps femoris; (n) Same as (j) but bilateral; (o) Tensor fasciae latae, gluteus (medius and minimus), piriformis; (p) Adductors, pectineus, and gracilis.

K5. (a) PF. (b) DF. (c) F. (d) In.

K6.

	DF	PF	In	Ev	F	Ex
Posterior compartment:						
a. Gastrocnemius and soleus		X				
b. Tibialis posterior		X	X			
c. Flexor digitorum longus		X	X		X	
Lateral compartment:						
d. Peroneus (longus and brevis)		X		X		
Anterior compartment:						
e. Tibialis anterior	X		X			
f. Extensor digitorum longus	X			X		X

K7. Figure LG 8.4(a): flexors: G, J, K, S, T (anterior portion), U, V, W, AA, BB, EE; extensors: BB, CC, DD. Figure LG 8.4(b): flexors: H, J, K, S; extensors: F, G, H, I, J, K, L, M (posterior fibers), N, O, R.

L3. (a) Acetylcholine (ACh); receptors. (b) Antibodies; cannot.

CRITICAL THINKING: CHAPTER 8

1. Explain why muscle contraction is described as the sliding filament mechanism.
2. Explain how the microscopic structure of cardiac muscle fibers accounts for its continuous, rhythmic activity.
3. Write a rationale for why endurance and strength training programs can be effective for maintaining healthy musculature in elderly persons.
4. Explain the kinds of information conveyed about muscles by examining muscle names. Give at least six examples of muscle names that provide clues to muscle locations, actions, size or shape, direction of fibers, numbers of origins, or points of attachment.
5. Identify types of movements required at shoulder, elbow, and wrist as you perform the action of tossing a tennis ball upward. Indicate groups of muscles that must contract and relax to accomplish this action.
6. Discuss the value of the following factors in increasing athletic performance: creatine supplements; stretching before exercising.

Questions 1–2: Arrange the answers in correct sequence.

_____ _____ _____ 1. Abdominal wall muscles, from superficial to deep:
   A. Transversus abdominis
   B. External oblique
   C. Internal oblique

_____ _____ _____ 2. From superior to inferior in location:
   A. Sternocleidomastoid
   B. Gastrocnemius
   C. Diaphragm and intercostal muscles

Questions 3–10: Choose the one best answer to each question.

3. All of these muscles are located on the anterior of the body except:
   A. Tibialis anterior
   D. Pectoralis major
   B. Rectus femoris
   E. Rectus abdominis
   C. Erector spinae

4. All of these muscles have attachments to the hip-bones *except:*
   A. Adductor muscles (longus and magnus)
   B. Biceps femoris
   C. Rectus femoris
   D. Vastus medialis
   E. Latissimus dorsi

5. The masseter and temporalis muscles are used for:
   A. Chewing
   C. Frowning
   B. Pouting
   D. Depressing tongue

6. All of these muscles are attached to ribs *except:*
   A. Serratus anterior
   D. Erector spinae
   B. Intercostals
   E. External oblique
   C. Trapezius

7. All of the following molecules are parts of thin filaments *except:*
   A. Actin
   C. Tropomyosin
   B. Myosin
   D. Troponin

8. Which statement about muscle physiology in the relaxed state is *false?*
   A. Myosin crossbridges are bound to ATP.
   B. Calcium ions are stored in sarcoplasmic reticulum.
   C. Myosin crossbridges are bound to actin.
   D. Tropomyosin–troponin complex is bound to actin.

9. Choose the one *false* statement.
   A. The hamstrings are antagonists to the quadriceps femoris.
   B. The name deltoid is based on the action of that muscle.
   C. The most important muscle used for normal breathing is the diaphragm.
   D. The insertion end of a muscle is the attachment to the bone that does move.

10. Choose the one *false* statement.
   A. In general, adductors (of the arm and thigh) are located more on the medial than on the lateral surface of the body.
   B. Both the pectoralis major and latissimus dorsi muscles extend the humerus.
   C. Muscle fibers remain relaxed if there are few calcium ions in the sarcoplasm.
   D. Atrophy means decrease in muscle mass.

Questions 11–15: Fill-ins. Refer to Figure LG 8.4(a) and (b). Write the word or phrase or key letters of muscles that best complete the statement or answer the questions.

_____ 11. Muscles G, S, U, V, and W on Figure LG 8.4(a) all have in common the fact that

they carry out the action of _____ .

_____ 12. Muscles G, N, O, and R on Figure LG 8.4(b) all have in common the fact that they

carry out the action of _____ .

_____ 13. If you colored all flexors red and all extensors blue on these two figures, the view

of the _____ surface of the body would appear more blue.

_____ 14. Choose the letters of all of the muscles listed below that would contract as you raise your left arm straight in front of you, as if to point toward a distant mountain: Figure LG 8.4(a): D T U V; Figure LG 8.4(b): N O P.

_____ 15. Choose the letters of all the muscles listed below that would contract as you raise your knee and extend your leg straight out in front of you, as if you are starting to march off to the distant mountain: Figure LG 8.4(a): AA BB CC DD; Figure LG 8.4(b): H I J K R.

ANSWERS TO MASTERY TEST: ■ CHAPTER 8

Arrange
1. B C A
2. A C B

Multiple Choice
3. C
4. D
5. A
6. C
7. B
8. C
9. B
10. B

Fill-ins
11. Flexion
12. Extension
13. Posterior
14. Figure LG 8.4(a): T (anterior fibers), U, V; Figure LG 8.4(b): none.
15. Figure LG 8.4(a): AA BB CC DD; Figure 8.4(b): H I.

FRAMEWORK 9
Nervous Tissue

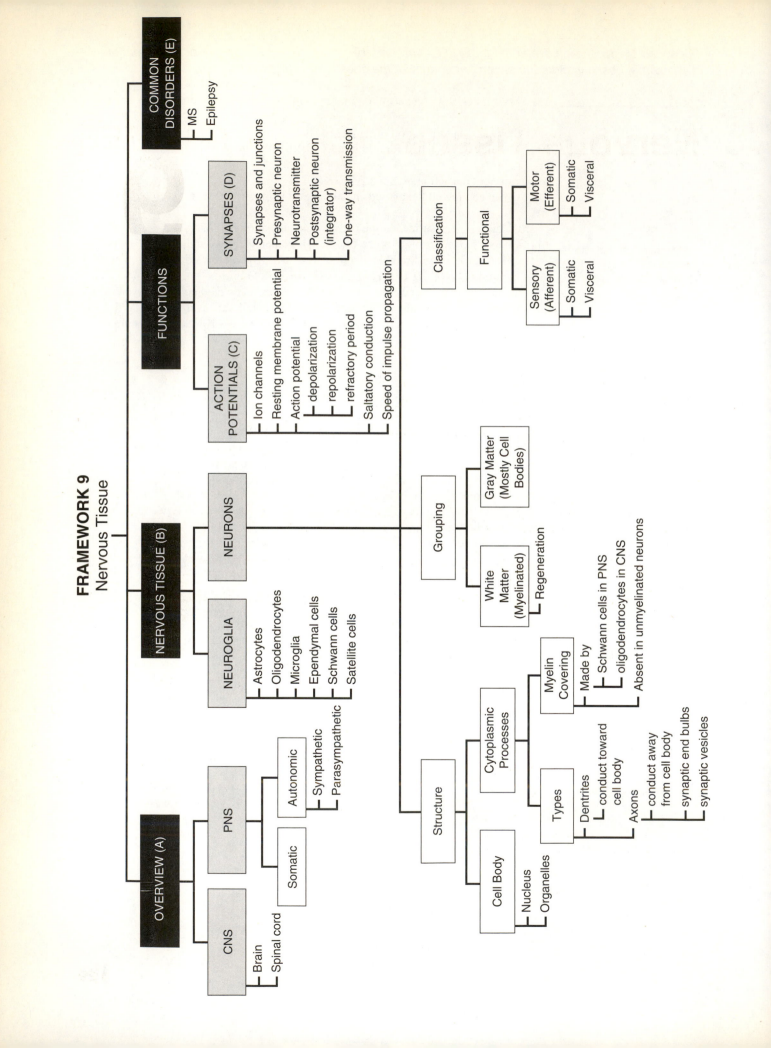

Nervous Tissue

Two systems—nervous and endocrine—are responsible for regulating our diverse body functions. Each system exerts its control with the help of specific chemicals, namely neurotransmitters (nervous system) and hormones (endocrine system). In the coming chapters, both systems of regulation will be considered, starting with the nervous system.

Nervous tissue consists of two types of cells: neurons and neuroglia. The name neuroglia (*glia* = glue) offers a clue as to the function of these cells: they bind, support, and protect neurons. Neurons perform the work of transmitting nerve impulses. These long and microscopically slender cells sometimes convey information several feet along a single neuron. Their function relies on an intricate balance between ions (Na^+ and K^+) found in and around nerve cells. Neurons release neurotransmitters that bridge the gaps between adjacent neurons and at nerve–muscle or nerve–gland junctions. Analogous to a complex global telephone system, the nervous tissue of the human body boasts a design and an organization that permit accurate communication, coordination, and integration of virtually all thoughts, sensations, and movements.

As you begin your study of nervous tissue, carefully examine the Chapter 9 Topic Outline and Objectives; check off each one as you complete it. To organize your study of nervous tissue, glance over the Chapter 9 Framework now. Be sure to refer to the Framework frequently and note relationships among key terms in each section.

TOPIC OUTLINE AND OBJECTIVES

A. Overview of the nervous system

☐ 1. List the structures and basic functions of the nervous system
☐ 2. Describe the organization of the nervous system

B. Nervous tissue

☐ 3. Contrast the histological characteristics and the functions of neurons and neuroglia.
☐ 4. Distinguish between gray matter and white matter.

C. Action potentials

☐ 5. Describe how a nerve impulse is generated and conducted.

D. Synaptic transmission

☐ 6. Explain the events of synaptic transmission and the types of neurotransmitters used.

E. Common disorders and medical terminology

WORDBYTES

Now become familiar with the language of this chapter by studying each wordbyte, its meaning, and an example of its use within a term. After you study the entire list, self-check your understanding by writing the meaning of each wordbyte on the line. As you continue through the *Learning Guide,* identify (and fill in) additional terms that contain the same wordbyte.

Wordbyte	Self-check	Meaning	Example(s)
af-	_____	toward	*af*ferent
astro-	_____	star	*astro*glia
dendr-	_____	tree	*dendr*ite
ef-	_____	away from	*ef*ferent
-ferent	_____	carried	af*ferent*
-glia	_____	glue	neuro*glia*
lemm-	_____	sheath	neuro*lemma*
neuro-	_____	nerve	*neuro*n
olig-	_____	few	*olig*odendrocytes
saltat-	_____	leaping	*saltat*ory
-soma-	_____	body	axo*soma*tic
syn-	_____	together	*syn*apse

CHECKPOINTS

A. Overview of the nervous system (pages 226–228)

A1. Contrast functions of the nervous system and the endocrine system in maintaining homeostasis.

■ **A2.** Check your understanding of the structures and organization of the nervous system by selecting answers that best fit descriptions below.

Aff. Afferent	ENS. Enteric nervous system
ANS. Autonomic nervous system	PNS. Peripheral nervous system
CNS. Central nervous system	SNS. Somatic nervous system
Eff. Efferent	

_____ a. Brain and spinal cord

_____ b. Sensory nerves

_____ c. Carry information from CNS to skeletal muscles

_____ d. Consists of sympathetic and parasympathetic divisions

_____ e. Nerves that convey impulses to smooth muscle, cardiac muscle, and glands; involuntary

_____ f. Cranial nerves, spinal nerves, ganglia, and sensory receptors

_____ g. "Brain of the gut"

■ **A3.** List several functions of the nervous system.

B. Nervous tissue (pages 228–230)

■ **B1.** Write *neurons* or *neuroglia* after descriptions of these cells.

a. Conduct impulses from one part of the nervous system to another: _____

b. Provide support and protection for the nervous system: _____

c. Bind nervous tissue to blood vessels, form myelin, and serve phagocytic functions: _____

d. Smaller in size, but more abundant in number: _____

e. Can form brain tumors known as gliomas: _____

B2. Complete the table describing neuroglia cells.

Type	Description	Functions	Location (CNS or PNS)
a.	Star-shaped cells		
b. Oligodendrocytes			
c.		Phagocytic	
d.		Line ventricles of brain and central canal of spinal cord	
e.		Produce myelin around axons of PNS neurons	
f. Satellite cells			PNS

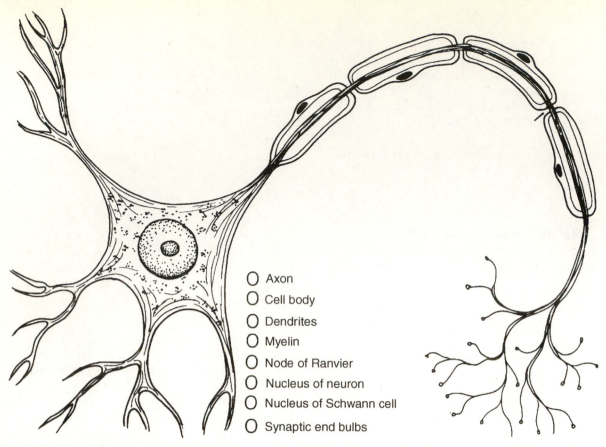

○ Axon
○ Cell body
○ Dendrites
○ Myelin
○ Node of Ranvier
○ Nucleus of neuron
○ Nucleus of Schwann cell
○ Synaptic end bulbs

Figure LG 9.1   Structure of a typical neuron as exemplified by an efferent (motor) neuron. Complete the figure as directed in Checkpoint B3.

B3.   On Figure LG 9.1, color structures with color code ovals.
Then draw arrows beside the figure to indicate direction of nerve impulses.

■ **B4.**   Match the parts of a neuron listed in the box with descriptions below.

A. Axon	D. Dendrite	My. Myelin
CB. Cell body	Mit. Mitochondria	

_____ a. Contains nucleus; cannot regenerate since lacks mitotic apparatus

_____ b. Lipid and protein covering that insulates many axons and increases speed of nerve impulse transmission

_____ c. Conducts impulses toward cell body

_____ d. Conducts impulses away from cell body; has synaptic end bulbs that secrete neurotransmitter

_____ e. Provide energy for neurons

■ **B5.**   Describe how myelin is laid down on nerve fibers in the:

a. PNS

b. CNS

132

■ **B6.** Describe the significance of the following parts of neurons. Comment on what you think might result from absence of that part of the neuron.

a. Myelin

b. Nodes of Ranvier

■ **B7.** Contrast key terms related to the nervous system in this exercise.

a. White matter consists of clusters of axons that *(are? are not?)* myelinated. For example, a nerve viewed in a lab

dissection will appear white (or slightly yellow) due to the presence of _____ .

b. Cell bodies of neurons *(are? are not?)* myelinated, and they comprise much of *(white? gray?)* matter.

c. The inner core (H- or butterfly-shaped) region of the spinal cord as seen in cross section is composed of *(gray? white?)* matter, whereas tracts that carry impulses "north" or "south" within the cord are made of *(gray? white?)* matter.

■ **B8.** _____ cells are required for regeneration of axons and dendrites. These cells are present in the *(CNS? PNS?)*, but not in the *(CNS? PNS?)*. Therefore regeneration of injured neurons occurs more readily in the *(CNS?*

PNS?) where these cells can form a _____ tube across the gap in the injured axon.

C. Action potentials (pages 230–234)

■ **C1.** Complete this Checkpoint about action potentials.

a. Nerve action potentials are also known as nerve _____ . List the two features in a nerve

cell membrane necessary for a nerve impulse to take place: _____ _____ .

b. Since the phospholipid bilayer normally permits *(little? much?)* passage of ions, two types of

_____ channels provide the main path for flow of current across the membrane. One type of such channels are *(leakage? voltage-gated?)* channels that are always open. The second type of channel opens

only in response to a change in membrane potential; such channels are known as _____ channels.

c. The second feature that is a prerequisite for starting a nerve impulse is the presence of a resting nerve cell membrane that is *(polarized? not polarized?)*, in this case with the inside of the cell more *(positive? negative?)* than the outside of the cell. In fact the resting membrane potential is about *(+70? 0? −70?)* mV.

Figure LG 9.2 Diagram of a nerve cell. Symbols for ions inside and outside cell are to be inserted, as well as arrows showing directions of movement of ions. Label and color as directed in Checkpoint C1.

d. Use Figure LG 9.2 to demonstrate three factors that account for the resting membrane potential. First, draw on the figure the major cations inside and outside of the cell. (Remember that *cations* bear positive charges.) The membrane has many more *(Na⁺? K⁺?)* leakage channels, causing much more K⁺ to leak out of the cell compared to the amount of Na⁺ that leaks in. Draw a thick arrow showing K⁺ exiting the cell, leading to negativity inside the plasma cell membrane.

e. Now draw the major anions within the cell of Figure LG 9.2. Proteins are *(large? small?)* chemicals. How does this factor contribute to the negative voltage inside the cell membrane?

f. A third factor involved with maintenance of resting potential is the _____/_____ pump. This form of

(active? passive?) transport pumps out the _____ ions that do leak into the cell. Nerve cells require a constant

supply of energy in the form of _____ to maintain this pump.

g. K⁺ is then attracted back into the cell as a result of _____ . Show this by adding a thick white arrow to the figure.

C2. Define these terms:

a. Excitability of neurons

b. Stimulus

c. Threshold

■ **C3.** Examine your understanding of how a nerve impulse occurs in this Checkpoint.

a. A stimulus causes the nerve cell membrane to become *(more? less?)* permeable to Na⁺. Na⁺ can then enter the

cell as Na⁺ _____ open.

b. At rest, the membrane had a potential of _____ mV. As Na⁺ enters the cell, the inside of the membrane becomes more *(positive? negative?)*. The potential will tend to go toward *(−80? −60?)* mV. The process of *(polarization? depolarization?)* is occurring. This process causes structural changes in more Na⁺ channels so that even more Na⁺ enters.

c. The membrane reaches threshold at about *(−55? 0? +30?)* mV. Na⁺ channels stay open until the inside of the membrane potential is *(reversed? repolarized?)* at +30 mV.

d. After a fraction of a second, K⁺ voltage-gated channels at the site of the original stimulus open. K⁺ is more concentrated *(outside? inside?)* the cell (as you showed on Figure LG 9.2); therefore K⁺ diffuses *(in? out?)*. This

causes the inside of the membrane to become more negative again and return to its resting potential of _____ mV.

The process is known as *(de? re?)*-polarization. In fact, an "overshoot" of outflow of K⁺ may cause _____ -polarization in which the membrane potential is *(−60? −80?)* mV.

134

e. The events just described (depolarization and repolarization) are called a *nerve impulse* or nerve

_____ _____ . This process occurs in two types of cells in

the body; these are _____ and _____ .

f. The impulse travels rapidly (or is _____) from the original point (site of the stimulus) along the nerve or muscle as adjacent areas of Na⁺ and K⁺ channels open and close.

Figure LG 9.3   Diagram for showing nerve action potentials. Identify letter labels in Checkpoint C4.

■ **C4.**   Write the correct label (A–E) of Figure LG 9.3 next to each description. One letter will be used twice.

_____ a. The stimulus is applied at this point.

_____ b. Resting membrane potential is at this level.

_____ c. Membrane becomes so permeable to K⁺ that K⁺ diffuses rapidly out of the cell.

_____ d. The membrane is becoming more positive inside as Na⁺ enters; its potential is −30 mV. The process of depolarization is occurring.

_____ e. The membrane is repolarizing at this point.

_____ f. Reversed polarization occurs; enough Na⁺ has entered so that this part of the cell is more positive inside than outside.

C5.   Define these terms:

a. All-or-nothing principle

b. Continuous conduction

■ **C6.**   Describe how myelination, fiber thickness, and temperature affect speed of impulse propagation.

a. Saltatory conduction occurs along *(myelinated? unmyelinated?)* nerve fibers. Saltatory transmission is *(faster? slower?)*. Explain why.

135

b. Which nerve fibers are likely to conduct impulses more rapidly? Fibers with *(large? small?)* diameter.

c. *A clinical challenge. (Warm? Cool?)* nerve fibers conduct impulses faster. How can this information be applied clinically?

d. Explain why Laurie feels no pain during dental work after Dr. Staubitz injects an anesthetic (Novocaine) near Laurie's teeth.

D. Synaptic transmission (pages 234–237)

■ **D1.**   Do this activity about synapses.

a. In Chapter 8, you studied the point at which a neuron comes close to contacting a muscle. This is known as a

_____ junction, shown in Figure LG 8.2, page 110. The minute space between two

neurons is known as a _____ .

The two neurons involved are known as a _____ -synaptic neuron and a

_____ -synaptic neuron.

b. Does a nerve impulse actually "jump" across the synaptic cleft? Explain.

c. Write these structures in order to summarize the anatomical pathway of a synapse.

EB. End bulb of presynaptic neuron	SC. Synaptic cleft
N. Neurotransmitter	SV. Synaptic vesicle
NR. Neurotransmitter receptor	VGCaC. Voltage-gated Ca^{2+} channels

_____ → _____ → _____ → _____ → _____ → _____
 (first)   (last)

d. Explain what causes Ca^{2+} to enter voltage-gated Ca^{2+} channels of presynaptic bulbs.

e. Describe the effect of entrance of Ca^{2+} into these channels.

f. An explanation for one-way impulse conduction at synapses is that only synaptic bulbs of *(pre? post?)*-synaptic neurons release neurotransmitter.

g. Excitatory neurotransmitters cause *(de? hyper?)*-polarization, but inhibitory neurotransmitters cause *(de? hyper?)*-polarization with a membrane more likely to exhibit a potential of *(−60? −80)* mV.

h. The *(pre? post?)*-synaptic neuron integrates excitatory and inhibitory effects in a process known as _____ .

D2.   Briefly describe the three main mechanisms by which neurotransmitters are normally removed from the synaptic cleft.

■ **D3.** Explain the mechanism of action of antidepressants such as Prozac and other SSRIs.

■ **D4.** Answer these questions about different types of neurotransmitters (NTs).

a. We have already discussed acetylcholine (_____), an NT released at synapses and neuromuscular junctions. ACh is *(excitatory? inhibitory?)* toward skeletal muscle, but ACh released from the vagus nerve is *(excitatory? inhibitory?)* toward cardiac muscle. So when the vagus nerve sends impulses to your heart, your pulse (heart rate) becomes *(faster? slower?)*.

b. GABA and glycine are both *(excitatory? inhibitory?)* NTs, whereas glutamate has powerful *(excitatory? inhibitory?)* effects

c. Name three neurotransmitters formed from amino acids.

d. Which NT is produced only as needed (not stored in synaptic vesicles) and plays a role in learning and

   memory? _____

e. _____ are the body's natural painkillers, and their production may be increased by acupuncture treatments, resulting in anesthesia and euphoria. As described in Checkpoint D3, the NT named

   _____ in mood control, and also helps induce sleep.

E. Common disorders and medical terminology (pages 237–238)

■ **E1.** Discuss multiple sclerosis (MS) in this Checkpoint.

a. This condition involves destruction of _____ that surrounds axons. This disorder is more common in

   (men? women?). An autoimmune disorder, it may be triggered by _____
b. The name multiple sclerosis refers to the many locations in which myelin is replaced by

   _____ . MS *(is? is not?)* progressive and *(does? does not?)* involve remissions and relapses.

E2. Check your understanding of epilepsy in this Checkpoint.

a. Define the term epileptic seizure.

b. Most epileptic seizures are *(caused by changes in blood chemistry? idiopathic [of unknown cause]?)*. Epilepsy *(usually? almost never?)* affects intelligence.

E3. Guillain-Barre syndrome exerts greater effects upon *(motor? sensory?)* neurons.

ANSWERS TO SELECTED CHECKPOINTS: CHAPTER 9

A2. (a) CNS. (b) Aff. (c) SNS or Eff (*Hint:* remember S A M E: Sensory = Afferent; Motor = Efferent). (d) ANS. (e) ANS. (f) PNS. (g) ENS.

A3. Sensation, integration, and motor or glandular response.

B1. (a) Neurons. (b–e) Neuroglia.

B4. (a) CB. (b) My. (c) D. (d) A. (e) Mit.

B5. (a) By Schwann cells wrapping their cell membranes around axons. (b) By a similar process involving oligodendrocytes.

B6. (a) Myelin increases the speed of nerve impulse conduction. Less myelin (as occurs in infants and in disorders such as multiple sclerosis) causes nerve impulses to occur at a slower rate. (b) These gaps in the myelin sheath cause the nerve impulse to be conducted rapidly from node to node.

B7. (a) Are; myelin. (b) Are not; gray. (c) Gray, white.

B8. Schwann; PNS, CNS; PNS, regeneration.

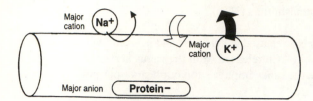

Figure LG 9.2A Diagram of a nerve cell.

C1. (a) Impulses; presence of ion channels and a resting membrane potential. (b) Little; ion; leakage; voltage-gated. (c) Polarized; negative, −70. (d) Inside the cell: K^+; outside the cell: Na^+; K^+. (e) Proteins; large; their size prohibits them from leaving the cell. (f) Na^+/K^+; active, Na^+; ATP. (g) Negativity inside the cell. See Figure LG 9.2A.

C3. (a) More; voltage-gated channels. (b) −70; positive; −60; depolarization. (c) −55; reversed. (d) Inside; out; −70; re; hyper, −80. (e) Action potential; neurons, muscle cells. (f) Propagated or conducted.

C4. (a) B. (b) A. (c) E. (d) C. (e) E. (f) D.

C6. (a) Myelinated; faster, voltage-gated Na^+ and K^+ channels are located primarily at nodes of Ranvier (gaps in myelin). Current carried by these ions at one node generates currents (triggers impulse) at the next node. (b) Large. (c) Warm; ice or other cold applications can slow conduction of pain impulses. (d) It blocks openings of voltage-gated Na^+ channels.

D1. (a) Neuromuscular; synaptic cleft; pre, post. (b) No; as described next, the nerve impulse triggers release of chemicals (neurotransmitters) that cross the synaptic cleft. (c) EB → VGCaC → SV → N → SC → NR. (d) Depolarization of the neuron causes Ca^{2+} channels to open. Because Ca^{2+} is more concentrated in interstitial fluid, it flows into channels and enters the neuron. (e) Ca^{2+} causes release of neurotransmitters into these synaptic clefts. (f) Pre. (g) De, hyper, −80. (h) Post, summation.

D3. The neurotransmitter serotonin is allowed to stay longer in the synapse (since its "uptake" is inhibited), thereby activating postsynaptic neurons that maintain a more positive mood.

D4. (a) ACh; excitatory, inhibitory; slower. (b) Inhibitory, excitatory. (c) Norepinephrine, dopamine, and serotonin. (d) Nitric oxide. (e) Endorphins; serotonin

E1. (a) Myelin; women, a virus. (b) Hardened scars or plaques; is, does.

CRITICAL THINKING: CHAPTER 9

1. Contrast the process of myelin formation in the CNS with that in the PNS.
2. Describe changes in impulse transmission, as well as signs or symptoms that occur with loss of myelin, as in multiple sclerosis (MS).
3. Contrast axons and dendrites with regard to structure and function.
4. If you cut a nerve in your finger, it can regenerate. However, severed axons in the brain or spinal cord cannot heal. Explain why.
5. Contrast effects of the following neurotransmitters: glutamate, glycine, and norepinephrine.
6. Describe the steps in repair of peripheral neurons.

MASTERY TEST: ■ CHAPTER 9

Questions 1 and 2: Arrange the answers in correct sequence.

_____ _____ _____ 1. In order of transmission across synapse, from first structure to last:
   A. Presynaptic end bulb
   B. Postsynaptic neuron
   C. Synaptic cleft

_____ _____ _____ 2. Membrane potential values, from most negative to most positive:
   A. Resting membrane potential (RMP)
   B. Reversed membrane potential
   C. Threshold potential

3. Choose the one *false* statement.
   A. The membrane of a resting neuron has a membrane potential of -70 mV.
   B. In a resting membrane, permeability to K^+ ions is less than permeability to Na^+ ions.
   C. Thin, unmyelinated fibers have a relatively slow rate of nerve transmission.

4. A term that means the same thing as *afferent* is:
   A. Autonomic   B. Somatic
   C. Peripheral   D. Motor
   E. Sensory

5. When an action potential reaches presynaptic end bulbs, the plasma membrane there becomes more permeable to _____ which then initiate release of neurotransmitter from synaptic vesicles.
   A. Na^+ B. K^+ C. Ca^{2+} D. Cl^-

Questions 6–10: Circle T (true) or F (false). If the statement is false, change the underlined word or phrase so that the statement is correct.

T   F   6. The concentration of potassium ions (K^+) is considerably <u>greater</u> inside a resting cell than outside it.

T   F   7. <u>Oligodendrocytes, astrocytes, ependymal cells, and neurons are all</u> neuroglial cells.

T   F   8. Neurotransmitter substances are released at <u>synapses and at neuromuscular junctions.</u>

T   F   9. The <u>brain and spinal nerves</u> are parts of the peripheral nervous system (PNS).

T   F   10. An excitatory nerve stimulus will temporarily <u>increase</u> permeability of the neuron's plasma membrane to <u>Na$^+$</u>.

Questions 11–15: Fill-ins. Complete each sentence with the word or phrase that best fits.

_____ 11. The _____ nervous system consists of the sympathetic and parasympathetic divisions.

_____ 12. Application of cold to a painful area can decrease pain in that area because _____ .

_____ 13. One-way nerve impulse transmission can be explained on the basis of release of transmitters only from the _____ of neurons.

_____ 14. Schwann cells are found only around fibers of the _____ nervous system, meaning that only these cells can _____ .

_____ 15. ACh is an abbreviation for the neurotransmitter named _____ .

ANSWERS TO MASTERY TEST: ■ CHAPTER 9

Arrange
1. A C B
2. A C B

Multiple Choice
3. B
4. E
5. C

True or False
6. T
7. F. Oligodendrocytes, astrocytes, ependymal cells, but not neurons
8. T
9. F. Spinal nerves (as well as some other structures, but not the brain)
10. T

Fill-ins
11. Autonomic
12. Cooling of neurons slows down the speed of nerve transmission, for example, of pain impulses
13. End bulbs of axons
14. Peripheral, regenerate
15. Acetylcholine

FRAMEWORK 10
Central Nervous System, Spinal Nerves, and Cranial Nerves

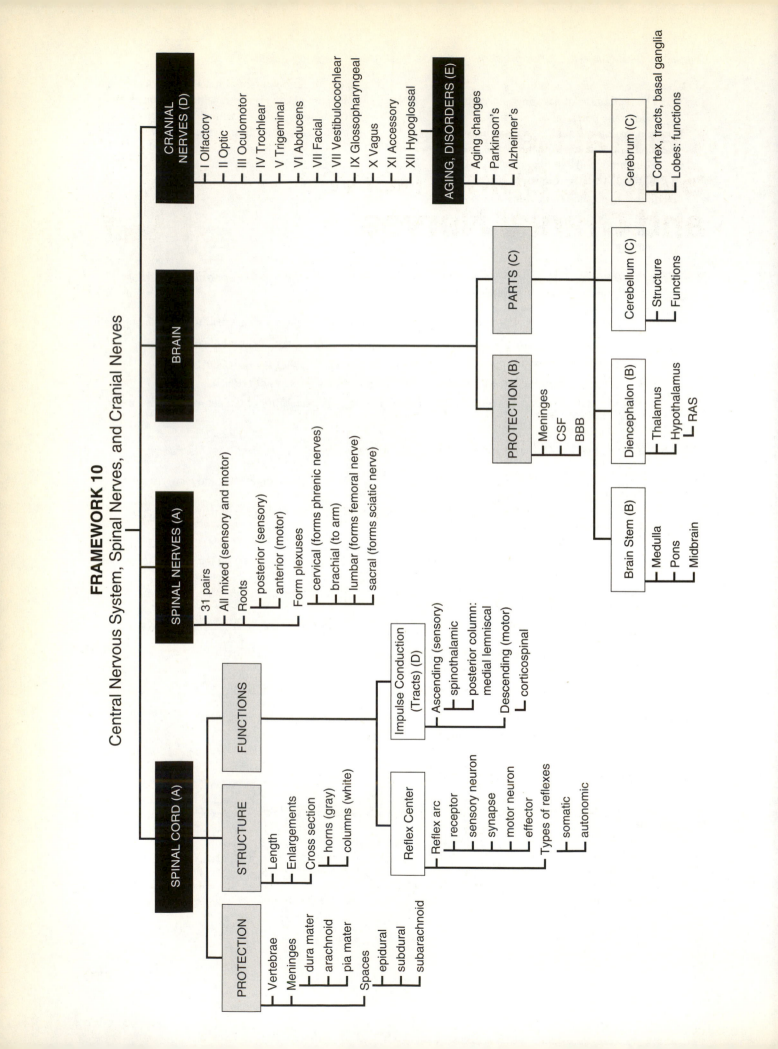

Central Nervous System, Spinal Nerves, and Cranial Nerves

In this chapter, we will tour the central nervous system (CNS)—the brain and spinal cord—as well as the spinal nerves and cranial nerves that permit communication, control, and integration between the CNS and distant body parts.

We begin with the spinal cord and spinal nerves. These serve as the major links in the communication pathways between the brain and all other parts of the body. Nerve impulses are conveyed along routes (or tracts) in the spinal cord, laid out much like train tracks: some head north to regions of the brain, and others carry nerve messages south from the brain toward specific body parts. Spinal nerves branch off from the spinal cord, perhaps like a series of bus lines that pick up passengers (nerve messages) to or from train depots (points along the spinal cord) en route to their final destinations. Organization is critical in this nerve impulse transportation network. Any structural breakdowns—by trauma, disease, or other disorders—can lead to interruption in service with resultant chaos (such as spasticity) or standstill (such as sensory loss or paralysis).

Next we move on to the brain—the major control center for the global communication system of the body. The brain requires round-the-clock protection and maintenance afforded by bones, meninges, cerebrospinal fluid, a special blood–brain barrier, along with a fail-safe blood supply. This vital control center consists of four major substructures: the brain stem, diencephalon, cerebrum, and cerebellum. Each brain part carries out specific functions and each releases specific chemical neurotransmitters. Twelve pairs of cranial nerves convey information to and from the brain. Just as a giant computer network may experience minor disruptions in service or major shutdowns, disorders in the brain or cranial nerves may lead to minor, temporary changes in nerve functions or profound and fatal outcomes.

As you begin your study of the central nervous system and many nerves attached to it, carefully examine the Chapter 10 Topic Outline and Objectives; check off each one as you complete it. To organize your study of the content, glance over the Chapter 10 Framework now. Be sure to refer to the Framework frequently and note relationships among key terms in each section.

TOPIC OUTLINE AND OBJECTIVES

A. Spinal cord and spinal nerves

☐ 1. Describe how the spinal cord is protected.
☐ 2. Describe the structure and function of the spinal cord.
☐ 3. Describe the composition, coverings, and distribution of spinal nerves.

B. Brain: protection; brainstem and diencephalon

☐ 4. Discuss how the brain is protected and supplied with blood.
☐ 5. Name the major parts of the brain and explain the function of each part.

C. Brain: cerebrum and cerebellum

☐ 6. Describe three somatic and sensory motor pathways.

D. Cranial nerves

☐ 7. Identify the 12 pairs of cranial nerves by name and number and give the functions of each.

E. Aging, common disorders, and medical terminology of the nervous system

☐ 8. Describe the effects of aging on the nervous system

WORDBYTES

Now study each wordbyte, its meaning, and an example of its use in a term. After you study the entire list, check your understanding by writing the meaning of each wordbyte in the margins. As you continue through the *Learning Guide,* identify (and fill in) additional terms that contain the same wordbyte.

Wordbyte	Meaning	Example(s)	Wordbyte	Meaning	Example(s)
a-, an	without	*an*esthesia	falx	sickle	*falx* cerebri
-algia	pain	neur*algia* *an*alge*sia*	glossi-	tongue	hypo*glossal*
			hemi-	half	*hemi*sphere
arachn-	spider	*arachn*oid	hypo-	under	*hypo*thalamus
cauda-	tail	*cauda*te	mater-	mother	pia *mater*
cephalo-	head	hydro*cephalic*	para-	abnormal	*par*esthesia
-ceptor	receiver	proprio*ceptor*	pia	delicate	*pia* mater
cortico-	bark	cerebral *cortex*	plexus	braid or network	cervical *plexus*
dura	hard	*dura* mater	pons	bridge	*pons*
-ellum	little	cerebe*llum*	soma-	body	*soma*tic
enceph-	brain	*enceph*alitis			
-esthesia	sensation	an*esthesia*, par*esthesia*			

CHECKPOINTS

A. Spinal cord and spinal nerves (pages 243–248)

■ **A1.** The spinal cord and brain form the *(central? peripheral?)* nervous system; spinal nerves and cranial nerves—

which are attached to the cord and brain—are parts of the _____ nervous system.

■ **A2.** Do the following exercise about protection for the CNS.

a. List several types of protection afforded by the body for the brain and spinal cord.

b. Write names of meninges and related spaces in sequence from superficial to deep. Choose answers from terms in the box and write on lines provided.

Meninges	Spaces
Arachnoid	Subarachnoid space
Dura mater	Subdural space
Pia mater	

Vertebrae (or cranial bone) → (epidural) space with fat and other connective tissue →

1. _____ → 2. _____ → 3. _____ → 4. _____ →

5. _____ → Spinal cord (or brain)

Now place an asterisk (*) in the space where cerebrospinal fluid (CSF) is located.

c. Inflammation of meninges is a condition called _____ .

■ **A3.** Do this activity about your own spinal cord.

a. Identify the location of your own spinal cord. It lies within the vertebral canal, extending from just inferior to the cranium to about the level of your *(waist? sacrum?)*. This level corresponds with about the level of *(L2? L4? S4?)* vertebrae. In other words, the *(spinal cord? vertebral column?)* reaches a lower (more inferior) level in your

body. The spinal cord is about 42 to 45 cm (_____ inches) in length.

b. Circle the two regions of your spinal cord that are notably enlarged where nerves exit to your upper and lower extremities:

cervical   thoracic   lumbar   sacral   coccygeal

c. The term *cauda equina* means _____ _____ . Describe the cauda equina.

d. A spinal tap (lumbar puncture) involves removal of fluid known as _____ fluid

from the *(epidural? subarachnoid?)* space. The puncture is made at the level of about the _____

to _____ vertebrae. Why is this location a relatively safe site for this procedure?

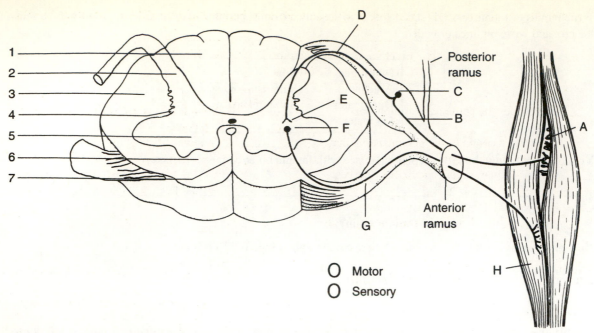

Figure LG 10.1 Outline of the spinal cord, roots, and nerves with reflex arc. Complete as directed in Checkpoints A4 and A7–A8.

■ **A4.** Refer to Figure LG 10.1 and do this activity about the spinal cord.

a. Gray matter is located *(within? outside?)* the H-shaped outline in regions of the cord known as gray *(columns? horns?)*. With a pencil, shade the gray matter portion on the right side of the figure only.

b. White matter is located *(within? outside?)* the H-shaped outline in regions of the spinal cord known as white *(columns? horns?)*. Tracts here are much like train "tracks"—sites of impulse conduction. Ascending tracts (those going "north") are *(sensory? motor?)*, whereas descending tracts are _____ . Tracts are white because they consist of axons that *(are? are not?)* myelinated.

c. Now label parts 1–7 on the left side of the figure.

d. Area 4 contains *(autonomic? somatic?)* cell bodies. Area 4 is located (bilaterally) only in regions of the spinal cord with autonomic functions . Circle those regions of the cord. *(Cervical? Thoracic? Upper lumbar? Lower lumbar? Sacral?)*. (More on this in Chapter 11.)

■ **A5.** Complete this exercise about spinal nerves. Spinal nerves are attached to the spinal cord.

a. There are _____ pairs of spinal nerves. Write the number of pairs in each region. _____ Cervical _____ Thoracic _____ Lumbar _____ Sacral _____ Coccygeal

b. Which of these spinal nerves form the cauda equina? _____

c. Spinal nerves are attached by two roots. As shown in Figure LG 10.1, the posterior root is *(sensory? motor? mixed?)*, the anterior root is _____ , and the spinal nerve is _____ .

d. Which term refers to the covering over an individual axon? *(Endoneurium? Epineurium? Perineurium?)*. Which term applies to the covering around an entire nerve? _____

e. Shortly after spinal nerves (with their connective tissue coverings) pass through intervertebral foramina, most form networks known as _____ . Refer to Figure 10.2, page 244 in the text. In which region do spinal nerves not group into plexuses, but instead form segmental arrangements? _____

■ **A6.** Match the plexus names in the box with descriptions. Refer to Figure 10.2 (page 244 in your text) for help.

B. Brachial	I. Intercostal	S. Sacral
C. Cervical	L. Lumbar	

_____ a. Provides the entire nerve supply for the arm

_____ b. Contains origin of the phrenic nerve that supplies diaphragm

_____ c. Forms ulnar, radial, and axillary nerves

_____ d. Forms the largest nerve in the body (the sciatic), which supplies posterior of thigh and the leg

_____ e. Supplies nerves to scalp, neck, and part of shoulder

_____ f. Supplies fibers to the femoral nerve, which innervates the quadriceps, so injury to this plexus would interfere with actions such as kicking a leg upward

■ **A7.** Each of the 31 pairs of spinal nerves is attached by *(1? 2? 3?)* roots. The *(anterior? posterior?)* root contains sensory nerve fibers, and the _____ root contains motor fibers. Color lightly the sensory and motor roots on Figure LG 10.1. Select colors according to the color code ovals there.

■ **A8.** Besides conveying nerve impulses "north and south," the spinal cord also serves as a _____ center. Complete this checkpoint on the function of the spinal cord as a reflex center. Label structures A–H on Figure LG 10.1 using the following terms: *effector, motor neuron axon, motor neuron cell body, receptor, sensory neuron axon, sensory neuron cell body, sensory neuron dendrite,* and *integrating center*. Note that these structures are lettered in alphabetical order along the conduction pathway of a reflex arc. Add arrows showing the direction of nerve transmission in the arc.

■ **A9.** *A clinical challenge.* Indicate the meaning of each of these diagnostic results for two members of the Adkins family who have been in an auto accident.

a. Elena, age 42, has multiple skull fractures. Her pupillary light reflex is absent bilaterally (both eyes remain dilated).

b. Jimmy, age 10, has no patellar reflex in his right leg.

■ **A10.** Do this exercise on reflexes.

a. How many neurons are contained in the reflex in Figure LG 10.1? *(1? 2? 3?)*. This is the type of reflex exemplified by the *(patellar reflex [knee jerk]? withdrawal reflex?)*.

b. The knee jerk and withdrawal reflexes are both *(autonomic? somatic?)* reflexes. The ability of your body to sense low blood pressure (BP) and then increase your heart rate as an attempt to raise your BP is an example of a(n) *(autonomic? somatic?)* reflex.

c. The photopupil is an example of a *(cranial? spinal?)* reflex.

B. Brain: protection; brainstem and diencephalon (pages 248–255)

■ **B1.** The human brain weighs about _____ lb (1300 g); it consists of *(billions? trillions?)* of cells, most of which are *(neurons? neuroglia?)*. Now identify numbered parts of the brain on Figure LG 10.2.

a. Structures 1–3 are parts of the _____ .

b. Structures 4 and 5 together form most of the _____ .

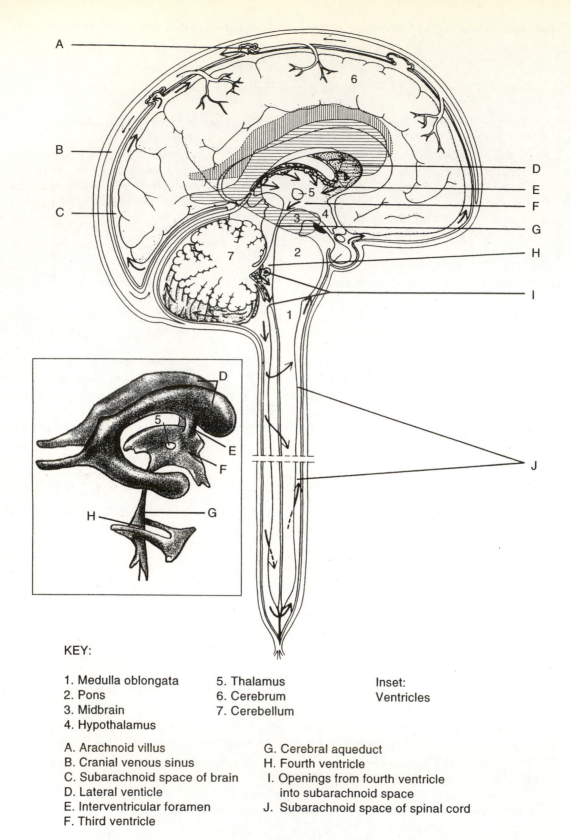

KEY:

1. Medulla oblongata
2. Pons
3. Midbrain
4. Hypothalamus

5. Thalamus
6. Cerebrum
7. Cerebellum

Inset:
Ventricles

A. Arachnoid villus
B. Cranial venous sinus
C. Subarachnoid space of brain
D. Lateral venticle
E. Interventricular foramen
F. Third ventricle

G. Cerebral aqueduct
H. Fourth ventricle
I. Openings from fourth ventricle
   into subarachnoid space
J. Subarachnoid space of spinal cord

Figure LG 10.2 Brain and meninges seen in sagittal section with lateral ventricles superimposed. Inset: entire ventricular system. Parts of the brain are numbered: refer to Checkpoint B1. Letters indicate pathway of cerebrospinal fluid (CSF); refer to Checkpoint B6.

146

c. Structure 6 is the largest part of the brain, the _____ .

d. The second largest part is structure 7, the _____ .

B2. Although the brain makes up about 2% of body weight, the brain normally receives about _____% of the

body's blood supply. This blood is conveyed by the arterial circle (of _____). Briefly state results of oxygen starvation of the brain for even about 4 minutes.

The brain stores *(much? little?)* glucose. List several effects of glucose deprivation of the brain.

■ **B3.** Describe the blood-brain barrier (BBB) in this exercise.
  a. Explain how the BBB helps to protect the brain.

  b. List two or more substances needed by the brain that normally cross the BBB.

  c. Besides the BBB, what other structures or fluids protect the brain?

B4. Review the layers of the meninges covering the brain and spinal cord by listing them here. *For extra review.* Label the layers on Figure LG 10.2 and review Checkpoint A2 (page LG 142).

■ **B5.** Circle all correct answers about CSF.
  a. The entire nervous system contains 80–150 mL of CSF. This amount is equal to approximately:
      A. 1 to 2 tablespoons   C. 1 to 2 cups
      B. ⅓ to ⅔ cup   D. 1 quart
  b. The color of CSF is:
      A. Yellow   C. Red
      B. Clear, colorless   D. Green
  c. Choose the function(s) of CSF.
      A. Serves as a shock absorber for brain and cord
      B. Contains red blood cells
      C. Contains wastes
      D. Contains nutrients
  d. How and where is CSF formed?
      a. By diffusion from neurons in the cerebrum
      B. By filtration and secretion from capillaries called choroid plexuses in the ventricles of the brain
      C. By active transport from arachnoid villi
  e. Which statement(s) describes its pathway?
      A. It is formed initially from blood, and finally flows back to blood.
      B. It circulates around the brain and the cord within the subarachnoid space.
      C. It bathes the brain by flowing through the subdural space.
      D. It passes via projections (villi) of the arachnoid into blood vessels (superior sagittal sinus) surrounding the brain.
  f. An excessive accumulation of CSF within the ventricles is a condition known as:
      A. Hydrarthrosis   C. Hydrocephalus
      B. Hydrophobia   D. Hydronephrosis

■ **B6.** To check your understanding of the pathway of cerebrospinal fluid (CSF), list in order the structures through which it passes. Use key letters on Figure LG 10.2. Start at the site of formation of CSF.

■ **B7.** Describe the principal functions of the medulla in this exercise.

a. The medulla serves as a pathway for all ascending and descending tracts. Its white matter therefore transmits *(sensory? motor? both sensory and motor?)* impulses.

b. A hard blow to the skull can be fatal because the medulla is the site of several vital centers. List these.

c. Some input to the medulla arrives by means of cranial nerves; these nerves may serve motor functions also.

Which cranial nerves are attached to the medulla? _____
(Functions of these nerves will be discussed later in this chapter.)

■ **B8.** Arrange in correct sequence from most inferior to most superior within the brainstem: _____ _____ _____

MED. Medulla   MID. Midbrain   P. Pons

■ **B9.** *A clinical challenge.* Answer these questions about the following patients (clients).

a. Elena (question A9) has an absent pupillary light reflex which specifically points to injury to the *(medulla? midbrain? pons?)*. *(Hint:* this brain part is the site of attachment of cranial nerve III which regulates pupil size).

b. Paula has a sneezing disorder that is attributed to a problem with the part of her brain that controls sneezing.

Which brain part? _____

c. Infant Olivia lacks a normal startle reflex. With which part of the brain is this reflex most associated? _____

d. Jonathan has a problem with his reticular activating system (RAS). What function is most likely to be

affected? _____

■ **B10.** Do this exercise on the diencephalon.

a. It consists mainly of the _____ and the _____ . Which part is

larger? _____ Which part is located directly above the pituitary gland?

_____

b. The thalamus is involved mostly with relay and interpretation of *(motor? sensory?)* impulses.

c. Write six words or phrases that describe key roles in the "job description" of the hypothalamus.

_____ _____ _____

_____ _____ _____

■ **B11.** *For extra review.* Check your understanding of these parts of the brain stem and diencephalon by matching them with the descriptions given below.

H. Hypothalamus	Mid. Midbrain	T. Thalamus
Med. Medulla	Pin. Pineal gland	
	Pons. Pons	

_____ a. It is the principal regulator of visceral activities because it acts as a liaison between cerebral cortex and autonomic nerves that control viscera.

_____ b. It is the site of the cerebral peduncles, which contain major sensory and motor tracts within the brain.

_____ c. Cranial nerves III–IV attach to this brain part.

_____ d. Cranial nerves V–VIII attach to this brain part.

_____ e. Cranial nerves VIII–XII attach to this brain part.

_____ f. Feelings of hunger, fullness, and thirst stimulate centers here so that you can respond accordingly.

_____ g. Almost all sensations are relayed through here, and these contribute to maintenance of consciousness.

_____ h. Regulation of heart, blood pressure, and respiration occurs by centers located here.

_____ i. It serves as the body's thermostat, regulating body temperature.

_____ j. Its name means "bridge"; it connects medulla and midbrain.

_____ k. Neuroendocrine structure since it helps to regulate the pituitary

_____ l. Site of the substantia nigra and red nucleus, both involved with muscle coordination

_____ m. Help to regulate the body's "biological clock" (2 answers)

C. Brain: cerebrum and cerebellum (pages 255–263)

C1. Describe the cerebellum.

a. Where is it located?

b. Describe its structure.

c. Describe its functions. Use these key terms: *coordinated movements, posture,* and *equilibrium.*

■ **C2.** *A clinical challenge.* Mr. Benson is diagnosed as having *ataxia* after he fell backwards down 10 steps. List three or more simple tests (requiring no equipment) that might have been used to identify his ataxia.

What part of his brain is likely to have been injured in this fall? _____

■ **C3.** Complete this exercise about cerebral structure.

a. The outer layer of the cerebrum is called the _____ . It is composed of *(white? gray?)* matter. This means that it contains mainly *(cell bodies? tracts?)*.

b. The surface of the cerebrum looks much like a view of tightly packed mountains or ridges, called

_____ . The parts where the cerebral cortex dips down into valleys are called

_____ (deep valleys) or _____ (shallow valleys).

c. The cerebrum is divided into halves called _____ . Connecting them is a band of *(white?*

gray?) matter called the _____ . Notice this structure in Figure LG 10.2 and in Figure 10.10 page 254 of your text.

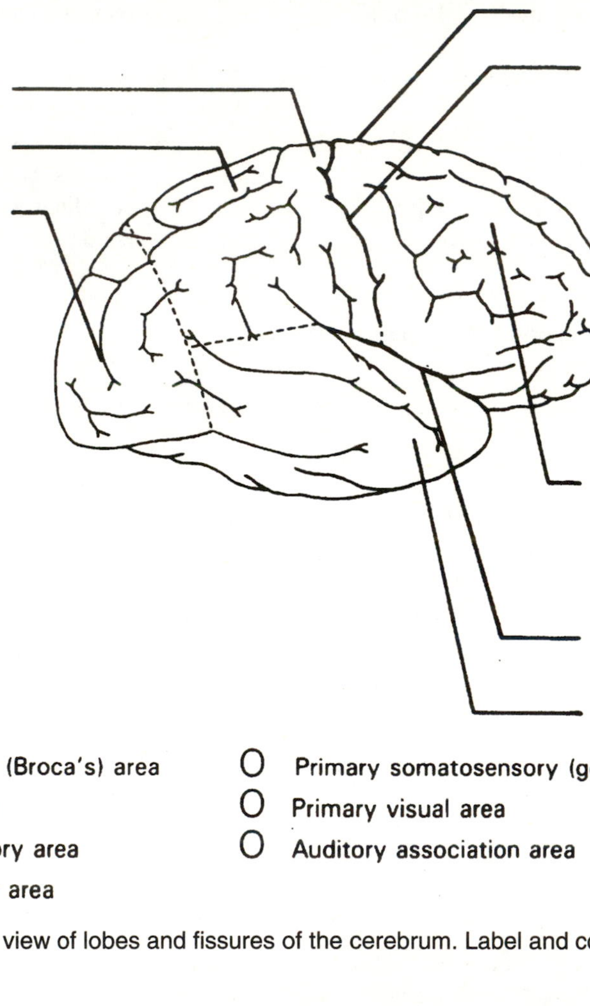

○ Motor speech (Broca's) area   ○ Primary somatosensory (general sensory) area
○ Premotor area   ○ Primary visual area
○ Primary auditory area   ○ Auditory association area
○ Primary motor area

Figure LG 10.3   Right lateral view of lobes and fissures of the cerebrum. Label and color as directed in Checkpoints C4 and C8.

■ **C4.** Label the following structures using leader lines on Figure LG 10.3: *frontal lobe, occipital lobe, parietal lobe, temporal lobe, central sulcus, lateral cerebral sulcus, precentral gyrus, postcentral gyrus.*

■ **C5.** List three functions of white matter (tracts) within the cerebrum.

■ **C6.** Refer to Figure 10.10 in the text, and complete this exercise about basal ganglia.

a. List the names of structures that are considered basal ganglia. _____ _____ _____
b. Basal ganglia are *(superficial? deep?)* in location within the cerebrum.
c. List three functions of basal ganglia.

d. Damage to the basal ganglia occurs in the condition known a _____ disease (PD). Since the client with PD lacks normal inhibition of some muscles (such as the triceps when biceps contracts), classic signs of

PD include rapidly alternately flexion and extension at joints, leading to _____ and rigidity.

C7. Describe the limbic system in this exercise.

a. List parts of the brain that are included in the limbic system.

b. Explain why the limbic system is sometimes called the "emotional" brain.

c. One other function of the limbic system is _____ . Forgetfulness, such as inability to recall recent events, results partly from impairment of this system.

■ **C8.** Color functional areas of the cerebral cortex listed with color code ovals on Figure LG 10.3.

■ **C9.** Check your understanding of functional areas of the cerebral cortex by doing this matching exercise. Answers may be used more than once.

PA. Primary auditory area	PS. Primary somatosensory area
PM. Primary motor area	PV. Primary visual area
PO. Primary olfactory arca	

_____ a. In the occipital lobe

_____ b. In the postcentral gyrus

_____ c. Receives sensations of pain, touch, itching tickling, pressure, and temperature

_____ d. In the parietal lobe

_____ e. In the temporal lobe; permits hearing

_____ f. In precentral gyrus of the frontal lobe; controls specific muscles or groups of muscles

_____ g. Controls smell

■ **C10.** Draw two important generalizations about brain functions in this activity.

a. In general, the anterior of the cerebrum is more involved with *(motor? sensory?)* control, whereas the posterior of the cerebrum is more involved with *(motor? sensory?)* functions. (Take a moment to visualize those activities taking place in the front and back of your own brain.)

b. As a general rule, *(primary? association?)* sensory areas receive sensations and *(primary? association?)* sensory areas are involved with interpretation and memory of sensations. For example, your ability to see the outline of the cerebrum, as in Figure LG 10.3, depends on your *(primary? association?)* visual areas. The fact that you can distinguish this diagram as a cerebrum (not a hand or heart), along with your memory of its structure for your next test, is based on the health of the neurons in your *(primary? association?)* visual areas.

■ **C11.** Do this exercise on the roles of the brain in verbal communication.

a. _____'s speech area is located in frontal lobes; in 97% of persons, this language control areas is located in the *(left? right?)* hemisphere.

b. Damage of Broca's speech areas results in ability to properly *(form? understand?)* words. This is known as *(fluent? nonfluent?)* aphasia.

c. Which area recognizes spoken words? _____ area

d. Which area directs a group of muscles to write in a specific sequence such as to write a sentence? _____

e. Which area of the brain controls scanning movements of eyes, such as searching for a name in a telephone

   book? _____

■ **C12.** Describe nerve pathways in this exercise.

a. Sensory and motor impulse conduction takes place mostly in tracts. These are located in *(gray horns? white*

   columns?). Ascending tracts are *(sensory? motor?)*. Descending tracts are _____ .

b. Write an arrow pointing north (↑) for ascending tracts or south (↓) for descending tracts in the table below.

Name of Tract	Direction of Impulses (↑ or ↓)	Functions
1. Spinothalamic		
2. Posterior column—medial lemnisus		
3. Corticospinal (direct)		

c. Now fill in the letter of the correct functions listed below for each tract in the table above.
   A. Control of precise, voluntary movements
   B. Sensations of precise touch, vibration, proprioception (position of muscles and joints), and stereognosis (size, shape, texture)
   C. Sensations of pain, temperature, and "crude" awareness of touch, pressure, tickle and itch

■ **C13.** Imagine that you have nicked your right thumb with a sharp scalpel in lab. Your response as you feel the pain is to move your right hand—and probably to make a mental note to be more careful in dissection work! Refer to Figures 10.14 and 10.15, pages 260 and 261 in your text, and describe pathways involved.

a. Sensory pathways from your thumb to the highest centers of your brain consist of a sequence of *(two? three?)* neurons. The first neuron carries impulses from pain receptors in your skin into the spinal nerves that enter

   the *(left? right?)* side of your _____ . (Note that in the case of sensory stimuli originating

   in the face or neck, impulses would pass into nerves that enter the _____ .) The second neuron then cross to the opposite side of the CNS transmits impulses to the *(left? right?)* side of the

   _____ region of your brain. The third neuron then transmits impulses to the

   (left? right?) side of the _____ region of your brain.

b. Integrative activities allowing thoughtful responses such as outbursts of pain as well as quick determinations to stop the bleeding and to exert more care in dissection are initiated by the *(cerebral cortex? cerebellum?)* of your brain.

c. Motor responses to muscles in your right hand consist of a *(two? three?)*-neuron sequence. The primary motor area of the left side of the cortex is the site of *(upper? lower?)* motor neurons in this pathway. Axons of these

neurons *(do? do not?)* cross to the opposite side of the CNS and descend in the _____ tracts. These axons then synapse with cell bodies of *(upper? lower?)* motor neurons located in the anterior gray horn of the *(left? right?)* side of your spinal cord. Axons of these neurons finally pass out within spinal nerves to your hand muscles, causing you, for example, to press your middle finger against your thumb to stop the bleeding.

d. Injury to *(upper? lower?)* motor neurons either within the brain or within *(posterior column? corticospinal?)* tracts, is more likely to lead to *(spastic? flaccid?)* paralysis. Explain.

e. Which other parts of the central nervous system help to regulate lower motor neurons so that smooth movements are normally produced?

■ **C14.** Circle the cerebral hemisphere that is likely to have been injured in each patient:

a. Mr. Lauren's brain damage has left him unable to speak or write. He is not able to learn sign language as a means of communication. *(Left? Right?)*

b. Ms. Ringold has lost the ability to recognize familiar songs, faces, tastes, or fragrances. *(Left? Right?)*

C15. Define the term *memory,* and give examples of parts of the brain involved with memory.

C16. Define these two terms: *brain waves, EEG.*

D. Cranial nerves (pages 263–265)

■ **D1.** Answer these questions about cranial nerves.

a. There are _____ pairs of cranial nerves. They are all attached to the _____ ; they leave

the _____ via foramina.

b. They are numbered by roman numerals in the order that they leave the cranium. Which is most anterior? *(I? XII?)* Which is the most posterior? *(I? XII?)*

c. All spinal nerves are *(purely sensory? purely motor? mixed?).* Are all cranial nerves mixed? *(Yes? No?)*

D2. Complete the table about cranial nerves.

Number	Name	Functions
a.	Olfactory	
b.		Vision (not pain or temperature of the eye)
c. III		
d.	Trochlear	
e. V		
f.		Stimulates lateral rectus muscle to abduct eye; proprioception of the lateral rectus
g.	Facial	
h.		Hearing; equilibrium
i. IX		
j.	Vagus	
k. XI		
l.		Supplies muscles of the tongue with motor and sensory fibers

■ **D3.** Check your understanding of cranial nerves by completing this exercise. Write the name and number of the correct cranial nerve following the related description.

a. Differs from all other cranial nerves in that it originates from the brain stem and from the spinal cord:

_____

b. Eighth cranial nerve (VIII): _____

c. It is widely distributed into neck, thorax, and abdomen: _____

d. Senses toothache, pain under a contact lens, wind on the face: _____

e. The largest cranial nerve; has three parts (ophthalmic, maxillary, and mandibular): _____

f. Controls contraction of muscle of the iris, causing constriction of pupil: _____

g. Innervates muscles of facial expression: _____

h. Two nerves that contain taste fibers and autonomic fibers to salivary glands: _____ ,

_____

154

i. Two purely sensory cranial nerves: _____ , _____

   which cranial nerve is almost entirely sensory? _____

j. Two cranial nerves that carry impulses to the brainstem regarding blood levels, of oxygen and carbon

   dioxide: _____ _____

E. Aging, common disorders, and medical terminology of the nervous system (pages 265-267)

■ **E1.** Describe brain development and aging in this exercise.

a. Brain growth occurs more rapidly during the *(first? last?)* few years of life. Which cells are more able to reproduce after birth? *(Neurons? Neuroglia?)*

b. With normal aging, the number of synapses between neurons is likely to _____-crease. Also conduction

   velocity _____-creases so reflex times _____-crease.

■ **E2.** Match the terms in the box with descriptions below.

> H. Hemiplegia   P. Paraplegia
> M. Monoplegia   Q. Quadriplegia

_____ a. Paralysis of one extremity only

_____ b. Paralysis of both legs

_____ c. Paralysis of both arms and both legs

_____ d. Paralysis of the arm, leg, and trunk on one side of the body

E3. A cerebrovascular accident (CVA) is commonly known as a _____. CVA is a *(common? rare?)* disorder in the U.S. Contrast a CVA with a TIA.

■ **E4.** Contrast two neural disorders by writing *ALS* after descriptions of amyotropic lateral sclerosis and *P* after descriptions of poliomyelitis.

a. Commonly known as "Lou Gehrig's disease": _____
b. Involves excessive buildup of glutamate, and excitatory neurotransmitter that can cause death of neurons: _____
c. Death typically occurs 2 to 5 years after onset (which is often about age 40): _____
d. Condition that can affect motor neurons, but typically not sensory neurons: _____
e. A vaccine had virtually eradicated this disease in the U.S, although recently, domestic incidence has increased:

   _____

f. Most cases involve only fever, headache, neck and back stiffness, deep muscle pain and weakness; but involvement of motor neurons can lead to serious and death: _____

■ **E5.** Complete this activity about Parkinson's disease (PD). (*Hint*: refer back to Checkpoint C6.)

a. PD typically affects persons about age *(10? 35? 60?)*. *(Most? Few?)* PD patients have a family history of the disease.

b. A common characteristic of PD patients is *(paralysis? shaking or tremor?)* and *(flaccid? rigid?)* muscles. This sign is caused by *(excessive? deficient?)* production of the neurotransmitter dopamine.

c. Describe facial changes that are typical of Parkinson's patients.

d. The PD patient is likely to walk with *(smaller? larger?)* than normal steps and have handwriting with *(smaller? larger?)* than normal characters.

E6. Describe Alzheimer's disease (AD) in this activity.

a. AD is *(common? rare?)* among very elderly persons. The early-onset form *(has? has not?)* been shown to involve genetic factors. Early-onset AD is *(the most common? a relatively rare?)* type of AD.

b. Early in the disease Alzheimer's patients are likely to forget *(recent? past?)* events. Over the course of the disease AD patients *(are? are not?)* likely to lose the ability to walk and talk.

c. AD involves a decrease in the number of neurons that release the neurotransmitter _____ . Other changes include increase of the abnormal protein known as _____ outside of neurons, and abnormal proteins known as _____ tangles inside of affected brain cells.

■ **E7.** Circle names of three disorders in this list that are known to be caused by viral infections.

Alzheimer disease	Parkinson disease
Amyotrophic lateral sclerosis (ALS)	Poliomyelitis
Shingles	Reye's syndrome
TIA	

■ **E8.** Select the name of the disorders in the box that fits each description below.

Dementia	Neuritis
Encephalitis	Sciatica
Neuralgia	Shingles

a. Acute inflammation of a nerve by the *herpes zoster* (chickenpox) virus: _____

b. Attack of pain along the largest nerve of the body, for example, caused by pressure of a "slipped disc" in the lumbar region: _____

c. Acute inflammation of the brain: _____

d. State of loss of intellectual disabilities, impaired memory and judgement, accompanied by change in personality; example: Alzheimer disease: _____

■ **E9.** Match the treatments of clinical conditions in the box with related descriptions below.

Analgesia	Anesthesia	Epidural block	Nerve block

a. Loss of sensation, for example, during surgery: _____
b. Pain relief, for example, by action of aspirin, ibuprofen, topical agents, or injected nerve block: _____
c. Injection of local anesthetic, for example, by a dentist, to control pain during dental procedures
d. Injection of anesthesia into the space between the dura mater and vertebrae, for example, to control pain during labor and delivery

ANSWERS TO SELECTED CHECKPOINTS: CHAPTER 10

A1. Central, peripheral.
A2. (a) Bones (cranial bones and vertebrae), meninges, and cerebrospinal fluid (CSF), as well as vertebral ligaments for spinal cord. (b) 1, Dura mater; 2, subdural space; 3, arachnoid; 4, subarachnoid space*; 5, pia mater. (c) Meningitis.
A3. (a) Waist; L2; vertebral column; 16–18. (b) Cervical, lumbar. (c) Horse's tail; lumbar and spinal nerves that droop downward (like a horse's tail) below the end of the spinal cord. These nerves then

exit through openings (foramina) between lumbar vertebrae or through the sacrum. (d) Cerebrospinal (CSF), subarachnoid; L3–4 or L4–5; because the spinal cord ends at a higher level (about L2) so it is not likely to be injured, and cauda equina nerves are much like cooked pasta noodles in a bowl of water (CSF) in that a needle inserted into the area is not likely to puncture a single nerve (noodle).
A4. (a) Within, horns; refer to Figure 10.3, page 249 in your text. (b) Outside, columns; sensory,

motor; are, (c) 1, posterior white column; 2, posterior gray horn; 3, lateral white column; 4, lateral gray horn; 5, anterior gray horn; 6, anterior white column; 7, anterior median fissure. (d) automonic thoracic, upper lumbar, sacral.

A5. (a) 31; 8, 12, 5, 5, 1. (b) Lumbar, sacral, and coccygeal. (c) Sensory, motor, mixed.
(d) Endoneurium; epineurium. (e) Plexuses; thoracic (T2–12) forming intercostal nerves that lie in grooves in the inferior of each rib.

A6. (a) B. (b) C. (c) B. (d) S. (e) C. (f) L.

A7. 2; posterior (or dorsal), anterior (or ventral).

A8. Reflex. A, receptor; B, sensory neuron dendrite; C, sensory neuron cell body; D, sensory neuron axon; E, integrating center (synapse or possibly an interneuron here; F, motor neuron cell body; G, motor neuron axon; H, effector (such as skeletal muscle fiber). Add arrows alphabetically from A → H, as shown on Figure 10.5, page 248 of the text.

A10. (a) 2; patellar reflex [knee jerk]. (b) Somatic; autonomic. (c) cranial

B1. Three (3); trillions, neuroglia. (a) Brain stem. (b) Diencephalon. (c) Cerebrum. (d) Cerebellum.

B3. (a) It protects the brain from many harmful chemicals and pathogens. (b) Oxygen, carbon dioxide, nutrients, and most anesthetics. (c) Cranial bones, meninges, and cerebrospinal fluid (CSF).

B5. (a) B. (b) B. (c) A, C, D. (d) B. (e) A, B, D. (f) C.

B6. D E F G H I J C A B.

B7. (a) Both sensory and motor. (b) Centers controlling heartbeat, diameter of blood vessels, and breathing; also important reflexes such as coughing and swallowing. (c) Part of VIII, as well as IX–XII.

B8. MED P MID.

B9. (a) Midbrain. (b) Medulla. (c) Midbrain. (d) Sleep.

B10. (a) Thalamus and hypothalamus; thalamus; hypothalamus. (b) Sensory. (c) Autonomic, pituitary, emotional/behavioral, thirst and eating, body temperature, circadian rhythm.

B11. (a) H. (b) Mid. (c) Mid. (d) Pons. (e) Med. (f) H. (g) T. (h) Med. (i) H. (j) Pons. (k) H. (l) Mid. (m) H, Pin.

C2. Ask him to touch his nose while his eyes are closed (cannot do), rapidly supinate and pronate his forearms on his thighs (cannot do), walk a straight line (staggering), and speak (slurring); cerebellum.

C3. (a) Cerebral cortex; gray; cell bodies. (b) Gyri; fissures, sulci. (c) Hemispheres; white, corpus callosum.

C4.

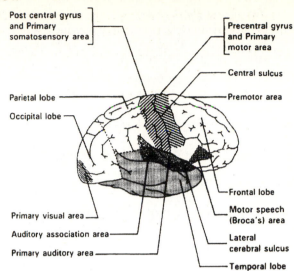

Figure LG 10.3A Right lateral view of lobes and fissures of the cerebrum.

C5. To transmit nerve impulses between (a) gyri within the same hemisphere of the cerebrum; (b) the right and left hemispheres; (c) the cerebrum and other levels of the brain or spinal cord.

C6. (a) Globus pallidus, putamen, and caudate nucleus. (b) Deep. (c) Help initiate and terminate movements, help regulate muscle tone and control subconscious contractions of skeletal muscles. (d) Parkinson's disease; tremor.

C8. See Figure LG 10.3A.

C9. (a) PV. (b) PS. (c) PS. (d) PS. (e) PA. (f) PM. (g) PO.

C10. (a) Motor, sensory. (b) Primary, association; primary; association.

C11. (a) Broca's; left. (b) Form, nonfluent. (c) Wernicke's. (d) Premotor. (e) Frontal eye field.

C12. (a) White columns; sensory; motor. (b, c)

Name of Tract	Direction of Impulses (↑ or ↓)	Functions
1. Spinothalamic	↑	C
2. Posterior column-medial lemniscus	↑	B
3. Corticospinal (direct)	↓	A

C13. (a) Three; right, spinal cord; brain stem; left, thalamus; left, cerebral cortex. (b) Cerebral cortex. (c) Two; upper; do, direct (corticospinal); lower, right. (d) Upper, corticospinal, spastic; lower motor neurons can still be stimulated by other neurons, for example, by sensory neurons in spinal cord reflexes such as the "knee jerk." (e) Neurons with cell bodies in basal ganglia or in the cerebellum, and sensory neurons (cell bodies in sensory ganglia) or local interneurons involved in spinal cord reflexes.

C14. (a) Left. (b) Right.

D1. (a) 12; brain; cranium. (b) I; XII. (c) Mixed; no (two are purely sensory).
D3. (a) Accessory (XI). (b) Vestibulocochlear. (c) Vagus (X). (d) Trigeminal (V). (e) Trigeminal (V). (f) Oculomotor (III). (g) Facial (VII). (h) Facial (VII) and glossopharyngeal (IX). (i) Olfactory (I), optic (II); vestibulocochlear (VIII). (j) Glossopharyngeal (IX) and vagus (X).
E1. (a) First; neuroglia. (b) De; de, in.
E2. (a) M. (b) P. (c) Q. (d) H.
E4. (a) ALS. (b) ALS. (c) ALS. (d) ALS and P. (e) P. (f) P.

E5. (a) 60; few. (b) Shaking or tremor, rigid; deficient. (c) Masklike expression and drooling. (d) Smaller; smaller.
E6. (a) Common; has; relatively rare. (b) Recent; are (c) Acetylcholine; beta-amyloid, neurofibrillary.
E7. Polio, Reye's syndrome (in children or teens usually after aspirin also), shingles.
E8. (a) Shingles. (b) Neuritis. (c) Neuralgia. (d) Encephalitis. (e) Anesthesia. (f) Analgesia.
E9. (a) Anesthesia. (b) Analgesia. (c) Nerve block. (d) Epidural block.

CRITICAL THINKING: CHAPTER 10

1. Describe the anatomy and physiology of the meninges, as well as the spaces formed between and surrounding meninges.
2. Describe a reflex arc, including the structure and function of its five components.
3. Define the term *plexus* and describe what is meant by *ventral rami* that form a plexus. State examples of two plexuses and the major nerves they form.
4. Describe the structure and functions of the following: (a) limbic system, (b) diencephalon, (c) reticular formation.

5. Contrast motor and sensory functions of cranial nerves V, VII, and IX.
6. Contrast a CVA with a TIA.
7. Consider what effects are likely to result from the following: (a) blockage of the cerebral aqueduct that connects the third and fourth ventricles of the brain by a tumor of the midbrain; (b) a severe blow to the medulla; (c) a stroke involving the occipital lobe.

MASTERY TEST: ■ CHAPTER 10

Questions 1–5: Arrange the answers in correct sequence.

_____ _____ _____ 1. From superficial to deep:
   A. Subarachnoid space
   B. Epidural space
   C. Dura mater

_____ _____ _____ 2. From superior to inferior:
   A. Thalamus
   B. Hypothalamus
   C. Corpus callosum

_____ _____ _____ _____ 3. Pathway of cerebrospinal fluid, from formation to final destination:
   A. Choroid plexus in ventricle
   B. Subarachnoid space
   C. Superior sagittal sinus
   D. Arachnoid villi

_____ _____ _____ _____ 4. The plexuses, from superior to inferior:
   A. Lumbar
   B. Brachial
   C. Cervical
   D. Sacral

_____ _____ _____ _____ _____ 5. Order of structures in a conduction pathway, from origin to termination:

   A. Motor neuron
   B. Sensory neuron
   C. Integrative center
   D. Receptor
   E. Effector

Questions 6–10: Circle the letter preceding the one best answer to each question.

6. Damage to the occipital lobe of the cerebrum would most likely cause:
   A. Loss of hearing
   B. Loss of vision
   C. Loss of ability to smell
   D. Paralysis
   E. Loss of feeling in muscles (proprioception)

7. Which of these is a function of the postcentral gyrus?
   A. Controls specific groups of muscles, causing their contraction
   B. Receives general sensations from skin, muscles, and viscera
   C. Receives olfactory impulses
   D. Primary visual reception area
   E. Primary auditory area

8. All of these are functions of the hypothalamus *except:*
   A. Control of body temperature
   B. Hunger and thirst center
   C. An integrating center for the autonomic nervous system (ANS)
   D. The principal relay center for sensory impulses
   E. Involved in maintaining sleeping or waking state

9. Herniation (or "slipping") of the disc between L4 and L5 vertebrae is most likely to result in damage to the _____ nerve.
   A. Femoral   C. Radial
   B. Sciatic   D. Musculocutaneous

10. Choose the *false* statement about the spinal cord.
   A. It has enlargements in the cervical and lumbar areas.
   B. It lies in the vertebral foramen.
   C. It extends from the medulla to the sacrum.
   D. It is surrounded by meninges.
   E. In cross section an H-shaped area of gray matter can be found.

Questions 11–15: Fill-ins. Complete each sentence with the word or phrase that best fits.

_____ 11. The phrenic nerve innervates the _____ .

_____ 12. The limbic system functions in control of _____ .

_____ 13. The lumbar and sacral nerves extending below the end of the cord and resembling a horse's tail are known as the _____ .

_____ 14. An inflammation of the dura mater, arachnoid, or pia mater is known as _____ .

_____ 15. The _____ controls wakefulness.

ANSWERS TO MASTERY TEST: ■ CHAPTER 10

Arrange
1. B C A
2. C A B
3. A B D C
4. C B A D
5. D B C A E

Multiple Choice
6. B
7. B
8. D
9. B
10. C

Fill-ins
11. Diaphragm
12. Emotional aspects of behavior
13. Cauda equina
14. Meningitis
15. Reticular activating system (RAS)

FRAMEWORK 11
The Autonomic Nervous System (ANS)

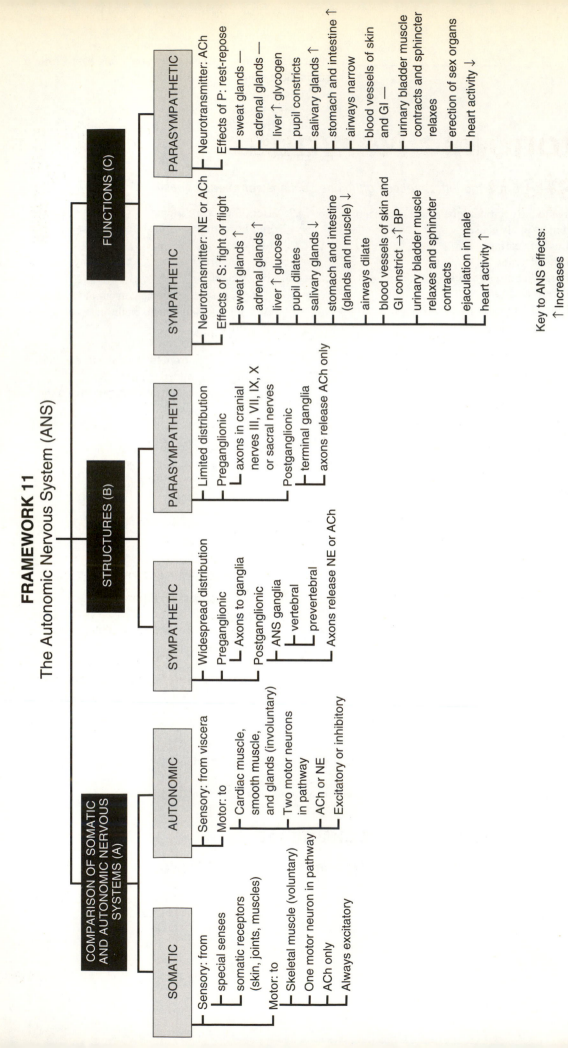

Autonomic Nervous System

The autonomic nervous system (ANS) consists of the special branch of the nervous system that exerts unconscious control over viscera. The ANS regulates activities such as heart rate and blood pressure, glandular secretion, and digestion. The two divisions of the ANS—the sympathetic and parasympathetic—carry out a continuous balancing act, readying the body for response to stress or facilitating rest and relaxation, according to momentary demands. ANS neurotransmitters hold particular significance clinically because they are often mimicked or inhibited by medications.

As you begin your study of ANS, carefully examine the Chapter 11 Topic Outline and Objectives; check off each one as you complete it. To organize your study of this content, glance over the Chapter 11 Framework now. Be sure to refer to the Framework frequently and note relationships among key terms in each section.

TOPIC OUTLINE AND OBJECTIVES

A. Comparison of somatic and autonomic nervous systems

☐ 1. Compare the main structural and functional differences between the somatic and autonomic parts of the nervous system.

B. Structure of the autonomic nervous system

☐ 2. Identify the structural features of the autonomic nervous system.

C. Functions of the autonomic nervous system

☐ 3. Describe the functions of the sympathetic and parasympathetic divisions of the autonomic nervous system.

D. Focus on wellness; common disorders

WORDBYTES

Now study each wordbyte, its meaning, and an example of its use in a term. After you study the entire list, check your understanding by writing the meaning of each wordbyte in the margin. As you continue through the *Learning Guide,* identify (and fill in) additional terms that contain the same wordbyte.

Wordbyte	Meaning	Example(s)	Wordbyte	Meaning	Example(s)
auto-	self	*auto*nomic	post-	after	*post*ganglionic
-nomos	law, governing	auto*nomic*	pre-	before	*pre*ganglionic
para-	near, beside	*para*vertebral, *para*thyroid			

CHECKPOINTS

A. Comparison of somatic and autonomic nervous systems (page 272-273)

■ **A1.** Why is the autonomic nervous system (ANS) so named? Is the ANS entirely independent of higher control centers? Explain.

■ **A2.** Contrast the somatic and autonomic nervous systems in this table.

	Somatic Motor	Autonomic Motor
a. Motor control is *(voluntary? involuntary?)*		
b. Types of tissue innervated by motor nerves	Skeletal muscle	
c. Motor neurons are excitatory (E), inhibitory (I), or both (E + I)		
d. Number of neurons in efferent pathway		
e. Neurotransmitter(s) released by motor neurons	Acetylcholine (ACh)	

■ **A3.** Circle all the sensations that are classified as *visceral sensations*.

A. Pain of a stomach ulcer

B. Awareness that you are contracting your quadriceps femoris muscle

C. Detection by the medulla of high levels of carbon dioxide (CO_2) in the blood

■ **A4.** Name the two divisions of the ANS. _____ _____

■ **A5.** Many visceral organs have dual innervation by the autonomic system.

a. What does dual innervation mean?

b. How do the sympathetic and parasympathetic divisions work in harmony to control viscera?

B. Structure of the autonomic nervous system (pages 273–277)

■ **B1.** Complete this exercise summarizing structural differences between autonomic motor and somatic motor pathways.

a. Somatic pathways begin at *(all? only certain?)* levels of the cord, whereas ANS routes begin at

_____ levels of the cord.

b. Between the spinal cord and effector, somatic pathways include _____ neuron(s), whereas autonomic pathways

require _____ neuron(s), known as the pre- _____ and _____ neurons.

■ **B2.** Contrast preganglionic and postganglionic neurons of the two divisions of the ANS in this exercise.

a. All preganglionic neurons have their cell bodies in the *(CNS? PNS?)*, specifically in *(brain stem or spinal cord?*

ganglia?). All postganglionic neurons have their cell bodies located in _____ , which are clusters of cell bodies that lie in the *(CNS? PNS?)*.

b. Sympathetic preganglionic neurons have their cell bodies located in the *(cervical? thoracic? lumbar? sacral?)* regions of the spinal cord. Axons of these neurons pass to cell bodies of postganglionic neurons located in one of two types of ganglia where synapsing occurs. These are known as *(circle two)*:

A. Sympathetic trunk ganglia   C. Terminal ganglia

B. Prevertebral ganglia   D. Posterior root ganglia

c. The superior cervical ganglion is a sympathetic *(trunk? prevertebral?)* ganglion that sends ANS axons mainly to the *(head? thorax?)*. Middle and inferior cervical ganglia primarily supply nerves to the *(head? heart? abdominal organs?)*. The celiac as well as superior and inferior mesenteric ganglia are sympathetic *(trunk? prevertebral?)* ganglia that innervate the *(head? heart? abdominal organs?)*.

d. Where are cell bodies of preganglionic neurons of the parasympathetic division located? _____

In which nerves are axons of these neurons contained? Cranial nerves _____ , _____ , _____ , _____ , and in spinal nerves from the _____ portion of the cord. Where do parasympathetic preganglionic axons synapse?

e. Most viscera *(do? do not?)* receive fibers from both the sympathetic and parasympathetic divisions of the ANS. However, the origins of these divisions in the CNS and the pathways taken to reach viscera *(are the same? differ?).*

■ **B3.** Refer to Figures 11.1 and 11.2, pages 272 and 274 in your text. Image walking along a sympathetic pathway. Now describe that route in this exercise.

a. First arrange in order the structures in the pathway common to all sympathetic preganglionic neurons:

_____ _____ _____

   A. Anterior root of spinal nerve and myelinated branch (ramus) off it
   B. Sympathetic trunk ganglion
   C. Gray matter in the thoracic or lumbar (T1 and L2) levels of the spinal cord

b. Where is the first place a sympathetic preganglionic neuron could end and synapse with a postganglionic neuron?

_____ Where else could that neuron end and synapse? _____

c. With how many neurons does a sympathetic preganglionic neuron typically synapse? *(1? 4 or 5? 20 or more?)* (Notice in Figure 11.2 that these blue axons can ascend or descend the trunk ganglia and give off as many as 20 branches, each of which can synapse with a different postganglionic neuron.) Does any one branch of a sympathetic preganglionic synapse more than once? *(Yes? No?)*

d. In Figure 11.2, axons of preganglionic neurons are shown as *(solid? broken?)* lines, and axons of postganglionic neurons are shown as *(solid? broken?)* lines.

■ **B4.** Contrast parasympathetic (P) and sympathetic (S) pathways in this exercise. (Refer to Figures 11.2 and 11.3 [pages 275 and 276] in the text.)

a. Which division has more preganglionic axons that are very long (right up to the effector organ) and has very short postganglionic axons (located within the effector organ)? _____ Explain why based on the anatomy of that division.

b. Which division of the ANS exerts a more widespread effect on the body? _____ Circle correct answers below to explain reasons for this fact:

   (1) Only the *(P? S?)* division has preganglionic neurons that branch out to trunk ganglia located on the entire length of the spinal column from cervical vertebrae through sacrum.
   (2) *(P? S?)* preganglionic neurons synapse with about 20 postganglionic neurons heading to a variety of effectors (See checkpoint B3c), whereas *(P? S?)* preganglionic neurons synapse with only about 4 or 5 postganglionic neurons leading to only one effector.
   (3) Some viscera, such as adrenal glands, sweat glands, hair muscles, and most blood vessels, receive only *(P? S?)* innervation.

C. Functions of the autonomic nervous system   (pages 277-280)

■ **C1.**   Complete this exercise about autonomic neurotransmitters.

a. Which neurotransmitter is released by all parasympathetic postganglionic neurons?

  ACh. Acetylcholine   NE. Norepinephrine

b. Which neurotransmitter is released by most sympathetic postganglionic neurons? *(ACh? NE?)*

c. During stress, the *(sympathetic? parasympathetic?)* neurons predominate. Therefore, stress responses primarily involve the neurotransmitter *(ACh? NE?)*.

d. The ANS neurotransmitter that "hangs around" longer (has more lasting effect) is *(ACh? NE?)*. This fact provides one explanation for the longer-lasting effects of *(sympathetic? parasympathetic?)* neurons.

■ **C2.**   After you study Table 11.2, page 279 of your text, use arrows to indicate effects of parasympathetic (P) and sympathetic (S) nerves on the activities listed below. The first one is done for you. Choose from the following answers:

  ↑. Stimulate or increase   ↓. Inhibit or decrease   —. No innervation

a. P ___↓___   S ___↑___   Dilation of pupil

b. P _____   S _____   Heart rate and blood flow to coronary arteries that supply heart muscle

c. P _____   S _____   Blood sugar level

d. P _____   S _____   Salivation and other digestive organ contractions and secretions

e. P _____   S _____   Erection of genitalia

f. P _____   S _____   Dilation of airways for easier breathing

g. P _____   S _____   Contraction of bladder and relaxation of internal urethral sphincter, causing urination

h. P _____   S _____   Contraction of smooth muscle of hair follicles, causing "goose bumps"

i. P _____   S _____   Release of epinephrine and norepinephrine from adrenal medulla

j. P _____   S _____   Coping with stress, fight-or-flight responses

k. P _____   S _____   Enhances "rest-and-digest" activities

l. P _____   S _____   Increases sweat to cool the body

m. P _____   S _____   Contracts spleen to release stored blood to the general circulation (which increases blood pressure)

n. P _____   S _____   Stimulates breakdown of adipose tissue to increase blood levels of fatty acids

o. P _____   S _____   Secretion of ADH by the posterior pituitary, causing increase in blood volume and blood pressure

■ **C3.** The acronymn "SLUDD" describes responses when *(sympathetic? parasympathetic?)* tone is increased.

Write the five functions described by SLUDD: S _____

L _____ U _____ D _____ D _____

D. Focus on wellness; common disorders (pages 278 and 280)

D1. List several examples of mind-body activities, and state several helpful effects for lowering stress.

■ **D2.** Review ANS disorders in this exercise. Fill in the blanks with names of disorders listed in the box, and also circle correct answers.

AD. Autonomic dysreflexia	RD. Raynaud's disease

a. _____ is a condition that occurs in *(most? few?)* persons with spinal cord injury (SCI) at or above level T6. A common trigger to an AD episode is overstretching of the urinary bladder (due to lack of sensation of fullness) so that this organ presses against the sympathetic trunk, overstimulating it. Blood vessels in regions below the injury then *(dilate? constrict?),* leading to pale, cool skin there. More significant is the consequence of diversion of blood to upper body parts so that blood pressure there becomes dangerously *(high? low?).*

b. _____ most commonly occurs in *(warm? cold?)* climates and among young *(women? men?).* This condition involves excessive *(sympathetic? parasympathetic?)* stimulation of smooth muscles of arterioles in fingers and

toes—which then appear _____. Digits may later appear red. Explain.

A1. At one time this system was thought to be autonomous (self-governing). However, it is regulated by brain centers in the hypothalamus and brain stem, with input from the cerebrum.

A2.

	Somatic Motor	Autonomic Motor
a. Motor control is (voluntary? involuntary?)	Voluntary	Involuntary
b. Types of tissue innervated by motor nerves	Skeletal muscle	Cardiac muscle, smooth muscle, and most glands
c. Motor neurons are excitatory (E), inhibitory (I), or both (E + I)	E	E + I
d. Number of neurons in efferent pathway	One (from spinal cord to effector)	Two (pre-and postganglionic)
e. Neurotransmitter(s) released by motor neurons	Acetylcholine (ACh)	ACh or norepinephrine (NE)

A3. A C

A4. Sympathetic, parasympathetic.

A5. (a) Innervation by both sympathetic and parasympathetic divisions. (b) One division excites, and the other division inhibits the organ's activity.

B1. (a) All, only certain. (b) 1, 2, ganglionic, postganglionic.

B2. (a) CNS, brain stem or spinal cord; ganglia, PNS. (b) Thoracic and lumbar; A, B. (c) Trunk, head; heart; prevertebral, abdominal organs. (d) Brain stem and sacral portions of the spinal cord; III, VII, IX, and X, sacral; cell bodies of postganglionic neurons located in terminal ganglia (within the walls of the effector organs themselves). (e) Do; differ.

B3. (a) C A B. (b) Sympathetic chain ganglion (at the same level as that pathway began, such as the chain ganglion at level T1 or T12); a pre-vertebral ganglion (such as celiac ganglion). (c) 20 or more; no. (d) Solid; broken.

B4. (a) P; because the preganglionic neurons end and synapse in terminal ganglia that are in the walls of viscera. (b) S. (b1) S. Note in Figure 11.3 that the parasympathetic division has no trunk ganglia. (b2) S, P. (b3) S. Refer to the left side of Figure 11.2 that shows one of millions of postganglionic neurons that pass from sympathetic ganglia back out to gain access to spinal nerves to reach organs such as blood vessels, sweat glands, and hair muscles. Anatomically, parasympathetic nerves have no pathways to allow them to pass into every spinal nerve, so they cannot innervate these "outlying organs". Arms, legs, and skin over thorax and abdomen, have no parasympathetic nerves whatsoever in them!

C1. (a) ACh. (b) NE. (c) NE. (d) NE; sympathetic.

C2. (b) P ↓, S ↑. (c) P ↓, S ↑. (d) P ↑, S ↓. (e) P ↑, S ↓. (f) P ↓, S ↑. (g) P ↑, S ↓. (h) P —, S ↑. (i) P—, S ↑. (j) P ↓, S ↑. (k) P ↑, S ↓. (l) P —, S ↑. (m) P —, S ↑. (n) P —, S ↑. (o) P —, S ↑.

C3. Parasympathetic; S, salivation; L, lacrimation (tear formation); U, urination; D, digestion; D, defecation.

D2. (a) AD; most; constrict; high. (b) RD, cold, women; sympathetic, more white or blue (cyanotic) than normal; erythema results from rewarming after cold exposure.

CRITICAL THINKING: CHAPTER 11

1. Identify ten specific organs in your own body that are supplied by autonomic neurons.

2. Explain why the names *craniosacral* and *thoracolumbar* are given to the two divisions of the ANS.

3. Contrast locations of autonomic ganglia in sympathetic and parasympathetic divisions of the ANS.

4. Describe a "sympathetic response" that you might exhibit when faced with a threatening situation. Include specific changes likely to occur in your viscera.

5. State three reasons why the parasympathetic division of the autonomic nervous system exerts a more limited effect than the sympathetic division.

Questions 1–3: Circle the letter preceding the one best answer to each question.

1. All of the following are parasympathetic effects *except:*
   A. Increase contractions of stomach and intestine
   B. Increase salivation
   C. Constrict the pupil
   D. Increase sweating and cause "goose pimples"

2. All of the following are activities characteristic of stress response, or fight-or-flight reaction, *except:*
   A. The liver breaks down glycogen to glucose.
   B. Blood vessels in skin and abdominal organs widen so that more blood flows to those regions.

   C. Hairs stand on end ("goose bumps") due to contraction of smooth muscles of hair follicles.
   D. The heart rate increases.

3. All of the following nerve fibers release the neurotransmitter acetylcholine *except:*
   A. All somatic neurons
   B. All preganglionic axons
   C. All postganglionic parasympathetic axons
   D. Most postganglionic sympathetic axons

Questions 4–10: Circle T (true) or F (false). If the statement is false, change the underlined word or phrase so that the statement is correct.

T F 4. Sympathetic cardiac nerves <u>stimulate</u> heart rate, and the vagus <u>slows down</u> heart rate.

T F 5. Preganglionic sympathetic nerve fibers are found in anterior roots of spinal nerves emerging from <u>cervical</u> and <u>lumbar</u> regions of the spinal cord.

T F 6. <u>Both sympathetic and parasympathetic</u> nerve fibers are found in all spinal nerves.

T F 7. Preganglionic sympathetic nerve fibers synapse in <u>terminal</u> ganglia.

T F 8. Preganglionic parasympathetic nerve fibers synapse in <u>sympathetic trunk ganglia and prevertebral ganglia.</u>

T F 9. Under stress conditions the <u>sympathetic</u> system dominates over the <u>parasympathetic</u> system.

T F 10. The sympathetic division of the ANS has a <u>more</u> widespread effect in the body than the parasympathetic division.

Questions 11–15: Fills-ins. Write the word or phrase that best fits the description.

_____ 11. The three main types of tissue (effectors) innervated by the ANS nerves

are _____ .

_____ 12. List the four "E situations" that trigger sympathetic responses.

_____ 13. The _____ is the part of the brain that serves as the major control and integration center of the ANS.

_____ 14. The cranial nerves that contain parasympathetic neurons are _____ , _____ ,

_____ , and _____ .

_____ 15. The nerves that carry almost all parasympathetic preganglionic axons in the body

are the _____ .

Multiple Choice

1. D
2. B
3. D

True or False

4. T
5. F. Thoracic and lumbar
6. F. Sympathetic
7. F. Sympathetic trunk ganglia and prevertebral ganglia
8. F. Terminal ganglia
9. T
10. T

Fill-ins

11. Cardiac muscle, smooth muscle, and glandular epithelium
12. Exercise, emergency, excitement, and embarrassment
13. Hypothalamus
14. III, VII, IX, and X
15. Vagus (cranial nerve X)

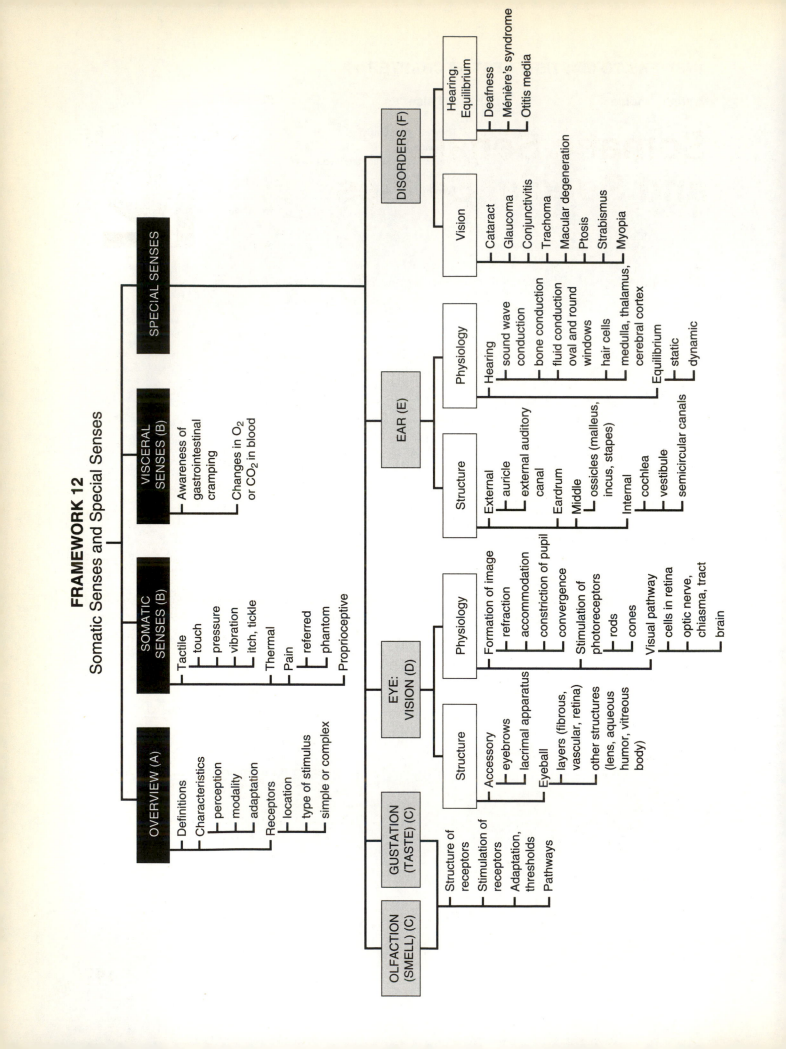

FRAMEWORK 12
Somatic Senses and Special Senses

Somatic Senses and Special Senses

The past two chapters have introduced the components of the nervous system and demonstrated the arrangement of neurons in the spinal cord, brain, spinal nerves, and cranial nerves. Chapter 12 integrates this information in the study of major nerve pathways including those for general senses (such as touch, pain, and temperature) and special senses including smell, taste, vision, hearing, and equilibrium. Special afferent pathways in many ways resemble general afferent pathways. However, a major point of differentiation is the arrangement of special sense receptors in complex sensory organs, specifically the nose, tongue, eyes, and ears.

As you begin your study of sensations, carefully examine the Chapter 12 Topic Outline and Objectives; check off each one as you complete it. To organize your study of sensations, glance over the Chapter 12 Framework now. Be sure to refer to the Framework frequently and note relationships among key terms in each section.

TOPIC OUTLINE AND OBJECTIVES

A. Overview of sensations

☐ 1. Define a sensation and describe the conditions necessary for a sensation to occur.

B. Somatic senses

☐ 2. Describe the location and function of the receptors for tactile, thermal, and pain sensations.

☐ 3. Identify the receptors for proprioception and describe their functions.

C. Olfaction: sense of smell; gustation: sense of taste

☐ 4. Describe the receptors for olfaction and the olfactory pathway to the brain.

☐ 5. Describe the receptors for gustation and the gustatory pathway to the brain.

D. Vision

☐ 6. Describe the accessory structures of the eye, the layers of the eyeball, the lens, the interior of the eyeball, image formation, and binocular vision.

☐ 7. Describe the receptors for vision and the visual pathway to the brain.

E. Hearing and equilibrium

☐ 8. Describe the structures of the outer, middle, and inner ear.

☐ 9. Describe the receptors for hearing and equilibrium and their pathways to the brain.

F. Focus on homeostasis; common disorders, medical terminology

WORDBYTES

Study each wordbyte, its meaning, and an example of its use in a term. Check your understanding by writing the meaning of each wordbyte in the margin. As you continue through the *Learning Guide,* identify (and fill in) additional terms that contain the same wordbyte.

Wordbyte	Meaning	Example(s)	Wordbyte	Meaning	Example(s)
aqua-	water	*aque*ous humor	olfact-	smell	*olfact*ory
-esthesia	sensation	an*esthesia,* par*esthesia*	ophthalm-	eye	*ophthalm*ologist
hemi-	half	*hemi*plegia, *hemi*sphere	opt-	eye	*opt*ic nerve, *opt*ician
kin-	motion	*kin*esthesia	orbit-	eye socket	supra*orbital*
macula	spot	*macula* lutea	ossi-	bone	*ossi*cle
med-	middle	otitis *med*ia	ot(o)	ear	*oto*lith
noci-	hurt, pain	*noci*ceptor	sclera	hard	*sclera*
ocul-	eye	*ocul*omotor nerve	tact-	touch	*tact*ile
			tympano-	drum	*tympan*ic membrane

CHECKPOINTS

A. Overview of sensations   (page 285-286)

A1. For a moment, visualize what your life would be like if you were unable to experience any sensations. List the three types of sensations that you believe you would miss most.

_____   _____   _____

Now write a sentence describing how your health and safety might be endangered by lack of ability to perceive sensations.

■ **A2.** Match each term in the box with the best description below. Use each answer only once.

GS. General sense	SpS. Special sense
SomS. Somatic sense	VS. Visceral sense

_____ a. Any sense that is not a special sense, for example touch or pressure

_____ b. Sense transmitted by receptors in your stomach that inform you that you are hungry

_____ c. Sense that allows you to smell and taste soup

_____ d. Sense of pain that you feel when you spill hot soup on your arm

■ **A3.** Arrange in correct order the conditions necessary for a sensation to occur by writing in the name of the term after its description below. Use the following list of terms: *conduction, receptor* or *sense organ, stimulus, translation.*

a. A change in the environment of a neuron that can alter its permeability to ions: _____

$\downarrow$

b. Picks up stimulus and converts it from resting potential to an electrical signal and nerve impulse when threshold

is reached: _____

$\downarrow$

c. Nerve impulse transmitted to the central nervous system: _____

$\downarrow$

d. Nerve impulse is converted to sensation; usually occurs in the cerebral cortex: _____

■ **A4.** Do this exercise on two characteristics of sensations.

a. Recall getting dressed this morning. The fact that you felt the shirt touching your back just after you put on that shirt, but that the awareness of that touch has dissipated over time is known as the characteristic of

_____ . Explain how relatively slow adaptation to pain is advantageous to you.

b. You actually see in your *(eyes? brain stem? cerebral cortex?)*, although you seem to see with your eyes. This

characteristic of sensation is known as _____ and is the result of integration in the *(thalamus? cerebral cortex?)*.

■ **A5.** Identify types of receptors listed below according to their structural classifications. Select from answers in the box.

> ENE. Encapsulated nerve endings
>
> FNE. Free nerve endings
>
> SC. Separate cells

_____ a. Receptors for in the retina of the eye for vision and in the inner ear for hearing

_____ b. Receptors for pain, itch, or tickle

_____ c. Receptors for touch or pressure

■ **A6.** Select terms from the box to correctly categorize different classes of sensory receptors according to function. Write one answer on each line provided.

Awareness of joint positions	Pain	Touch
Change of blood pressure	Pressure	Vibration
Cold	Smell	Vision
Hearing	Taste	Warm

a. Thermoreceptor: _____ _____

b. Chemoreceptor: _____ _____

c. Nocireceptor: _____

d. Mechanoreceptor: _____ _____ _____

_____ _____ _____

e. Photoreceptor: _____

B.   Somatic senses   (pages 286–290)

■ **B1.**   Circle the locations of receptors below that are classified as *somatic* receptors. (The other answers are all sites of *visceral* receptors). Receptors that sense:

A. Injury to your tongue after you accidentally bite down on it

B. Joint structures that allow you to know the exact position of your right knee

C. Cramp in the gastrocnemius muscle

D. Cramping of the intestine that accompanies a gastrointestinal infection

E. Changes in oxygen level in blood passing through your aorta

F. Itching on your neck from a mosquito bite

■ **B2.**   Do the following exercise about tactile sensations.
a. List five types of tactile sensation.

b. Receptors for these sensations are all (*evenly? unevenly?*) distributed throughout skin.

■ **B3.**   Match names of receptors with their descriptions. Answers may be used more than once.

C. Corpuscles of touch (Meissner's corpuscles)	N. Nociceptors
HRC. Hair root complexes	T. Tendon organs
J. Joint kinesthetic receptors	Type I. Type I cutaneous mechanoreceptors (tactile or Merkel discs)
L. Lamellated (Pacinian) corpuscles	Type II. Type II cutaneous mechanoreceptors (end organs of Ruffini)
M. Muscle spindles	

_____ a. Egg-shaped receptors located in dermal papillae, especially in fingertips, palms of hands, and soles of feet; rapidly adapting

_____ b. Onion-shaped structures sensitive to pressure and high-frequency vibration

_____ c. Free nerve endings that sense pain

_____ d. Touch receptors (two answers); slowly adapting

_____ e. Proprioceptors (three answers)

_____ f. May respond to any type of stimulus if stimulus is strong enough to cause tissue damage

_____ g. Respond to chemicals released from injured tissue, such as prostaglandins

_____ h. Receptors between skeletal muscle cells that are sensitive to stretch

_____ i. Receptors that protect muscles and their tendons from excessive tension

_____ j. Receptors stimulated when you feel an ant walking on your skin

B4.   *For extra review.* Refer to Figure LG 5.1 (page 59). Identify and label types of receptors on that figure. Add your own drawings of three other types of receptors in appropriate layers of skin. Label those receptors also.

■ **B5.** Complete this activity about different types of somatic sensations in yourself.

   a. Explain why it is not possible to tickle yourself.

   b. Explain why your skin itches at the site of a mosquito bite.

   c. In which part of your skin are cold receptors located? _____ What about receptors for warm stimuli? _____

   d. Identify which sense allows you to distinguish between a paperclip and a key while your eyes are

      closed. _____

■ **B6.** Complete this activity about two different types of pain.

   a. (*Fast? Slow?*) pain is known as acute, sharp, or prickly pain. Write one or more causes of this type of pain.

      _____

   b. (*Fast? Slow?*) pain can be classified as chronic, aching, or throbbing pain. This type of pain (*does? does not?*) occur in deep tissues or internal organs. Slow pain is likely to be localized to a (*precise? diffuse?*) area.

■ **B7.** *A clinical challenge.* Do this exercise about pain in two patients.

   a. Explain why patients experiencing visceral pain may feel pain in locations quite distant from the heart. Pain impulses that originate in the heart, as during a "heart attack," enter the spinal cord at the same level as do sensory

      fibers from skin covering the _____ . This level of the cord is about _____ to _____ .

   b. Refer to Figure 12.2, page 288 in the text. Which are most likely the organs causing referred pain in Ms. Benson's right shoulder and the right side of her neck?

   A. Pancreas and stomach   B. Liver and gallbladder   C. Colon and bladder   D. Kidneys

B8. Complete this checkpoint on pain.

   a. Defend or dispute these statements:

   (1) "Pain is a useful sensation."

   (2) "A single type of pain treatment should work for everyone."

   (3) "A 'runner's high' is truly a chemical phenomenon."

b. Define chronic pain, and list several examples of causes of chronic pain.

c. List five or more possible approaches to pain management.

C. Olfaction: sense of smell; gustation: sense of taste (pages 290–293)

■ **C1.** What cells undergo division to produce new olfactory receptors? _____ Explain why this fact is remarkable.

■ **C2.** Describe olfactory receptors and pathways in this exercise.

a. Receptors for smell are located in the *(superior? inferior?)* portion of the nasal cavity. Receptor neurons have

several cilia known as olfactory _____ that are reactive to odors.

b. What is the function of olfactory glands in the nose?

c. Individual olfactory receptors appear to respond to about *(four? twelve? hundreds of?)* different odors. Adaptation to odors occurs *(slowly? rapidly?).*

d. Upon stimulation of olfactory hairs, impulses pass to cell bodies and axons; the axons of these olfactory cells form cranial nerves *(I? II? III?),* the olfactory nerves. These pass from the nasal cavity to the cranium to

terminate in the olfactory _____ located just inferior to the _____ lobes of the cerebrum.

e. Neurons in the olfactory bulb then convey impulses along the olfactory _____ directly to

the cerebrum, specifically to the _____ and _____

lobes, as well as the _____ system and hypothalamus which evoke memories and

_____ responses.

■ **C3.** Describe receptors for taste in this exercise.

a. Receptors for taste, or _____ sensation, are located in taste buds with their sensitive

gustatory _____ projecting through the taste bud pores.

b. Taste buds are located on projections known as _____ which give the tongue its rough appearance. *(Filiform? Fungiform? Vallate?)* papillae are mushroom-shaped and are found over the entire tongue. Besides the tongue, where else are tastebuds located?

c. Name the five primary tastes: _____, _____, _____, _____, and _____. Explain how a wide variety of tastes can be detected.

d. Receptors for smell are much *(more? less?)* sensitive than those for taste, meaning that *(a large? only a small?)* amount of chemical substance must be present for olfaction to occur. If you have a cold (with a "stuffy nose"), you

may not be able to discriminate the usual variety of tastes, largely because of the loss of the sense of _____.

e. Are gustatory receptor cells neurons (as olfactory receptors are)?

f. Taste impulses are conveyed to the brain by cranial nerves _____ , _____ , and _____ . Cranial nerves enter the brainstem and then *(some do? none?)* pass to the limbic system. Pathways through the thalamus termi-

nate in area *(17? 42? 43?)* of the _____ lobe of the cerebral cortex. Only then does the perception of taste occur.

C4. Explain why certain foods can evoke aversion or even disgust.

D. Vision (pages 293–301)

D1. Examine your own eye structure in a mirror. With the help of Figure 12.5 (page 294 in your text), identify each of these accessory structures of the eye: *conjunctiva, eyebrow, eyelashes*. Identify the sites of *lacrimal glands* and *nasolacrimal ducts*.

D2. Write a paragraph discussing the functions of tears. Include the following terms in your paragraph: *lacrimal apparatus, lysozyme, irritants, parasympathetic emotions*.

■ **D3.** On Figure LG 12.1, color the three layers (or tunics) of the eye using color code ovals. Next to the ovals write letters of labeled structures that form each layer. One is done for you. Then label all structures in the eye and check your answers against the key.

■ **D4.** Complete this exercise about blood vessels of the eye.
a. Retinal blood vessels *(can? cannot?)* be viewed through an ophthalmoscope. Of what advantage is this?

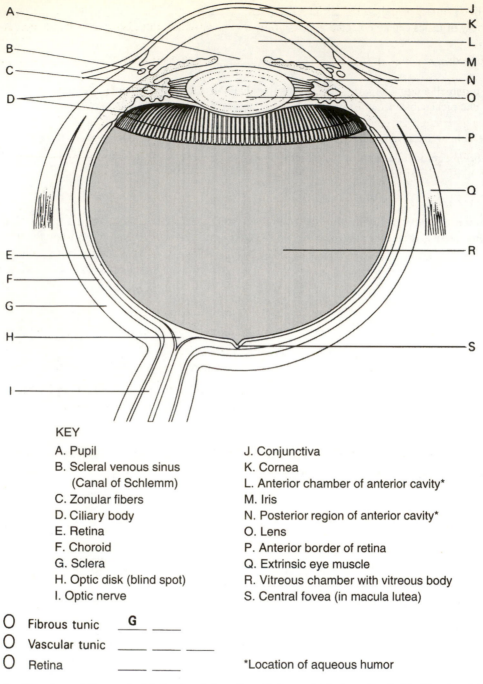

KEY

A. Pupil
B. Scleral venous sinus
   (Canal of Schlemm)
C. Zonular fibers
D. Ciliary body
E. Retina
F. Choroid
G. Sclera
H. Optic disk (blind spot)
I. Optic nerve

J. Conjunctiva
K. Cornea
L. Anterior chamber of anterior cavity*
M. Iris
N. Posterior region of anterior cavity*
O. Lens
P. Anterior border of retina
Q. Extrinsic eye muscle
R. Vitreous chamber with vitreous body
S. Central fovea (in macula lutea)

O Fibrous tunic   __G__ ____

O Vascular tunic   ____ ____ ____

O Retina   ____ ____   *Location of aqueous humor

Figure LG 12.1 Structure of the eyeball in horizontal section. Color and label as directed in Checkpoint D3.

b. The central retinal artery enters the eye at the *(optic disk? central fovea?)*. (Hint: Refer to Figure 12.6, page 295 in your text.)

■ **D5.** Contrast the two types of photoreceptor cells by writing R for rods or C for cones before the related descriptions.

_____ a. About 6 million in each eye; most concentrated in the central fovea of the macula lutea.

_____ b. Over 120 million in each eye; located mainly in peripheral regions of the eye.

_____ c. Sense color and acute (sharp) vision.

_____ d. Used for night vision.

■ **D6.** Do this exercise that describes the retina.

a. After light strikes the back of the retina, nerve impulses pass anteriorly through three zones of neurons. First is

the region of rods and cones, or _____ zone; next is the _____

layer; most anterior is the _____ layer.

b. Axons of the ganglion layer converge to form the optic _____ . This nerve exits from the

eye at the point known as the _____ disk, or _____ . The name indicates that no image formation can occur here because no rods or cones are present.

■ **D7.** Describe the lens and the cavities of the eye in this exercise.

a. The lens divides the eye into anterior and posterior regions. The posterior region is filled with

_____ body, which has a *(watery? jellylike?)* consistency. This body helps to hold the

_____ in place. Vitreous body is formed during embryonic life and *(is? is not?)* replaced
in later life.

b. A watery fluid called _____ humor is present in the anterior cavity. This fluid is formed

by the _____ processes, and it is normally replaced every 90 *(mintues? days?)*.

c. How are the formation and final destination of aqueous fluid similar to that of cerebrospinal fluid (CSF)?

d. Aqueous fluid creates pressure in the eye; in glaucoma this pressure *(increases? decreases?)*. What effects may
occur? (*Hint:* see text, p. 310.)

■ **D8.** *For extra review.* Check your understanding by matching eye structures in the box with descriptions below.
Use each answer once.

CF. Central fovea	Cor. Cornea	P. Pupil	SVS. Scleral venous sinus
CM. Ciliary muscle	I. Iris	S. Sclera	(canal of Schlemm)

_____ a. "White of the eye"

_____ b. Area of sharpest vision; area of densest
concentration of cones

_____ c. Most anterior portion of the eyeball;
contains no blood vessels; important in
focusing images on retina

_____ d. A hole; appears black, like a circular
doorway leading into a dark room

_____ e. Colored part of the eye; regulates whether
pupil is constricted or dilated, that is, the
amount of light that enters the eye

_____ f. Attaches to the lens by means of radially
arranged fibers called the zonular fibers

_____ g. Located at the junction of iris and cornea;
drains aqueous humor

D9. Define each of these processes involved in *image formation* on the retina.

a. *Refraction* of light

b. *Accommodation* of the lens

c. *Constriction* of the pupil

d. *Convergence* of the eyes

■ **D10.** Answer these questions about formation of an image on the retina.

a. Bending of light rays so that images focus exactly on the retina is a process known as

_____ . Which two eye structures accomplish most refraction?

_____ and _____

b. Draw how a letter "e" would look as it is focused on the retina: _____

■ **D11.** Explain how *accommodation* enables your eyes to focus clearly.

a. A *(concave? convex?)* lens bends light rays so that they converge and finally intersect. The lens in each of your eyes is *(biconcave? biconvex?)*.

b. When you bring your finger toward your eye, light rays from your finger must converge *(more? less?)* so that they focus on your retina. Thus, for near vision, the anterior surface of your lens must become *(more? less?)* convex. This occurs by a thickening and bulging forward of the lens caused by *(contraction? relaxation?)* of the

_____ muscle. In fact, after long periods of close work (such as reading), you may experience eye strain.

c. People who are far-sighted or *(myopic? hypermetropic?)* can see distant objects but cannot focus well on nearby objects. The inability to focus close up, for example, to read, is a normal aging change resulting from loss of

elasticity of the _____ . The condition is known as _____ -opia.

d. Near-sighted people can see *(near, but not far? far, but not near?)*. This condition is known as

_____ , and is corrected by lenses that are *(concave? convex?)*. (*Hint:* If you wear glasses to correct near-sightedness, note that the outer edges of your lenses are much thicker and the center more hollow, like a "cave.")

■ **D12.** Choose the correct answers to complete each statement.

a. Both accommodation of the lens and constriction of the pupil involve *(extrinsic? intrinsic?)* muscles, whereas convergence of the eyes involves *(extrinsic? intrinsic?)* muscles.

b. The pupil *(dilates? constricts?)* in the presence of bright light; the pupil *(dilates? constricts?)* in stressful situations when sympathetic nerves stimulate the iris. (*Hint:* Refer to Table 11.2, page 279 in your text.)

D13. Define these terms: *binocular vision, convergence.*

■ **D14.** List three principal processes that are necessary for vision to occur.

_____   _____

_____

180

■ **D15.** Do this exercise on photoreceptors and photopigments.

a. Rods contain the photopigment named _____ . Breakdown of this chemical allows you

to see in *(bright? dim?)* light. Night blindness is most often caused by lack of vitamin _____ because that

vitamin is required for formation of _____ .

b. Cones contain pigments that require *(bright? dim?)* light for breakdown. Three different pigments are responsive to different *(colors? intensities of light?)*. A person who is red-green color-blind lacks one of the photopigments in *(rod? cone?)* photoreceptors. This condition *(is? is not?)* hereditary and is more common in *(females? males?)*.

■ **D16.** Describe the conduction pathway for vision by arranging these structures in sequence. Write the letters in correct order on the lines provided.

C. Cerebral cortex (visual areas)	ON. Optic nerve	T. Thalamus
OC. Optic chiasm	OT. Optic tract	

_____ _____ _____ _____ _____

■ **D17.** After studying the visual pathways in Figure 12.11, page 301 of your text, complete this exercise.

a. Like most sensory pathways, the visual pathway *(does? does not?)* pass through the thalamus. Visual pathways

terminate in the _____ lobes of the cerebral cortex.

b. The primary visual area on the right side of the brain receives nerve impulses that originated in *(the right? the left? both the right and left)* eye(s). What advantage does this arrangement provide?

E. Hearing and equilibrium   (pages 301–308)

■ **E1.** Refer to Figure LG 12.2 and do the following exercise.

a. Color the three parts of the ear using color code ovals. Then write on lines next to ovals the letters of structures that are located in each part of the ear. One is done for you.

b. Label each lettered structure using leader lines on the figure.

■ **E2.** *For extra review.* Select the ear structures in the box that fit the descriptions below. Not all answers will be used.

AT. Auditory tube	M. Malleus	S. Stapes
AUR. Auricle	OW. Oval window	TM. Tympanic membrane
I. Incus	RW. Round window	

_____ a. Tube used to equalize pressure on either side of tympanic membrane

_____ b. Eardrum

_____ c. Structure on which stapes exerts piston-like action

_____ d. Ossicle adjacent to eardrum

_____ e. Anvil-shaped ear bone

E3. Contrast locations, anatomy, and physiology of the following structures:

a. Bony labyrinth/membranous labyrinth

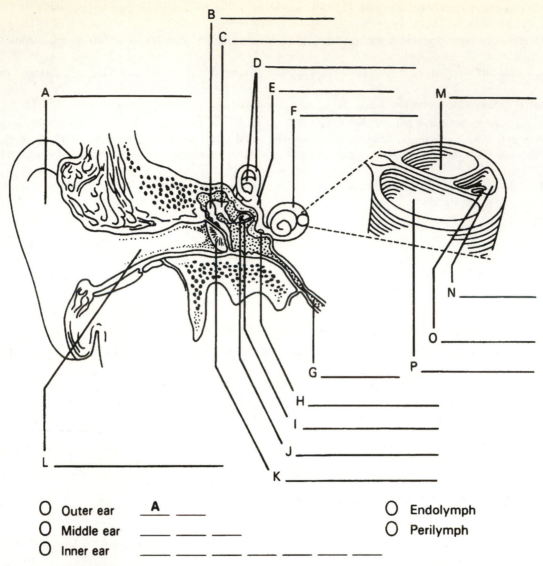

B _____
C _____
D _____
E _____
F _____
A _____
M _____
N _____
O _____
P _____
G _____
H _____
I _____
J _____
L _____
K _____

O Outer ear   __A__ ____
O Middle ear ____ ____ ____
O Inner ear   ____ ____ ____ ____ ____ ____ ____

O Endolymph
O Perilymph

Figure LG 12.2   Diagram of the ear in frontal section. Color and label as directed in Checkpoints E1 and E4.

b. Basilar membrane/tectorial membrane

■ **E4.**   Color *endolymph* and *perilymph* and related color code ovals in Figure LG 12.2.

■ **E5.**   Summarize events in the process of hearing in this activity. It may help to refer to Figure 12.14 (page 304 in the text).

a. Sound waves travel through the _____ and strike the _____
    membrane. A musical high note has a *(higher? lower?)* frequency of sound waves than a low note. So frequency
    is *(directly? indirectly?)* related to pitch. Sound waves are magnified by the action of the three

    _____ in the middle ear.

182

b. The ear bone named *(malleus? incus? stapes?)* strikes the *(round? oval?)* window, setting up waves in *(endo-? peri-?)* lymph. This pushes on the floor of the upper *scala (vestibuli? tympani?)*. As a result the cochlear duct is moved, and so is the perilymph in the lower canal, the *scala (vestibuli? tympani?)*. The pressure of the perilymph is finally expended by bulging out the *(round? oval?)* window.

c. As the cochlear duct moves, tiny hair cells embedded in the floor of the duct are stimulated. These hair cells are

part of the _____ organ; its name is based on its spiral arrangement on the

_____ membrane all the way around the coils of the cochlear duct.

The *(apex? base?)* of this duct (which is close to the helicotrema) is stimulated by low frequency sounds (as from a cello), whereas the *(apex? base?)* which is close to the oval window detects high frequency sounds such as those from a violin.

d. As spiral organ hair cells are moved by waves in endolymph, hairs move against the

_____ membrane. This movement stimulates release of neurotransmitter hair cells that

excite nearby sensory neurons of the _____ branch of cranial nerve _____ .

The pathway continues to the brain stem (medulla), _____ (relay center), and finally to the

_____ lobe of the cerebral cortex (areas _____ and _____). Injury to the primary auditory area of the right side of the brain is likely to cause hearing loss in *(the right ear? the left ear? both ears?)*.

■ **E6.** Contrast receptors for hearing and for equilibrium in this summary of the inner ear.

a. Receptors for hearing and equilibrium are all located in the *(middle? inner?)* ear. All consist of supporting cells

and _____ cells that are covered by the delicate membrane.

b. In the spiral organ, which senses _____ , the membrane is called the

_____ membrane. Hair cells move against this membrane as a result of *(sound waves? change in body position?)*.

c. In the macula, located in the *(semicircular canals? vestibule?)*, the membrane is embedded with calcium carbon-

ate crystals called _____ . These respond to gravity in such a way that the macula is the main receptor for *(static? dynamic?)* equilibrium. An example of such equilibrium occurs as you are aware of your *(position while lying down? change in position on a careening roller coaster?)*.

d. In the semicircular canals the membrane is called the _____ . Its shape is *(flat? like an inverted cup?)*. The cupula is part of the *(crista? saccule?)* located in the ampulla. Change in direction (as in a

roller coaster) causes _____ to bend hairs in the cupula. Cristae in semicircular canals are therefore receptors primarily for *(static? dynamic?)* equilibrium.

E7. Describe vestibular pathways, including the role of the cerebellum in maintaining equilibrium.

■ **F1.** Now that you have studied major aspects of the nervous system, including sensations, relate the impact of nerves and senses on other systems by writing in the name of the affected system.

a. Bumping your right tibial crest sets up pain in nociceptors there, alerting you of this injury:

_____ system.

b. Catching a glimpse of a kitten in your pathway, you quickly move to the side to avoid stepping on

the kitten: _____ system.

c. Your hypothalamus sends nerve impulses to the posterior pituitary to cause release of hormones:

_____ system. (More about this in the next chapter.)

d. The olfactory experience of dinner cooking causes your gastric glands to begin secreting:

_____ system.

e. Hearing your loved one's voice and recalling how much you care for one another, your erectile

tissues respond: _____ system.

■ **F2.** Match the name of the disorder with the description.

A. Anosmia	Mac. Macular degeneration	Pto. Ptosis
Cat. Cataract	Mén. Ménière's disease	Str. Strabismus
Con. Conjunctivitis	Myo. Myopia	Tra. Trachoma
G. Glaucoma	OM. Otitis media	

_____ a. Condition requiring corrective lenses to focus distant objects

_____ b. Excessive intraocular pressure (IOP) resulting in blindness

_____ c. Irreversible deterioration of the point of sharpest vision in the retina; much more common in elderly

_____ d. Pinkeye

_____ e. Cross-eyedness

_____ f. Inflammation of the middle ear

_____ g. Inability to smell

_____ h. Disturbance of the inner ear with excessive endolymph

_____ i. Loss of transparency of the lens

_____ j. The greatest single cause of blindness in the world; caused by bacterial infection

_____ k. Drooping of an eyelid

F3. Contrast these two types of deafness: *sensorineural/conduction.*

F4. Explain why middle ear infection (otitis media) is more common than inner ear infection.

A2. (a) GS. (b) VS. (c) SpS. (d) SomS (also GS).

A3. (a) Stimulus → (b) Receptor → (c) Conduction → (d) Integration.

A4. (a) Adaptation; protects you by continuing to provide warning signals of painful stimuli. (b) Cerebral cortex; perception; cerebral cortex.

A5. (a) SC. (b) FNE. (c) ENE.

A6. (a) Cold, warm. (b) Smell, taste. (c) Pain. (d) Touch, pressure, vibration, hearing, awareness of joint position (proprioception), change of blood pressure. (e) Vision.

B1. A B C F (Note that D and E are visceral).

B2. (a) Touch, pressure, vibration, itch, and tickle. (b) Unevenly.

B3. (a) C. (b) L. (c) N. (d) Type I and Type II. (e) J, M, T. (f) N. (g) N. (h) M. (i) T. (j) HRC.

B5. (a) Impulses to the cerebellum alert the brain that your own fingers are doing the tickling. (b) Free nerve endings are stimulated by chemicals such as bradykinin. (c) Epidermis; dermis. (d) Kinesthesia and proprioception.

B6. (a)Fast; knife cut or needle puncture. (b) Slow; does; diffuse.

B7. (a) Medial aspects of the left arm; T1, T5. (b) Liver and gallbladder.

C1. Basal cells; mature neurons are not normally replaced during adult life, yet these olfactory neurons die after only a month of "service," and then new recruits are formed.

C2. (a) Superior; hairs. (b) Produce mucus that acts as a solvent for odorants. (c) Hundreds of; rapidly. (d) I; bulbs, frontal. (e) Tracts, temporal, frontal, limbic, emotional.

C3. (a) Gustatory, hairs. (b) Papillae; fungiform; roof of the mouth, throat (pharynx), and epiglottis. (c) Sweet, sour, salty, bitter, and umami (meaty); combinations of the five tastes result in different patterns of impulses, added to the smell and touch of foods. (d) More, only a small; smell. (e) No, but they release a neurotransmitter that triggers impulses in first-order neurons. (f) VII, IX, and X; some do; 43, parietal.

D3. Fibrous tunic: G, K; vascular tunic: D, F, M; retina: E, S.

D4. (a) Can; health of blood vessels can be readily assessed, for example, in people with diabetes. (b) Optic disk.

D5. (a) C. (b) R. (c) C. (d) R.

D6. (a) Photoreceptor; bipolar; ganglion. (b) Nerve; optic, blind spot.

D7. (a) Vitreous, jellylike; retina; is not. (b) Aqueous; ciliary; minutes. (c) Both are formed from blood vessels (choroid plexuses) and the fluid finally returns to venous blood. (d) Increases; damage to the retina with possible blindness.

D8. (a) S. (b) CF. (c) Cor. (d) P. (e) I. (f) CM. (g) SVS.

D10. (a) Refraction; cornea, lens. (b) "∂" ("e" inverted 180° and much smaller).

D11. (a) Convex; biconvex. (b) More; more; contraction, ciliary. (c) Hypermetropic; lens; presby. (d) Near, but not far; myopia; concave.

D12. (a) Intrinsic; extrinsic. (b) Constricts; dilates.

D14. Formation of an image on the retina, stimulation of photoreceptors leading to a nerve impulse, and transmission along nerve pathways to the thalamus and visual cortex.

D15. (a) Rhodopsin; dim; A, the retinal portion of rhodopsin. (b) Bright; colors; cone; is, males.

D16. ON OC OT T C.

D17. (a) Does; occipital. (b) Both the right and left; injury or loss of blood flow to one side of the brain would still allow at least partial vision with *both* eyes.

E1.

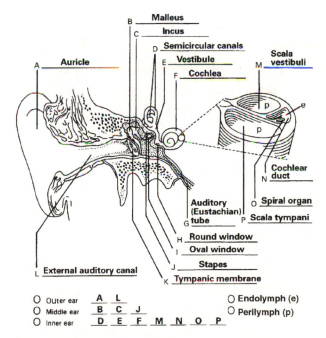

Figure LG 12.2A   Diagram of the ear in frontal section.

E2. (a) AT. (b) TM. (c) OW. (d) M. (e) I.

E4. See Figure LG 12.2A.

E5. (a) External auditory canal, tympanic; higher; directly; ossicles. (b) Stapes, oval, peri-; vestibuli; tympani; round. (c) Spiral; basilar; apex, base.

(d) Tectorial; cochlear, VIII; thalamus, temporal, 41 and 42; both ears.

E6. (a) Inner; hair. (b) Hearing, tectorial; sound waves. (c) Vestibule, otoliths; static; position while lying down. (d) Cupula; like an inverted cup; crista; endolymph; dynamic.

F1. (a) Skeletal. (b) Muscular. (c) Endocrine. (d) Digestive. (e) Reproductive (also cardio-vascular since engorgement of erectile tissues is due to dilation of blood vessels in genitalia).

F2. (a) Myo. (b) G. (c) Mac. (d) Con. (e) Str. (f) OM. (g) A. (h) Mén. (i) Cat. (j) Tra. (k) Pto.

CRITICAL THINKING: CHAPTER 12

1. Describe structures and mechanisms that help prevent overstretching of your muscles, tendons, and joints. Be sure to include roles of receptors and describe nerve pathways.

2. Contrast the sensations of proprioception and kinesthesia. Tell what information these senses provide to you.

3. How does the anatomy of olfactory pathways explain how certain odors can trigger intense memories and emotional responses?

4. Note the location of the pituitary gland in Figure 10.8, page 252, and in Figure 12.3, page 291 of your textbook. Determine which special sense is likely to be partially or totally lost by each of the following:
    a. A tumor (adenoma) of the pituitary gland.
    b. Surgery to remove the pituitary gland.

5. Contrast effects of damage to the following structures: *otoliths/ossicles.*

6. Contrast smell and taste according to the following criteria: (a) threshold, (b) cranial nerve pathways for each. Also describe how you could demonstrate that much of what most people think of as taste is actually smell.

7. Discuss causes and potential damage to the eye associated with the following conditions: (a) detachment of the retina, (b) glaucoma, (c) cataracts, (d) trachoma.

MASTERY TEST: ■ CHAPTER 12

Questions 1–4: Circle T (true) or F (false). If the statement is false, change the underlined word or phrase so that the statement is correct.

T   F   1. Intrinsic eye muscles are responsible for changes in pupil size, accommodation, and movement of eyes up and down and from right to left.

T   F   2. Crista, macula, otolith, and spiral organ are all structures located in the inner ear.

T   F   3. The receptor organs for special senses are less complex structurally than those for general senses.

T   F   4. The anterior cavity is filled with vitreous body.

Questions 5–7: Arrange the answers in correct sequence.

_____ _____ _____   5. Layers of the eye, from superficial to deep:
                              A. Sclera
                              B. Retina
                              C. Choroid

_____ _____ _____ _____   6. From anterior to posterior:
                              A. Aqueous humor
                              B. Optic nerve
                              C. Cornea
                              D. Lens
                              E. Vitreous body

_____ _____ _____ _____ _____ 7. Pathway of sound waves and resulting mechanical action:
   A. External auditory canal
   B. Stapes
   C. Malleus and incus
   D. Oval window
   E. Tympanic membrane

Questions 8–10: Circle the letter preceding the one best answer to each question.

8. Infections in the throat (pharynx) are most likely to lead to ear infections in the following manner. Bacteria spread through the:
   A. External auditory meatus to the external ear
   B. Auditory (Eustachian) tube to the middle ear
   C. Oval window to the inner ear
   D. Round window to the inner ear

9. Choose the *false* statement about rods.
   A. There are more rods than cones in the eye.
   B. Rods are concentrated in the central fovea and are less dense around the periphery.
   C. Rods enable you to see in dim (not bright) light.
   D. Rods contain rhodopsin.
   E. No rods are present at the optic disk.

10. Select the one answer that provides a term with a correct description.
    A. Cornea: known as the "whites of their eyes"
    B. Optic disk: point of sharpest vision
    C. Retinal blood vessels: the only blood vessels in the body that can be viewed directly
    D. Iris: contracts to change shape of the lens during accommodation
    E. Pupil: a hole in the lens

Questions 11–15: Fill-ins. Write the word or phrase that best fits the description.

_____ 11. Name the four processes necessary for formation of an image on the retina:

_____ of light rays, _____ of the lens, _____ of the pupil, and _____ of the eyes.

_____ 12. More than half of the sensory receptors in the human body are located in the

_____ .

_____ 13. The vestibular apparatus consists of the _____ and functions in regulation of

_____ .

_____ 14. The names of the three ossicles are _____ , _____ , and _____ .

_____ 15. Two functions of the ciliary body are _____ and _____ .

ANSWERS TO MASTERY TEST: ■ CHAPTER 12

True or False
1. F. Pupil size and accommodation
2. T
3. F. More
4. F. Aqueous humor

Arrange
5. A C B
6. C A D E B
7. A E C B D

Multiple Choice
8. B
9. B
10. C

Fill-ins
11. Refraction, accommodation, constriction, convergence
12. Eye
13. Utricle, saccule, and semicircular canals; equilibrium and balance
14. Malleus, incus, and stapes
15. Production of aqueous humor and alteration of lens shape for accommodation

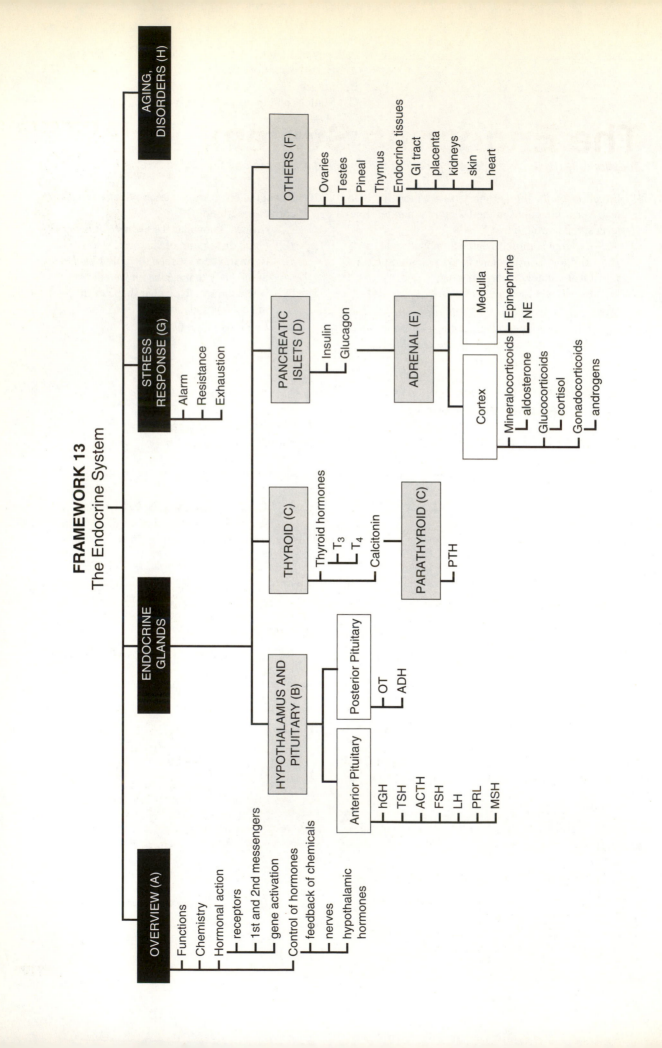

FRAMEWORK 13
The Endocrine System

The Endocrine System CHAPTER 13

Hormones produced by at least 18 organs exert widespread effects on just about every body tissue. Hormones are released into the bloodstream, which serves as the vehicle for distribution throughout the body. In fact, analysis of blood levels of hormones can provide information about the function of specific endocrine glands. How do hormones "know" which cells to affect? And how does the body "know" when to release more or less of a hormone? Mechanisms of action and regulation of hormone levels are discussed in this chapter. The roles of all major hormones and effects of excesses or deficiencies in those hormones are also included. The human body is continually exposed to stressors, such as rapid environmental temperature change, a piercing sound, or serious viral infection. Adaptations to stressors are vital to survival; however, certain adaptations lead to negative consequences. A discussion of stress and adaptation concludes Chapter 13.

As you begin your study of the endocrine system, carefully examine the Chapter 13 Topic Outline and Objectives; check off each one as you complete it. To organize your study of this content, glance over the Chapter 13 Framework now. Be sure to refer to the Framework frequently and note relationships among key terms in each section.

TOPIC OUTLINE AND OBJECTIVES

A. Overview: introduction; hormone action

- [ ] 1. List the components of the endocrine system.
- [ ] 2. Define target cells and describe the role of hormone receptors.
- [ ] 3. Describe the two general mechanisms of the action of hormones.

B. Hypothalamus and pituitary gland

- [ ] 4. Describe the locations of and relationship between the hypothalamus and pituitary gland.
- [ ] 5. Describe the functions of each hormone secreted by the pituitary gland.

C. Thyroid and parathyroid glands

- [ ] 6. Describe the location, hormones, and functions of the thyroid gland.
- [ ] 7. Describe the location, hormones, and functions of the parathyroid glands.

D. Pancreatic islets

- [ ] 8. Describe the locations, hormones, and functions of the pancreatic islets.

E. Adrenal glands

- [ ] 9. Describe the locations, hormones, and functions of the adrenal glands.

F. Other endocrine glands and tissues

- [ ] 10. Describe the locations, hormones, and functions of the ovaries and testes.
- [ ] 11. Describe the locations, hormone, and functions of the pineal gland.
- [ ] 12. List the hormones secreted by cells in tissues and organs other than endocrine glands, and describe their functions.

G. The stress response

☐ 13. Describe how the body responds to stress.

H. Aging and the endocrine system; common disorders; homeostasis

☐ 14. Describe the effects of aging on the endocrine system.

WORDBYTES

Study each wordbyte, its meaning, and an example of its use in a term. Then check your understanding by writing the meaning of each wordbyte in the margins. As you continue through the *Learning Guide,* identify (and fill in) additional terms that contain the same wordbyte.

Wordbyte	Meaning	Example(s)	Wordbyte	Meaning	Example(s)
adeno-	gland	*adeno*hypophysis	insipid-	without taste	diabetes *insipid*us
andro-	man	*andro*gens	mellit-	sweet	diabetes *mellit*us
crin-	to secrete	endo*crine*	oxy(s)-	swift	*oxy*tocin
endo-	within	*endo*crine	para-	around	*para*thyroid, *para*crine
exo-	outside	*exo*crine	somato-	body	*somato*tropin
gen-	to create	diabeto*genic*	-tocin	childbirth	pi*tocin*
hormon-	excite, get moving	*hormon*e	trop-	turn	thyro*tropin*

CHECKPOINTS

A. Overview (pages 316–318)

■ **A1.** Compare the ways in which the nervous and endocrine systems exert control over the body. Write N (nervous) or E (endocrine) next to descriptions that fit each system.

_____ a. Sends messages at synapses to muscles, glands, and neurons only.

_____ b. Sends messages via the cardiovascular system to virtually any part of the body.

_____ c. Effects are generally faster and shorter-lived.

■ **A2.** Complete the table contrasting exocrine and endocrine glands.

	Secretions Transported In	Examples
a. Exocrine		
b. Endocrine		

A3. Refer to Figure LG 13.1 and do this exercise.

■ a. Color and label each endocrine gland next to letters A–G.

■ b. Next to numbers 1–13, label each organ containing endocrine tissue. *For extra review,* identify the location of each organ in your own body.

c. *For extra review.* As you proceed through this chapter, write names (or abbreviations) of each hormone next to the endocrine organ or tissue that produces it.

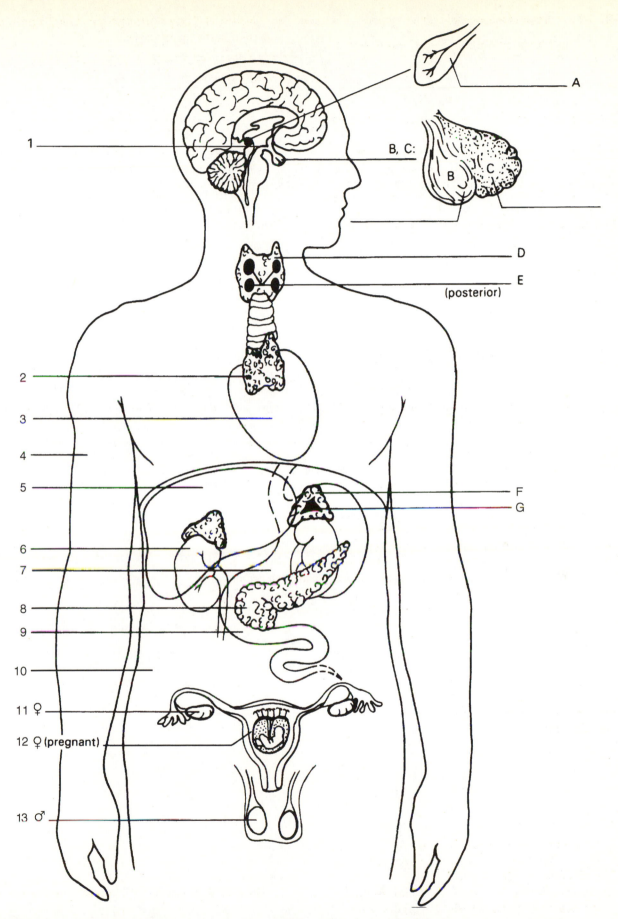

1

2

3

4

5

6

7

8

9

10

11 ♀

12 ♀ (pregnant)

13 ♂

A

B, C:

B

C

D

E

(posterior)

F

G

Figure LG 13.1 Diagram of endocrine glands on right side of figure (A–G) and organs containing endocrine tissue (1–13) on left side of figure. Label and color as directed in Checkpoints A3 and B1.

191

A4. In a word or two each, list seven functions of hormones.

1. _____   5. _____

2. _____   6. _____

3. _____   7. _____

4. _____

■ **A5.** Imagine yourself as a body cell watching dozens of different hormones passing by in the "street" (blood plasma) next to you. How would you know which hormones to "invite in" to interact with you?

■ **A6.** Certain hormones can readily enter cells because the hormones are *(water? lipid?)*-soluble. Name several of

these hormones: _____
Once inside the target cell, the hormone acts by binding to receptors there. The hormone-receptor complex then direct triggers genes *(DNA? mRNA?)* to "turn on" (or "turn off"). An activated gene is transcribed to form *(DNA? mRNA?)* which directs synthesis of new proteins. The new proteins form the major parts of

_____ that alter cell function.

■ **A7.** Refer to Figure 13.3, page 318 in your text, and complete this Checkpoint.

a. A number of hormones, including peptide and protein hormones, interact with plasma membrane receptors because these hormones are *(water? lipid?)*-soluble, and therefore *(can? cannot?)* penetrate the phospholipid bi-layer of the plasma membrane. Name several of these hormones.

b. One such hormone is human growth hormone (hGH). hGH acts as the *(first? second?)* messenger in the process of triggering body cells to respond to hormones, because hGH carries the message from the endocrine gland, namely

the _____ _____ gland, where hGH is made to the _____ cell where hGH acts, in this case, a muscle or bone cell.

c. The hormone then binds to a receptor on the *(inner? outer?)* surface of the plasma membrane. (Why not the inner

surface? _____) This starts a reaction that catalyzes

conversion of ATP to _____ which is known as the *(first? second?)* messenger.

d. cAMP then activates enzymes that cause the target cell to respond. In the case of hGH, activated enzymes in the

bone or muscle cell carry out activities to stimulate _____ of the bone or muscle.

e. Since most hormones are *(lipid? water?)*-soluble, most hormones act by *(alteration of gene expression? the second messenger mechanism?)*.

■ **A8.** Precise regulation of hormone levels is critical to homeostasis. Identify mechanisms that control release of each hormone listed below. Use answers in the box.

> C. Blood level of a chemical (such as calcium or glucose) that is controlled by that hormone
> H. Blood level of another hormone
> N. Nerve impulses

_____ a. Epinephrine   _____ c. Insulin

_____ b. Parathyroid hormone (PTH) or   _____ d. ACTH, which is controlled by hypo-
                calcitonin (CT)   thalamic releasing hormones

■ **A9.** *(Negative? Positive?)* feedback mechanisms are used for most hormonal regulation. For example, a low level of Ca^{2+} circulating through the parathyroid gland causes a(n) *(increase? decrease?)* in release of parathyroid hormone (PTH) and in blood level of Ca^{2+}. State one example of hormone control by *positive* feedback.

B. Hypothalamus and pituitary gland   (pages 319–323)

■ **B1.** Complete this exercise about the pituitary gland.

a. The pituitary gland has been nicknamed the "_____ ." It is attached by a stalk, known as

the _____ to the part of the brain called the _____ .

b. Most of the gland consists of the *(anterior? posterior?)* lobe. The _____ lobe consists of

endocrine gland tissue, whereas the _____ lobe contains ends of neurons that release hor-

mones made above in the hypothalamus. Hormones from the _____ lobe are regulated by
releasing and inhibiting hormones that are made by the hypothalamus and pass through blood directly to the
pituitary.

c. The anterior pituitary is known to secrete *(2? 5? 7?)* different hormones. Write abbreviations for each of these
hormones next to the diagram of the anterior pituitary on Figure LG 13.1.

■ **B2.** Do this Checkpoint about growth hormone.

a. A main function of human growth hormone (_____) is to stimulate growth and maintain size of

_____ and _____ . hGH stimulates growth by triggering secretion

of IGFs (or i_____ g_____ f_____) by many body cells. Name
several cells that respond to hGH in this way.

b. Effects of hGH (via IGFs) are to _____-crease protein synthesis in bones and muscles, and provide energy for

growth by _____-creasing fat breakdown and _____-creasing blood glucose.

c. hGH secretion is controlled by two regulating hormones made by the _____ ; these are

known as a releasing hormone and a(n) _____ hormone.

d. One stimulus that promotes release of hGH is low blood sugar (for example, during growth when cells require
much energy from glucose). This condition, *(hyper? hypo?)*-glycemia, stimulates secretion of the hGH-

_____ hormone, which then causes release of *(high? low?)* levels of hGH and *(high?
low?)* levels of blood glucose. The response to low levels of blood sugar by release of high levels of a hormone
(hGH) that raises blood sugar is an example of a *(positive? negative?)* feedback system.

■ **B3.** Match the seven anterior pituitary hormones with descriptions below. Two answers will be used twice.

ACTH.	Adrenocorticotropic hormone	MSH.	Melanocyte-stimulating hormone
FSH.	Follicle-stimulating hormone	PRL.	Prolactin
hGH.	Human growth hormone	TSH.	Thyroid-stimulating hormone
LH.	Luteinizing hormone		

_____ a. Stimulates ovaries to secrete estrogens and develop future eggs; stimulates sperm production in testes

_____ b. In females, stimulates ovaries to release secondary oocytes (ovulation) and to form corpus luteum, which then releases both estrogens and progesterone

_____ c. In males, stimulates testes to secrete testosterone

_____ d. Stimulates secretion of thyroid hormones

_____ e. Stimulates secretion of hormones from the adrenal cortex

_____ f. Excessive amounts cause skin to darken

_____ g. Stimulates milk production by mammary glands in breast

_____ h. Stress (including physical trauma or inter-leukin-1) triggers release of this hormone

■ **B4.** Define *tropic hormones*.

Write T next to the four hormones in the box above (Checkpoint B3) that are tropic hormones.

■ **B5.** Do this exercise about hormonal control of mammary glands.

a. Milk is produced in mammary glands following stimulation by the hormone _____

(_____), which is secreted by the *(anterior? posterior?)* pituitary. Release of PRL *(is? is not?)* regulated by hypothalamic hormones.

b. Prolactin levels increase due to _____-crease of PRH (prolactin-releasing hormone) or _____-crease of PIH (prolactin-inhibiting hormone). *(PRH? PIH?)* stimulates a rise in prolactin during pregnancy, whereas *(PRH? PIH?)* suppresses release of prolactin in women who are not pregnant or lactating.

c. Name two other hormones that prepare breasts during pregnancy for lactation: _____

_____ .

d. A different hormone causes ejection of milk from glands into ducts each time the baby breastfeeds. This

hormone, named _____ , is released by the *(anterior? posterior?)* pituitary.

e. What is the function of prolactin in males?

B6. Defend or dispute this statement: "The posterior pituitary is really not an endocrine gland."

■ **B7.** Answer these questions about posterior pituitary hormones.

a. Stretching (distension) of the uterus during labor _____-creases synthesis of oxytocin (OT). OT then

_____-creases uterine contractions, which then _____-crease release of more OT. This is an example of a

(positive? negative?) feedback control mechanism. Oxytocin is also known clinically as _____ .

b. Oxytocin causes *(formation of milk in? ejection or let-down of milk from?)* mammary glands. Suckling *(stimulates? inhibits?)* release of OT and milk.

c. On a day when your body becomes dehydrated by loss of sweat, your ADH production is likely to

_____-crease. As a result, your body with _____-crease urinary output, thereby _____-creasing blood

volume. This *(positive? negative?)* feedback mechanism attempts to bring body fluid level back to normal.

d. ADH is also known as vaso-_____ , indicating the effect of this hormone to *(in? de?)*-crease blood pressure by *(dilating? constricting?)* blood vessels (and therefore blood flow) leading into specific body tissues.

e. Alcohol *(stimulates? inhibits?)* ADH secretion, which contributes to _____-creased urine production after alcohol intake. *For extra review* of other factors affecting ADH production, refer to page 323 in the text.

C. Thyroid and parathyroid glands (pages 323–327)

■ **C1.** Identify the location of your own thyroid gland in this exercise. Place the tip of your index finger on the front of your throat. Now swallow. The part that moves up is your "Adam's apple," also known as your thyroid cartilage. It is visible in Figure 13.7, on page 324 in your text. The thyroid gland lies just *(anterior? posterior?)* and *(superior?*

inferior?) to the thyroid cartilage. The thyroid gland is shaped much like a(n) _____ .

■ **C2.** Describe the hormone secretion of the thyroid gland in this exercise.

a. The thyroid is composed of two types of glandular cells. *(Follicular? Parafollicular?)* cells produce the two

hormones _____ and _____ . Parafollicular cells manufacture the

hormone _____ .

b. The ion *(Cl^-? I^-? Br^-?)* is highly concentrated in the thyroid because it is an essential component of T_3 and T_4.

Each molecule of thyroxine (T_4) contains four of the I^- ions, whereas T_3 contains _____ ions of I^-.

■ **C3.** Describe functions of thyroid hormone in this Checkpoint.

a. Like growth hormone, thyroid hormones tend to _____-crease protein synthesis. Thyroid hormone especially

affects growth of the nervous tissue, so _____-crease of thyroid hormones during fetal development may lead to mental retardation. (See text page 337.)

b. Also like growth hormone, thyroid hormones tend to _____-crease breakdown of fats as energy sources. How-

ever, unlike hGH, thyroid hormones _____-crease use of glucose as an energy (ATP) source. Overall, thyroid

hormone _____-creases metabolic rate.

■ **C4.** Check your understanding of regulation of thyroid hormones (T_3 and T_4) in this Checkpoint.

a. T_3 and T_4 production decreases when blood levels of iodine are abnormally *(high? low?)*.

b. Like other releasing hormones, thyroid-releasing hormone (TRH) is made by the *(anterior pituitary? hypothalamus?)*. High levels of TRH stimulate *(high? low?)* levels of the thyroid-stimulating hormone (TSH).

c. TSH is a *(releasing? tropic? target?)* hormone made by the *(anterior pituitary? hypothalamus?)*. TSH *(stimulates? inhibits?)* thyroid production of T_3 and T_4.

d. In other words, high levels of TRH → *(high? low?)* levels of TSH → *(high? low?)* levels of T_3 and T_4.

e. A high level of the target hormone T_3 and T_4 *(stimulates? inhibits?)* TRH and TSH production. This mechanism is an example of a *(positive? negative?)* feedback system which normally keeps T_3 and T_4 levels in balance.

f. Circle factors that are likely to increase anterior pituitary secretion of TSH, and therefore increase thyroid hormone: *(warm environment? high altitude? pregnancy? aging?)*.

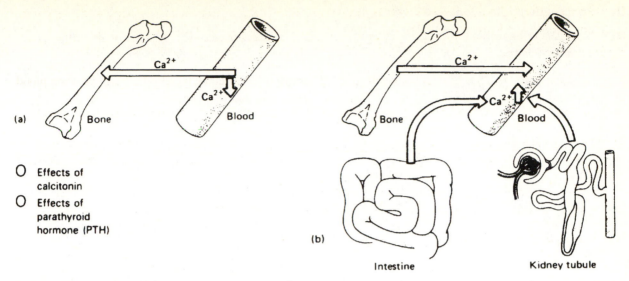

Figure LG 13.2 Hormonal control of calcium ion (Ca^{2+}). (a) Increased calcium storage in bones lowers blood calcium (hypocalcemic effect). (b) Calcium from three sources increases blood calcium level (hypercalcemic effect). Arrows show direction of calcium flow. Color arrows as directed in Checkpoint C5.

■ **C5.** Color arrows on Figure LG 13.2 to show hormonal regulation of calcium. This figure emphasizes that CT and PTH are *(synergists? antagonists?)* with regard to effects on calcium. *(CT? PTH?)* draws Ca^{2+} from bones and urine

and returns Ca^{2+} to blood. PTH also stimulates kidney production of the active form of vitamin _____ ; this vita-

min facilitates absorption of _____ in the intestine. Blood level of

Ca^{2+} increases as release of PTH _____-creases.

■ **C6.** Summarize the effects of CT and PTH on blood levels of different ions by drawing arrows. One is done for you, indicating that calcitonin (CT) decreases blood calcium.

a. CT: $\downarrow$ **Ca²⁺** _____ $PO_4{}^{3-}$

b. PTH: _____ Ca^{2+} _____ $PO_4{}^{3-}$ _____ Mg^{2+}

■ **C7.** Regulation of hormones PTH and calcitonin are controlled by blood levels of _____ .

A. Anterior pituitary hormones

B. Hypothalamic hormones

C. Blood levels of Ca^{2+}

Complete the arrows to show this regulation which maintains homeostasis by negative feedback:

$\downarrow$ blood Ca^{2+} → | secretion of PTH → | blood Ca^{2+} → | secretion of calcitonin → | blood Ca^{2+}

D. Pancreatic islets (pages 327–329)

■ **D1.** The islets of Langerhans are located in the _____ . Do this exercise describing hormones produced there.

a. The name *islets of Langerhans* suggests that these clusters of _____-crine cells lie amidst a "sea" of exocrine cells within the pancreas. Name two major types of islet cells.

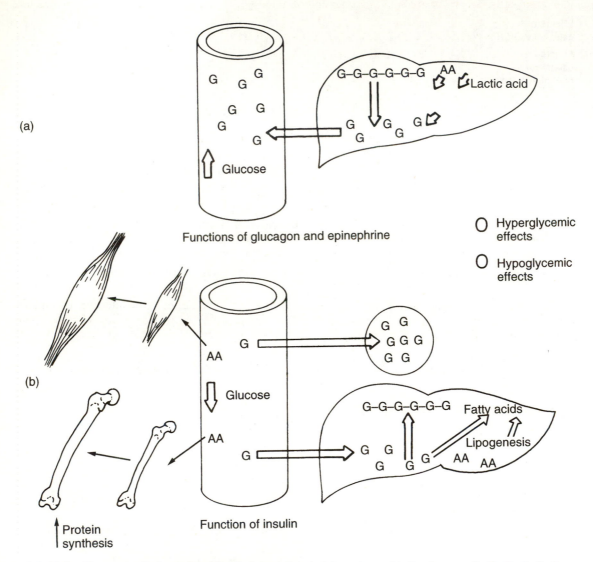

Figure LG 13.3 Hormones that regulate blood glucose level. AA, amino acid; G, glucose; G–G–G–G–G–G, glycogen. Color according to directions in Checkpoint D1.

b. Refer to Figure LG 13.3(a). Glucagon, produced by *(alpha? beta?)* cells, *(increases? decreases?)* blood

sugar. How? First it stimulates the breakdown of _____ to glucose. Second, it stimulates the conversion of amino acids and other compounds to glucose. This process stimulates release of glucose from the liver into the blood.

c. Is glucagon controlled directly by an anterior pituitary *tropic* hormone? *(Yes? No?)* In fact, control is by effect of

blood _____ level directly on the pancreas. When blood glucose is low, a *(high? low?)* level of glucagon is produced. This raises blood sugar. This is a *(positive? negative?)* feedback mechanism.

d. *For extra review.* Define the terms *glycogen* and *glucagon.*

e. Now look at Figure LG 13.3(b). The action of insulin is *(the same as? opposite?)* that of glucagon. In other words, *insulin decreases blood sugar* when it is above normal. (Repeat that statement three times; it is important!) Insulin acts by several mechanisms shown in the figure. First, insulin *(helps? hinders?)* transport of glucose

from blood into cells. Second, it accelerates conversion of glucose to _____ and to fats.

Insulin also facilitates entrance of _____ into cells, as in muscle or bone.

f. Identify two major factors that stimulate the pancreas to secrete insulin:

A. High blood sugar (hyperglycemia)

B. Low blood sugar (hypoglycemia)

C. Sympathetic nerves

D. Parasympathetic nerves

g. Hormones that raise blood glucose levels are called *(hypo? hyper?)*-glycemic hormones. Insulin, the one

hormone that lowers blood glucose level, is said to be _____-glycemic. To show this, color arrows in Figure LG 13.3.

■ **D2.** Do this exercise on Type II diabetes. (It may help to refer to text pages 332 and 338.)

a. This disorder is more likely to occur in persons who are: (circle all correct answers)

A. Underweight   B. Obese   C. Active   D. Sedentary

b. In type II diabetes, blood levels of insulin are *(higher? lower?)* than normal (exactly opposite to insulin levels in type I diabetes). The problem is that *(insulin is not produced by the pancreas? body cells do not respond to insulin?),* a condition known as insulin _____ .

c. People with type II diabetes tend to have more fat around the *(waist? hips?).* This "visceral fat" leads to *(elevated? low?)* levels of triglycerides and LDL's with _____-creased risk for arteriosclerosis and *(high? low?)* blood pressure.

d. Write three health tips that can reduce risk for type 2 diabetes.

E. Adrenal glands   (pages 329–333)

■ **E1.** Contrast the two regions of the adrenal gland by matching answers in the box with descriptions below.

AC. Adrenal cortex	AM. Adrenal medulla

_____ a. The outer portion of the adrenal gland

_____ b. Secretes steroid hormones such as aldosterone, cortisol, progesterone, and testosterone

_____ c. Regulated by ACTH from the anterior pituitary

_____ d. Secretes epinephrine and norepinephrine

■ **E2.** Injury to the outer zone of the adrenal cortex would be most likely to affect production of *(gluco? mineralo?)*-corticoids, whereas injury to the inner zone (adjacent to the adrenal medulla) would be likely to affect production of

_____ .

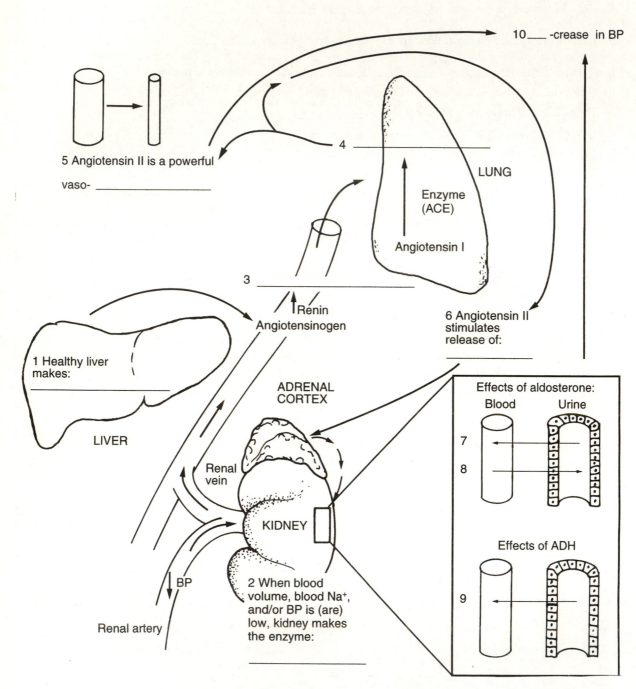

Figure LG 13.4 Hormonal control of fluid and electrolyte balance and blood pressure (BP) by renin, angiotensin, aldosterone, and ADH. Fill in blanks 1–10 as directed in Checkpoint E3. Write answers to 7, 8, and 9 on arrows next to those numbers to show movement of different chemicals.

■ **E3.** Fill in blank lines on Figure LG13.4 to show mechanisms that control fluid balance and blood pressure.

■ **E4.** The major mineralocorticoid is _____ , as shown in Figure LG 13.4; the major

glucocorticoid is _____ .

■ **E5.** In general, glucocorticoids such as cortisol help to regulate metabolism and response to stress. Describe effects of cortisol in this exercise.

   a. Cortisol promotes *(synthesis? breakdown?)* of proteins. In this respect, its effects are *(similar? opposite?)* to those of both hGH and thyroid hormone.

   b. However, cortisol may stimulate conversion of amino acid, lactic acid, and parts of fats to glucose. In this way

      cortisol (like hGH) _____-creases blood glucose level—which helps during stress. In fact, both cortisol and

      hGH levels _____-crease in response to stress (See page 335 of your text).

   c. Cortisol also acts to _____-crease blood pressure and to *(enhance? limit?)* the inflammatory process. (Cortisol is anti-inflammatory.)

E6. Androgens are produced by the adrenal *(cortex? medulla?)* in *(men? women? both men and women?)*. Effects of adrenal androgens are typically greater in *(females? males?)*. Describe effects of androgens.

E7. Do this exercise on the adrenal medulla.

   a. Name the two principal hormones secreted by the adrenal medulla.

      _____ _____

      Circle the one that accounts for 80 percent of adrenal medulla secretions.

   b. In general, effects of these hormones mimic the *(sympathetic? parasympathetic?)* nervous system. List three or more effects. *(Hint:* Refer to Figure LG 13.3(a).)

F.   Other endocrine glands and tissues   (pages 333–335)

■ **F1.** Do this exercise on sex hormones. (More details on these hormones in Chapter 23.)

   a. The two female sex hormones _____ and _____ are produced by

      organs named _____ . Levels of these hormones increase under the influence of *(releasing? tropic? target?)* hormones FSH and LH made by the *(hypothalamus? anterior pituitary?)*.

   b. Where is the hormone relaxin made? _____ Write one or more function(s) of relaxin.

   c. Name the primary androgen (male sex hormone). _____

   d. Inhibin is produced by *(females only? males only? both females and males?)*. Which hormone is inhibited by

      inhibin? _____

■ **F2.** Where is the pineal gland located? _____ Name a hormone produced by this gland.

_____ Secretion of the hormone increases in *(lightness? darkness?)*. SAD is an

acronym for s_____ a_____ d_____ , which
is more likely to occur in winter months when melatonin levels are *(elevated? low?)*.

■ **F3.** Where is the thymus located? (Hint: refer to Figure LG 13.1.) *(In the neck? Anterior to the heart?)* Name one

hormone made by the thymus gland. _____

■ **F4.** As you continue your study of anatomy and physiology, you will consider hormones produced in other parts

of the body: those from the _____ tract (Chapter 19), hormones made in the

_____ during pregnancy (Chapter 24), and also the kidney hormone erythropoietin, which

affects _____ production (Chapter 14). The heart produces a hormone called ANP

(_____ _____ _____) that helps to

_____-crease blood pressure.

F5. Write a brief description of functions of each of the following categories of chemicals produced by the body.
a. Leukotrienes (LTs)

b. Prostaglandins (PGs)

■ **F6.** Certain medications such as aspirin and ibuprofen (Advil) inhibit synthesis of one of the categories of chem-

icals listed in Checkpoint E7. Which one? _____ . Write three helpful effects of these
medications.

G. The stress response   (pages 335–336)

G1. Define *stressor* and list five examples of stressors.

■ **G2.** Summarize the body's response to stressors in this exercise.

a. The part of the brain that senses the stress and initiates response is the _____ . It responds
by two main mechanisms.
b. First is the *(alarm? resistance?)* reaction involving the adrenal *(medulla? cortex?)* and the

_____ division of the autonomic nervous system.

c. Second is the _____ reaction. Playing key roles in this response are three anterior pitui-

tary hormones: _____ , _____ , and _____
and their target hormones. Therefore the adrenal *(medulla? cortex?)* hormones are activated.

■ **G3.** Now describe events that characterize the first (or alarm) stage of response to stress. Complete this exercise.

a. The alarm reaction is sometimes called the _____ response.

b. During this stage blood glucose *(increases? decreases?)* by a number of hormonal mechanisms, including those

of adrenal medulla hormones _____ and _____ . Glucose must be
available for cells to have energy for the stress response.

c. Oxygen must also be available to tissues; the respiratory system *(increases? decreases?)* its activity. Heart rate
and blood pressure *(increase? decrease?)*. Blood is shunted to vital tissues such as the

_____ , and _____ .

G4. Describe the effects of each of these hormones during stress responses.

a. Mineralocorticoid (aldosterone)

b. Glucocorticoid (cortisol)

c. Thyroid hormones

d. Human growth hormone

■ **G5.** When the resistance stage fails to combat the stressor, the body moves into the _____ stage. Explain
how this stage is related to prolonged effects of:

a. Glucocorticoids

b. The renin-angiotensin-aldosterone (RAA) mechanism

H. Aging and the endocrine system; common disorders; homeostasis (pages 336–340)

H1. Defend or dispute this statement: "With aging, the endocrine system is usually compromised."

■ **H2.** With age, the physiologic changes that occur in various glands may or may not affect production of each gland's designated hormone. Identify aging changes in the following exercise.
   a. Adrenal glands secrete (*more? the same amounts of? less?*) cortisol and aldosterone, and they produce (*more? the same amounts of? less?*) epinephrine and norepinephrine.
   b. The thymus is largest in (*infancy? early adulthood? old age?*). T lymphocytes (*are? are not?*) still produced in old age.
   c. Aging ovaries (*increase? decrease?*) in size, and they (*do? do not?*) respond to gonadotropins. Because ovaries produce (*more? less?*) estrogen, FSH and LH levels are (*higher? lower?*) after menopause.
   d. Elderly males (*can? cannot?*) produce active sperm in normal numbers.

■ **H3.**   Now that you have studied major aspects of the endocrine system, identify the systems affected by hormones in each case.

   a. Jimmy's erythropoietin level is considerably lower than normal: _____ system.

   b. Mr. Israel's LH levels are low: _____ system.

   c. Ms. Benton is taking cortisone shots for her arthritis and she is more vulnerable to respiratory infections:

   _____ system.

   d. Androgens secreted as Alex reaches puberty are activating facial sebaceous glands, resulting in acne:

   _____ system.

   e. Sarah's parathyroid glands are overactive, altering her blood calcium levels (3 systems):

   _____ system, _____ system, _____ system.

■ **H4.**   Match the disorder with the hormonal imbalance.

Ac. Acromegaly	DM. Diabetes mellitus	PD. Pituitary dwarfism
Ad. Addison's disease	Go. Goiter	T. Tetany
Cr. Cretinism	Gr. Graves' disease	
Cu. Cushing's syndrome	M. Myxedema	
DI. Diabetes insipidus	PC. Pheochromocytoma	

_____ a. Deficiency of hGH in child; slow bone growth

_____ b. Excess of hGH in adult; enlargement of hands, feet, and jawbones

_____ c. Deficiency of ADH; production of enormous quantities of "insipid" (nonsugary) urine

_____ d. Deficiency of effective insulin; hyperglycemia and glycosuria (sugary urine)

_____ e. Deficiency of thyroid hormones in child; short stature and mental retardation

_____ f. Deficiency of thyroid hormones in adult; edematous facial tissues

_____ g. Overactive thyroid with protruding eyes and high metabolic rate

_____ h. Enlarged thyroid gland

_____ i. Result of deficiency of PTH; decreased calcium in blood and fluids around muscles, resulting in abnormal muscle contraction

_____ j. Deficiency of adrenocorticoids; increased K^+ and decreased Na^+, resulting in low blood pressure and dehydration

_____ k. Oversecretion of adrenal cortex; "moon face" and "buffalo hump"

_____ l. Due to a benign tumor resulting in hypersecretion of epinephrine and norepinephrine

■ **H5.** Contrast type 1 and type 2 diabetes mellitus (DM) by writing 1 or 2 next to the related descriptions.

_____ a. Also known as maturity-onset DM

_____ b. The more common type of DM

_____ c. Not due to deficiency of insulin, but to loss of sensitivity of cells to insulin

_____ d. Can be treated by diet, exercise, weight loss, and drugs that stimulate secretion of insulin

_____ e. More likely to lead to serious complications such as ketoacidosis

_____ f. Pancreatic beta cells are destroyed so replacement insulin is required

■ **H6.** Answer these questions about diabetes.

a. List the three "poly's" of diabetes.

b. Which condition is more likely to occur when the body has an excessive amount of insulin for current needs?

(Hyper? Hypo?)-glycemia. This condition which deprives the brain of glucose is known as insulin _____.

ANSWERS TO SELECTED CHECKPOINTS: CHAPTER 13

A1. (a) N. (b) E. (c) N.

A2.

	Secretions Transported In	Examples
a. Exocrine	Ducts	Sweat, oil, mucus, digestive juices
b. Endocrine	Interstitial fluid, then into blood	Hormones

A3. (a) A, pineal; B, posterior pituitary; C, anterior pituitary; D, thyroid; E, parathyroid; F, adrenal cortex; G, adrenal medulla. (b) 1, hypothalamus; 2, thymus; 3, heart; 4, skin; 5, liver; 6, kidney; 7, stomach; 8, pancreas; 9, small intestine; 10, adipose tissue; 11, ovaries; 12, placenta; 13, testes.

A5. You (the cell) would have only certain receptors on your surface (in plasma membrane) or within you (intracellular receptors). Any given hormone can affect only a cell with specific receptors that "fit" that hormone, just as only certain hormonal "keys" fit the "locks" on your doors. Many other hormones float by, never entering because they do not present the chemical "keys" that fit your "locks."

A6. Lipid; estrogens, progesterone, testosterone, and aldosterone; DNA, mRNA; enzymes.

A7. (a) Water, cannot; releasing and inhibiting hormones and OT (and ADH) made by the hypothalamus, FSH, LH, TSH, ACTH, and hGH from the anterior pituitary, insulin, glucagon, PTH, and CT. (b) First, anterior pituitary, target. (c) Outer; because water-soluble hormones cannot easily pass across the lipid bilayer portion of the plasma membrane; cyclic AMP (cAMP), second. (d) Growth. (e) Water, the second messenger mechanism.

A8. (a) N. (b, c) C. (d) H.

A9. Negative; increase. Example of positive feedback: uterine contractions stimulate more oxytocin.

B1. (a) Master gland; infundibulum, hypothalamus. (b) Anterior; anterior, posterior; anterior. (c) 7; hGH, PRL, ACTH, MSH, TSH, FSH, LH. (Refer to Checkpoint B3.)

B2. (a) hGH, bones, skeletal muscles; insulinlike growth factors; liver, skeletal muscle, cartilage, and bone. (b) In, in, in. (c) Hypothalamus, inhibiting. (d) Hypo, releasing, high, high; negative.

B3. (a) FSH. (b) LH. (c) LH. (d) TSH. (e) ACTH. (f) MSH. (g) PRL. (h) ACTH.

B4. Hormones that influence (stimulate production of) other hormones; ACTH, FSH, LH, and TSH.

B5. (a) Prolactin (PRL), anterior; is. (b) In, de; PRH, PIH. (c) Estrogens and progesterone. (d) Oxytocin (OT), posterior. (e) No known function, but hypersecretion can lead to erectile dysfunction.

B7. (a) In; in, in; positive; Pitocin. (b) Ejection or letdown of milk from; stimulates. (c) In; de, in; negative. (d) Pressin, in, constricting. (e) Inhibits, in.

C1. Anterior, inferior; butterfly or letter "H."

C2. (a) Follicular; thyroxine (T_4), triiodothyronine (T_3); calcitonin (CT). (b) I^-; 3.

C3. (a) In; de. (b) In; in; in.

C4. (a) High. (b) Hypothalamus; high. (c) Tropic, anterior pituitary; stimulates. (d) High, high. (e) Inhibits, negative. (f) High altitude, pregnancy.

C5. Labels on Figure LG 13.2: (a) Effects of calcitonin (CT). (b) Effects of parathyroid hormone (PTH). Antagonists; PTH; D; calcium, magnesium, and phosphate; in.

C6. (a) ↓ CA^{2+} ↓ PO_4^{3-}. (b) ↑ Ca^{2+} ↓ PO_4^{3-} ↑ Mg^{2+}.

C7. C; ↓ blood Ca^{2+} → ↑ secretion of PTH → ↑ blood Ca^{2+} → ↑ secretion of calcitonin → ↓ blood Ca^{2+}.

D1. Pancreas. (a) Endo; alpha, beta. (b) Alpha, increases; glycogen. (c) No; glucose; high; negative. (d) *Glycogen* is a storage form of glucose; *glucagon* is a hormone that breaks down glycogen to glucose. (e) Opposite; helps; glycogen; amino acids. (f) A D. (g) Hyper; hypo; glucagons and epinephrine are hyperglycemic; insulin is hypoglycemic.

D2. (a) B D. (b) Higher; body cells do not respond to insulin, resistance. (c) Waist; elevated, in, high. (d) Control weight and alcohol intake, exercise, and do not smoke.

E1. (a–c) AC. (d) AM.

E2. Mineralo, androgens.

E3. 1, The protein angiotensinogen; 2, renin; 3, angiotensin I; 4, angiotensin II; 5, constrictor; 6, aldosterone; 7, Na^+ (and H_2O by osmosis); 8, K^+ and H^+; 9, H_2O; 10, In.

E4. Aldosterone; cortisone.

E5. (a) Breakdown; opposite. (b) In; in. (c) In; limit.

F1. (a) Estrogens, progesterone, ovaries; tropic, anterior pituitary. (b) Ovaries and placenta; relaxes cartilage between hipbones (pubic symphysis) and dilates cervix of the uterus toward the end of pregnancy. (c) Testosterone. (d) Both females and males; FSH.

F2. In the brain near the hypothalamus; melatonin; darkness; seasonal affect disorder, elevated.

F3. Anterior to the heart; thymosin.

F4. Gastrointestinal, placenta, red blood cell; atrial natriuretic peptide, de.

F6. Prostaglandins (PGs); reduce fever, pain, and inflammation.

G2. (a) Hypothalamus. (b) Alarm, medulla, sympathetic. (c) Resistance; ACTH, hGH, TSH; cortex.

G3. (a) Fight-or-flight. (b) Increases, epinephrine and norepinephrine. (c) Increases; increase; skeletal muscles, heart.

G5. Exhaustion. (a) Cortisol leads to muscle wasting and suppression of the immune system which can lead to infection. (b) The RAA mechanism reduces blood flow to organs such as the GI tract and kidneys and can lead to stomach ulcers (and reduced eating and nutrition) and kidney failure.

H2. (a) Less; the same amount of. (b) Infancy; are. (c) Decrease, do not; less, higher. (d) Can.

H3. (a) Cardiovascular (blood). (b) Reproductive (ovaries). (c) Lymphatic and immune. (d) Integumentary. (e) Skeletal (loss of bone mass), muscular, and nervous.

H4. (a) PD. (b) Ac. (c) DI. (d) DM. (e) Cr. (f) M. (g) Gr. (h) Go. (i) T. (j) Ad. (k) Cu. (l) PC.

H5. (a–d) 2 (e–f) 1.

H6. (a) Polyuria, polydipsia, and polyphagia. (b) Hypo; shock.

CRITICAL THINKING: CHAPTER 13

1. Contrast sources and functions of releasing hormones and tropic hormones.
2. Contrast hormone production of the anterior and posterior lobes of the pituitary gland.
3. Describe hormonal regulation of blood levels of the following ions; (a) calcium, (b) potassium.
4. Describe the renin–angiotensin–aldosterone mechanism for controlling blood pressure.
5. Discuss the hormone imbalance (whether excessive or deficient) and symptoms of each of the following conditions: acromegaly, Addison's disease, Cushing's syndrome, diabetes insipidus, diabetes mellitus, and myxedema.

MASTERY TEST: ■ CHAPTER 13

Question 1: Arrange the answers in correct sequence.

_____ _____ _____ _____   1. Steps in renin–angiotensin mechanism to increase blood pressure:
   A. Angiotensin I is converted to angiotensin II.
   B. Renin converts angiotensinogen, a plasma protein, to angiotensin I.
   C. Angiotensin II causes vasoconstriction and stimulates aldosterone production, which causes water conservation and raises blood pressure.
   D. Low blood pressure or low blood Na^+ level causes secretion of enzyme renin from cells in kidneys.

Questions 2–5: Circle the letter preceding the one best answer to each question.

2. All of the following correctly match hormonal imbalance with related signs or symptoms *except:*
   A. Deficiency of ADH—excessive urinary output
   B. Excessive aldosterone—high blood pressure
   C. Excess of aldosterone—potassium depletion, muscle weakness
   D. Excess of PTH—low blood level of calcium, tetany (muscle spasms)

3. All of the following hormones are secreted by the anterior pituitary *except:*
   A. ACTH   D. Oxytocin
   B. FSH   E. hGH
   C. Prolactin

4. All of these hormones lead to increased blood glucose *except:*
   A. ACTH   D. Growth hormone
   B. Insulin   E. Epinephrine
   C. Glucagon

5. All of these compounds are synthesized in the hypothalamus *except:*
   A. ADH   D. PIH
   B. GHRH   E. Oxytocin
   C. Calcitonin

Questions 6–10: Circle T (true) or F (false). If the statement is false, change the underlined word or phrase so that the statement is correct.

T F 6. The hormones of the adrenal medulla mimic the action of <u>parasympathetic</u> nerves.

T F 7. Secretion of <u>growth hormone, insulin, and glucagon</u> is controlled (directly or indirectly) by blood glucose level.

T F 8. Most of the feedback mechanisms that regulate hormones are <u>positive (rather than negative)</u> feedback mechanisms.

T F 9. The alarm reaction occurs <u>before</u> the resistance reaction to stress.

T F 10. Most hormones studied so far appear to act by a mechanism involving <u>cyclic AMP</u> rather than by the <u>gene activation</u> mechanism.

Questions 11–15: Fill-ins. Write the word or phrase that best fits the description.

_____ 11. _____ is a hormone used to induce labor because it stimulates contractions of smooth muscle of the uterus.

_____ 12. A high level of thyroid hormones circulating in the blood stream will tend to lead to a low level of the releasing hormone _____ and the tropic hormone _____ .

_____ 13. Type _____ diabetes mellitus is a non-insulin-dependent condition in which insulin level is adequate but cells have decreased sensitivity to insulin.

_____ 14. "Releasing hormones" and "inhibiting hormones" are all made in the _____ and they affect the _____ .

_____ 15. _____ is a gland that secretes anti-inflammatory hormones similar to cortisone.

ANSWERS TO MASTERY TEST: ■ CHAPTER 13

Arrange
1. D B A C

Multiple Choice
2. D
3. D
4. B
5. C

True or False
6. F. Sympathetic
7. T
8. F. Negative (rather than positive)
9. T
10. T

Fill-ins
11. Oxytocin (OT or pitocin)
12. TRH, TSH
13. 2
14. Hypothalamus, anterior pituitary
15. Adrenal cortex

FRAMEWORK 14
The Cardiovascular System: Blood

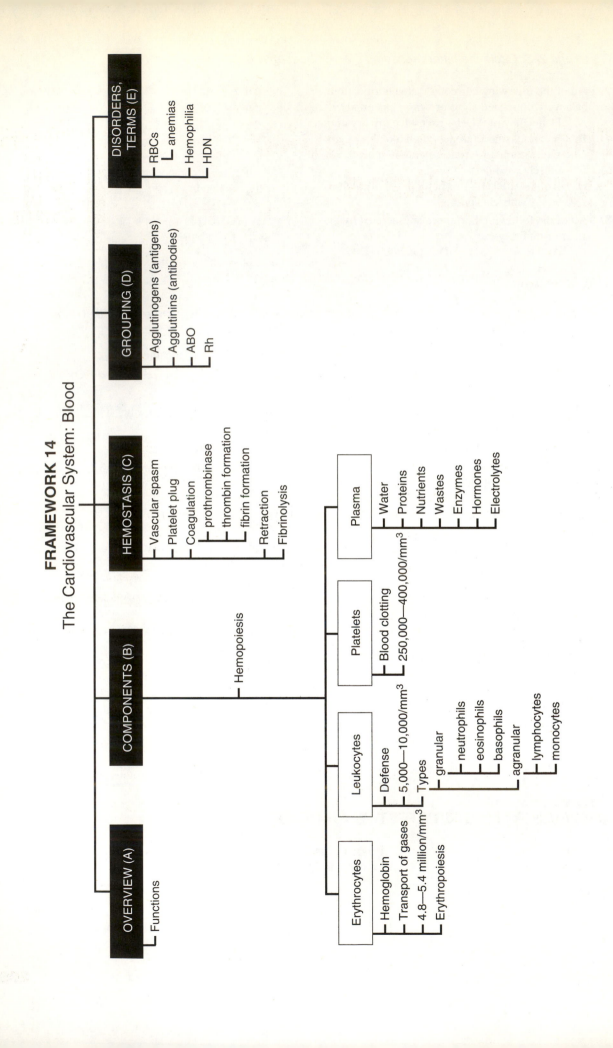

The Cardiovascular System: Blood

The major role of the blood is transportation through an intricate system of channels—blood vessels—that reach virtually every part of the body. Blood constantly courses through the vessels, carrying gases, fluids, nutrients, electrolytes, hormones, and wastes to and from body cells. Most of these chemicals "ride" in the fluid portion of blood (plasma). Some, such as oxygen, piggyback on red blood cells. White blood cell "passengers" take short trips in the blood and depart from the vessels at sites calling for defense. Platelets and other clotting factors congregate to plug holes in the network of vessels.

As you begin your study of the cardiovascular system, carefully examine the Chapter 14 Topic Outline and Objectives; check off each one as you complete it. To organize your study of blood, glance over the Chapter 14 Framework now. Be sure to refer to the Framework frequently and note relationships among key terms in each section.

TOPIC OUTLINE AND OBJECTIVES

A. Overview: functions of blood

☐ 1. List and describe the functions of blood.

B. Components of whole blood

☐ 2. Discuss the formation, components and functions of whole blood.

C. Hemostasis

☐ 3. Describe the various mechanisms that prevent blood loss.

D. Blood groups and blood types

☐ 4. Describe the ABO and Rh blood groups.

E. Common disorders and medical terminology

WORDBYTES

Study each wordbyte, its meaning, and an example of its use in a term. Check your understanding by jotting meanings of wordbytes in margins. Identify other examples of terms that contain these wordbytes as you continue through the text and *Learning Guide*.

Wordbyte	Meaning	Example(s)	Wordbyte	Meaning	Example(s)
a-, an-	not, without	*an*emia	-osis	condition	leukocyt*osis*
-crit	to separate	hemato*crit*	-pedesis	moving	dia*pedesis*
-cyte	cell	leuko*cyte*	-penia	want, lack	leukocyto*penia*
dia-	through	*dia*phragm	-phil-	love	hemo*phil*iac
-emia	blood	leuk*emia*	-poiesis	formation	hemo*poiesis*
erythro-	red	*erythro*poiesis	poly-	many	*poly*cythemia
heme-, hemo-	blood	*hemo*stasis	-rhag-	burst forth	hemor*rhage*
leuko-	white	*leuko*cytosis	-stasis	standing still	hemo*stasis*
mega-	large	*mega*karyocyte	thromb-	clot	*thromb*us, pro*thromb*in

CHECKPOINTS

A. Overview: functions of blood   (page 346)

A1. Write a description of the components and functions of the cardiovascular system.

■ **A2.** Describe functions of blood in this Checkpoint.

a. Blood transports many substances. List six or more.

_____ _____ _____

_____ _____ _____

b. List four aspects of homeostasis regulated by blood. One is done for you.

_____**pH**_____ _____

_____ _____

c. Explain how your blood protects you.

B. Components of whole blood (pages 346–354)

■ **B1.** Describe components of blood in this exercise.

a. Blood consists of about _____ percent plasma and _____ percent formed elements, which include

_____ and _____ .

b. The three types of formed elements (or cells) are: red blood cells (_____ -cytes), white

blood cells (_____ -cytes), and platelets (_____ -cytes).

■ **B2.** *A clinical challenge.* Answer these questions about hematocrit.

a. A hematocrit consists of the percentage of centrifuged blood that consists of _____ blood
cells. Which is a normal value for a hematocrit? *(15%? 45%? 70%?)* A hematocrit of 15 indicates *(mild? severe?)*
anemia. Which value indicates polycythemia? A hematocrit of *(14? 45? 60?)*

b. Which components of centrifuged blood are found in the buffy coat? _____ blood cells.
These make up less than *(1%? 20%? 42%? 70%?)* of blood.

■ **B3.** Match the names of components of plasma with their descriptions. (*Hint:* Refer to Figure 14.1, page 347 in
your text.)

A. Albumins	GAF. Glucose, amino acids, and fats
E. Electrolytes	HE. Hormones and enzymes
F. Fibrinogen	W. Water
G. Globulins	

_____ a. Makes up about 92 percent of plasma   _____ e. A protein used in clotting

_____ b. Regulatory substances carried in blood   _____ f. Antibody proteins

_____ c. Cations and anions carried in plasma   _____ g. Food substances carried in plasma

_____ d. Constitute about 54 percent of plasma
protein

■ **B4.** Do this exercise on blood formation.

a. Blood formation is a process known as _____ . All blood cells arise from

_____ cells.

b. After birth most blood cell formation takes place in red bone marrow. Name several bones in which this process
occurs after birth.

c. Name the two types of stem cells: _____ and _____ stem cells.
Which of these form most types of blood cells?

_____

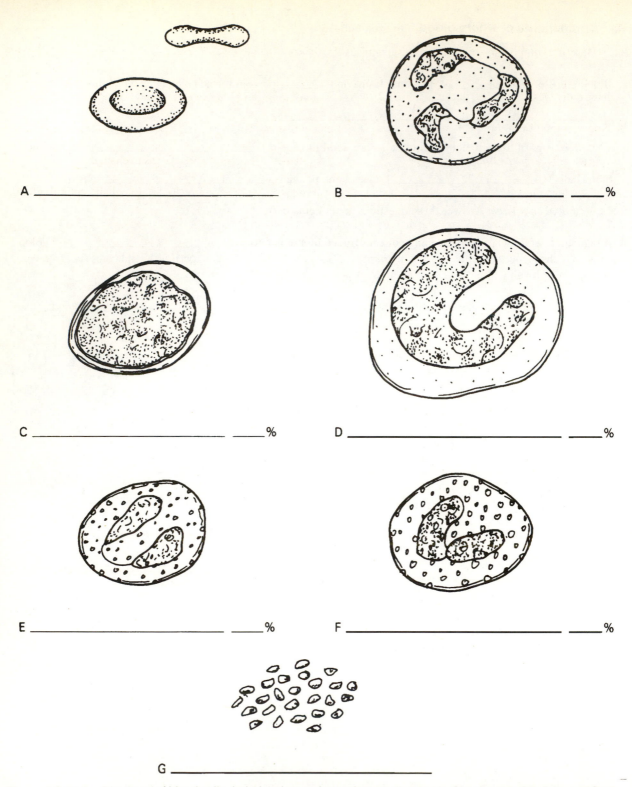

A _____

B _____ ____ %

C _____ ____ %

D _____ ____ %

E _____ ____ %

F _____ ____ %

G _____

Figure LG 14.1   Diagrams of blood cells. Label, color, and complete as directed in Checkpoints B5, B10, and B17.

■ **B5.** Do this exercise on red blood cell (RBC) structure.

a. Refer to Figure 14.2, p. 348 in your text. Name the type of cell that is the last stage in development before mature

red blood cells (RBCs) move into the blood stream. _____ This cell *(does? does not?)* have a nucleus. Typically, about *(1%? 5%? 10%?)* of circulating red blood cells are reticulocytes.

b. Now refer to Figure LG 14.1. Which diagram represents a mature RBC? _____ Note that it *(does? does not?)*

contain a nucleus. Its biconcave shape _____ -creases the surface area for gas exchange.

The chemical named _____ accounts for the red color of RBCs. Label and color the RBC.

■ **B6.** Circle the most normal blood values. (Note that values vary slightly according to age and sex.)

a. Average life of a red blood cell: 4 hours   4 days   4 months   4 years

b. RBC count in cube this size: □ μL (microliter)
500   5000   250,000   5 million   250 million

■ **B7.** Explain the roles of the terms in the box in red blood cell production and destruction by matching terms with descriptions.

AA. Amino acid	H. Hypoxia
BB. Bilirubin and biliverdin	IF. Intrinsic factor
B_{12}. Vitamin B_{12}	T. Transferrin
E. Erythropoietin	US. Urobilinogen and stercobilin
Fe. Iron	

_____ a. Decrease of oxygen in cells; serves as a signal that erythropoiesis is needed (to help provide more oxygen to tissues)

_____ b. Hormone produced by kidneys when they are hypoxic; stimulates erythropoiesis in red bone marrow

_____ c. Vitamin necessary for normal erythropoiesis

_____ d. Substance produced by the stomach lining and necessary for normal vitamin B_{12} absorption

_____ e. Component of the globin portion of hemoglobin

_____ f. Component of the heme portion of hemoglobin

_____ g. Protein transporter of iron within blood

_____ h. Products of breakdown of the heme portion of hemoglobin

_____ i. Waste products of bilirubin eliminated in urine or feces

B8. Define *induced polycythemia* and explain pro's and con's of using this procedure.

B9. Define these terms: *anemia, cyanosis.*

■ **B10.** Refer to Figure LG 14.1 and do this exercise about leukocytes.

a. Leukocytes *(have? lack?)* hemoglobin, and so these cells are known as *(red blood cells or RBCs? white blood cells or WBCs?).*

b. Label each leukocyte (WBC) on Figure LG 14.1 and indicate what percentage of the total WBC count is accounted for by each type of WBC. Such a breakdown of white blood cells is known as a

_____ WBC count.

c. Each white blood cell (WBC) *(has? lacks?)* a nucleus. Which type of WBC has a large kidney-shaped nucleus?

_____ Which WBC has a nucleus that occupies most of the cell?

_____ .

d. Which three WBCs are granular? _____ _____ _____

e. *For extra review.* Color nucleus, cytoplasm, and granules of all WBCs.

■ **B11.** Check your understanding of types of WBCs by matching names of WBCs with descriptions. Answers may be used more than once.

B. Basophils	L. Lymphocytes	N. Neutrophils
E. Eosinophils	M. Monocytes	

_____ a. Constitute the largest percentage of WBCs

_____ b. Largest of the leukocytes, they are 12–20 μm in diameter

_____ c. Involved in immunity: some develop into plasma cells that produce antibodies

_____ d. Involved in allergic response: release histamine, heparin, and serotonin

_____ e. Involved in allergic reactions, combat histamines and provide protection against parasitic worms

_____ f. Form wandering macrophages that clean up sites of infection

_____ g. Important in phagocytosis (two answers)

_____ h. Classfied as agranular leukocytes (two answers)

_____ i. Includes three types of cells: B, T, and natural killer

B12. Contrast these three methods for obtaining blood samples regarding purposes and risk factors: *venipuncture/fingerstick/arterial stick.*

■ **B13.** What are MHC antigens and how do they affect the success rate of transplants?

■ **B14.** A normal leukocyte count is _____/μL. In other words, a typical ratio of RBCs to WBCs is about:

A. 700:1   C. 2:1

B. 30:1   D. 1:1

■ **B15.** *A clinical challenge.* Mrs. Doud arrives at a health clinic with a suspected acute infection. During infection, it is likely that Mrs. Doud's leukocyte count will *(in? de?)*-crease. A count of *(4000? 8000? 15,000?)* leukocytes/μL blood is most likely. This condition is known as *(leukocytosis? leukopenia?)*.

■ **B16.** Circle correct answers related to thrombocytes.

a. Thrombocytes are also known as:
   A. Antibodies   C. Red blood cells
   B. Platelets   D. White blood cells
b. Thrombocytes are formed in:
   A. Bone marrow
   B. Tonsils and lymph nodes
   C. Spleen
c. Platelets are:
   A. Entire cells
   B. Chips off the old megakaryocytes

d. A normal range for platelet count is __/μL.
   A. 5,000–10,000   C. 4.5–5.5 million
   B. 150,000–400,000
e. The primary function of platelets is related to:
   A. O_2 and CO_2 transport   C. Blood clotting
   B. Defense   D. Blood typing

■ **B17.** Label platelets on Figure LG 14.1.

B18. List the components of a *complete blood count (CBC)* (*Hint:* Refer to Table 14.1, page 352 of the text).

C. Hemostasis   (pages 354–357)

■ **C1.** Define hemostasis. _____

List the three basic mechanisms of hemostasis:

a. Vascular _____

b. _____ plug

c. _____ (clotting)

■ **C2.** Check your understanding of the first two steps of hemostasis in this Checkpoint.

a. What typically triggers vascular spasm at the start of hemostasis?

For how long does the vascular spasm persist?

Platelets *(diminish? enhance?)* vasoconstriction.

b. Arrange in correct sequence these events in platelet plug formation: _____ _____ _____

PAD. Platelet adhesion

PAG. Platelet aggregation

PPF. Platelet plug formation

c. State two factors that result in a "snowball" effect, causing the platelet plug to enlarge.

d. The resulting platelet plug is useful in preventing blood loss in *(large? small?)* vessels.

■ **C3.** Consider the third step in hemostasis, coagulation (or blood clotting), in this Checkpoint.

a. Exactly what forms a clot? *(A "snowball" of red blood cells? An insoluble protein named fibrin plus trapped blood cells?)*

b. In a test tube with blood clotted at the bottom, the liquid found above the clot is called *(plasma? serum?)*. In other words, serum is the liquid part of blood *(containing? minus?)* clotting factors, whereas plasma is the liquid part of blood *(containing? minus?)* clotting factors.

c. Explain why clotting might be considered "good news" or "bad news," depending on the circumstances.

■ **C4.** Refer to Figure LG 14.2 and do this exercise on formation of blood clots. We will start from the final product, the clot itself (letter E), and work backward to see what leads to formation of the clot.

a. As we noted in Checkpoint C3, a clot (or _____) is made of the insoluble protein named

_____ (E). The term *insoluble protein* means that the strands of fibrin *(dissolve well? do not dissolve?)* in plasma.

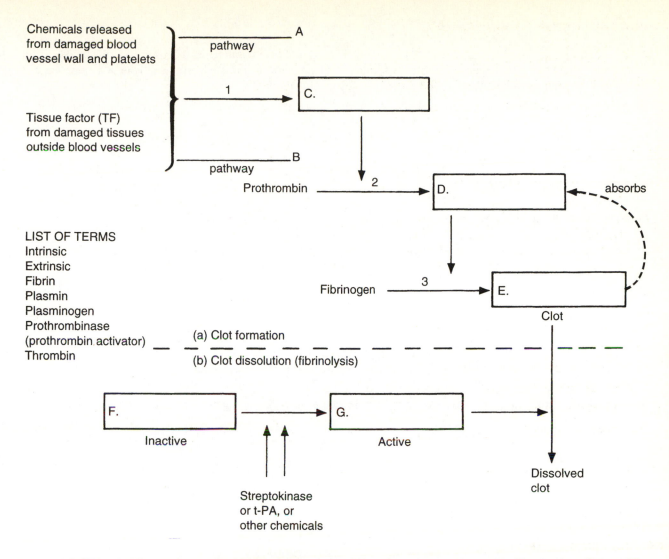

Figure LG 14.2 Summary of steps in clot formation and dissolution. Complete as directed in Checkpoints C4 and C6.

b. If fibrin were always present in blood, blood would clot continuously and excessively; blood flow would be inhibited. Fortunately, the coagulation mechanism is designed to form clots only when clotting is desirable (although it may happen when unwanted). Fibrin is formed initially in a "soluble" (dissolved) state known as

_____ , which is always ready and waiting in blood plasma. Only when the enzyme

_____ (letter _____ on the figure) is present is fibrinogen converted to insoluble fibrin.

c. Similarly, if thrombin were always present, the final step (fibrinogen → fibrin) would happen continually. However, a checks-and-balance mechanism is present by having the inactive enzyme named

_____ constantly present in plasma, with activation to thrombin occurring only in certain situations.

d. What enzyme triggers conversion of prothrombin → thrombin? _____ (letter _____). A cascade of events (in A and B) leads to formation of this enzyme.

e. Prothrombinase may be formed by either of two pathways, both of which involve release of chemicals from dam-

aged cells. In one case (letter _____), the cells are lining blood vessels, so the damage is inside (or _____- trinsic to) the cardiovascular system. This happens, unfortunately, when blood vessels have (*smooth? rough?*) insides in atherosclerosis.

f. The second, or _____-trinsic, pathway (letter _____) involves release of a chemical known as

_____ _____ (TF) from traumatized cells outside the blood vessels. When you bump against a corner of furniture and get a bruise, you are likely to activate both pathways (A and B) because you damage blood vessels and surrounding tissues.

g. Note that both pathways (intrinsic and extrinsic) have steps *(1 and 2? 1 and 3? 2 and 3?)* in common; only step

_____ differs. However, the *(intrinsic? extrinsic?)* pathway occurs more rapidly (within seconds), whereas

the _____ pathway requires several minutes. Write the word *faster* next to B (extrinsic pathway) on Figure LG 14.2.

h. Note the arrows in the upper part of the figure that point to the "snowball" effects, or

_____ feedback mechanisms. *Platelets,* sticking to the damaged blood vessel lining, "recruit" even more platelets to enhance the intrinsic pathway. Explain how platelets do this.

i. In order for this cascade of events (steps 1, 2, and 3 on the figure) to occur, the ion *(Na^+? Cl^-? Ca^{2+}?)* is required.

j. Patients who experience excessive clot formation may take anticoagulant medications such as those that interfere

with thrombin formation. These include _____ and _____ .

■ **C5.** Although there are many forms of hemophilia, all involve deficiency of _____ . Hemophilia *(is? is not?)* hereditary. Write three symptoms of hemophilia. (Refer to page 360 in your text.)

■ **C6.** Describe mechanisms that normally prevent clots from becoming excessively large.

a. Fibrin can absorb and inactivate up to 90% of _____ to reduce the amount of this chemical in blood.

b. Clot _____ occurs, in which the clot becomes *(tighter and smaller? looser and larger?)*.

c. The lower portion of Figure LG 14.2 shows the next step: clot dissolution or _____ . An inactive enzyme (F on the figure), which is part of a clot, is activated to (G), which dissolves clots. Label F and G on the figure. F is activated by chemicals produced within the body as shown on the figure.

■ **C7.** *A clinical challenge.* Match the correct term with the description.

Embolus	Pulmonary embolism	Thrombus	Thrombosis

_____ a. A blood clot

_____ b. A blood clot in an unbroken vessel, such as

in an atherosclerosed vessel: _____

_____ c. A "clot-on-the-run" dislodged from the site at which it formed (usually a

deep vein of the leg); also fat from broken bone, bubble of air, or amniotic fluid traveling through blood, possibly to lung (pulmonary) vessels

_____ d. Blood clot (or fat or air) that travels from veins to heart to blood vessels in lungs:

_____

■ **C8.** Mr. Richardson is rushed by ambulance to an emergency room where tissue plasminogen activator (t-PA) is administered. What is his probable diagnosis?

C9. Describe health practices that can reduce risk for unwanted blood clots.

D. Blood groups and blood types (pages 357–359)

■ **D1.** Contrast agglutination with coagulation.

a. Coagulation (or _____) involves coagulation factors found in platelets, plasma, or other tissue fluids. The process *(requires? can occur in absence of?)* red blood cells.

b. Agglutination (or "clumping") of erythrocytes (RBCs) is an example of an

antigen-_____ process that *(does? does not?)* require RBCs because these are sites of

_____ used in the agglutination reaction. In the hours following the clumping, RBCs

swell and burst, a process called _____ .

■ **D2.** Do this exercise about factors responsible for blood groups (types).

a. Antigens are located *(in plasma? on surface of RBCs?)*. Type A blood has *(A? B?)* antigens on RBCs.

b. Several months after birth, infants with type A blood normally begin producing *(anti-A? anti-B? both anti-A and anti-B? neither anti-A nor anti-B?)* antibodies in response to exposures to even minute amounts of type B blood. In other words, people with type A blood produce antibodies that attack type *(A? B? A and B? O?)* blood because this is "foreign" (antigenic) to them.

D3. Complete the table contrasting ABO types.

Blood Group	Percentage of White Population	Percentage of Black Population	Percentage of Native American Population	Sketch of Blood Showing Correct Antigens and Antibodies	Can Donate Safely to	Can Receive Blood Safely from
a. Type A						A, O
b.	11					
c. Type AB		4		A B B A neither A B anti-A B A nor anti-B		
d.					A, B, AB, O	

■ **D4.** Type O is known as the universal *(donor? recipient?)* with regard to the ABO group because type O blood

lacks _____ of the ABO group. Type _____ is known as the universal recipient.
Explain why.

■ **D5.** Complete this exercise about the *Rh system.*

a. The Rh *(+? −?)* group is more common. Rh *(+? −?)* blood has Rh antigens on the surfaces of RBCs.

b. Under normal circumstances plasma of *(Rh⁺ blood? Rh⁻ blood? both Rh groups? neither Rh group?)* contains anti-Rh antibodies.

c. Rh *(+? −?)* people can develop these antibodies when they are exposed to Rh *(+? −?)* blood.

d. The most common example of this occurs in fetal–maternal incompatibility when a mother who is Rh *(+? −?)* has a baby who is Rh *(+? −?)* and some of the baby's blood enters the mother's bloodstream. The mother develops anti-Rh antibodies, which may cross the placenta in future pregnancies and hemolyze the RBCs of Rh

(+? −?) babies. Such a condition is known as _____ .

e. *A clinical challenge.* Discuss precautions taken with Rh⁻ mothers during pregnancy or soon after delivery, miscarriage, or abortion of an RH⁺ baby to prevent future problems with Rh incompatibility.

(Hint: see text page 360.)

■ **E1.** Match names of types of anemia with descriptions below.

A. Aplastic	Hr. Hemorrhagic	P. Pernicious
Hl. Hemolytic	I. Iron-deficiency	S. Sickle cell

_____ a. Most prevalent anemia in the world

_____ b. Condition in which intrinsic factor is not produced, so absorption of vitamin B_{12} is inadequate

_____ c. Inherited condition in which hemoglobin forms stiff rodlike structures, causing erythrocytes to assume sickle shape and rupture, reducing oxygen supply to tissues

_____ d. Rupture of red blood cell membranes due to variety of causes, such as parasites, toxins, or antibodies

_____ e. Condition due to excessive bleeding, as from wounds, gastric ulcers, heavy menstrual flow

_____ f. Inadequate erythropoiesis as a result of destruction or inhibition of red bone marrow

■ **E2.** _A clinical challenge._ Amy (35 years old) has a red blood count of 7 million/µL. She has the condition known as _(anemia? polycythemia?)_ Amy has a hematocrit done. It is more likely to be _(under 32? over 46?)._ Her blood is _(more? less?)_ viscous than normal, which is likely to cause _(high? low?)_ blood pressure.

E3. Define these terms:
a. Leukemia

b. Septicemia

c. Thrombocytopenia

d. Jaundice

ANSWERS TO SELECTED CHECKPOINTS: CHAPTER 14

A2. (a) Oxygen, carbon dioxide and other wastes, nutrients, hormones, enzymes, heat. (b) pH, temperature, water (fluids), and dissolved chemicals (such as ions and proteins). (c) Contains phagocytic cells, blood-clotting proteins, and defense proteins such as antibodies.

B1. (a) 55, 45, cells, cell fragments. (b) Erythro, leuko, thrombo.

B2. (a) Red, 45%; severe; 60. (b) White, 1%.

B3. (a) W. (b) HE. (c) E. (d) A. (e) F. (f) G. (g) GAF.

B4. (a) Hemopoiesis; pluripotent stem cells. (b) Femurs and humeri, flat bones of skull, sternum, ribs, vertebrae, and pelvis. (c) Myeloid, lymphoid; myeloid.

B5. (a) Reticulocyte; does not; in;1%. (b) A; does not; hemoglobin; see Figure 14.2, page 348 of your text.

B6. (a) 4 months. (b) 5 million.

B7. (a) H. (b) E. (c) B_{12}. (d) IF. (e) AA. (f) Fe. (g) T. (h) BB. (i) US.

B10. (a) Lack, white blood cells or WBCs. (b) B, neutrophil (60–70); C, lymphocyte (20–25); D, monocyte (3–8); E, eosinophil (2–4); F, basophil (0.5–1.0); differential. (c) Has; monocyte (D); lymphocyte (C). (d) B E F. (e) See Table 14.2, page 355 of your text.

B11. (a) N. (b) M. (c) L. (d) B. (e) E. (f) M. (g) M, N. (h) L, M. (i) L.

B13. MHC antigens are proteins on cell surfaces; they are unique for each person. The greater the similarity between major histocompatibility (MHC) antigens of donor and recipient, the less likely is rejection of the transplant.

B14. 5000 to 10,000; A (for example, 4,900,000 RBCs: 7000 WBCs = 700:1).

B15. In; 15,000; leukocytosis.

B16. (a) B. (b) A. (c) B. (d) B. (e) C.

B17. G.

C1. Stoppage of bleeding. (a) Spasm. (b) Platelet. (c) Coagulation.

C2. (a) Damage to the wall of a blood vessel; minutes to hours; enhance. (b) PAD PPF PAG. (c) Platelets become sticky, and they produce chemicals that activate more platelets. (d) Small.

C3. (a) An insoluble protein named fibrin plus trapped blood cells. (b) Serum; minus, containing. (c) "Good news": clotting is necessary to avoid excessive bleeding; "bad news": excessive clot formation (thrombosis) blocks blood flow, for example, through vessels that supply blood to the heart (a "coronary thrombosis").

C4. (a) Thrombus, fibrin; do not dissolve. (b) Fibrinogen; thrombin, D. (c) Prothrombin. (d) Prothrombinase, C. (e) A, in; rough. (f) Ex, B, tissue factor. (g) 2 and 3; 1; extrinsic, intrinsic. (h) Positive; damaged platelets are "sticky," causing more platelets to accumulate, and platelets release chemicals that trigger step 1 of clot formation; (i) Ca^{2+}. (j) Heparin and warfarin (coumadin).

C5. Some clotting (coagulation) factor; is; hemorrhaging, nosebleeds, blood in urine, and joint damage due to bleeding.

C6. (a) Thrombin. (b) Retraction; tighter and smaller. (c) Fibrinolysis; F, plasminogen; G, plasmin.

C7. (a) Thrombus. (b) Thrombosis. (c) Embolus. (d) Pulmonary embolism.

C8. Heart attack or stroke; t-PA can activate his own plasminogen to form plasmin which can break down blood clots in vessels of the heart or brain.

D1. (a) Clotting; can occur in absence of. (b) Antibody, does, antigens; hemolysis.

D2. (a) On the surface of RBCs; A. (b) Anti-B; B.

D4. Donor; A and B antigens; AB; people with type AB blood lack both anti-A and anti-B antibodies.

D5. (a) +; +. (b) Neither Rh group. (c) −, +. (d) −, +; +; hemolytic disease of the newborn (HDN) or erythroblastosis fetalis. (e) Anti-Rh antibody therapy is given to the mother to bind to any fetal Rh antigens, destroying them. Therefore, the Rh^- mother will not produce antibodies that would attack antigens of future Rh^+ fetuses.

E1. (a) I. (b) P. (c) S. (d) Hl. (e) Hr. (f) A.

E2. Polycythemia; over 46; more, high.

CRITICAL THINKING: CHAPTER 14

1. Describe mechanisms for assuring homeostasis of red blood cell count.
2. Explain why athletes might be likely to train in a high-altitude city such as Denver for several weeks immediately before competing in that city.
3. Describe roles of the following types of cells in inflammation (including phagocytosis) and immune responses: neutrophils, monocytes, and lymphocytes.
4. Contrast the role of eosinophils and basophils in body defenses.
5. Contrast advantages and disadvantages of blood clotting in the human body. Then describe control mechanisms that provide checks and balances so that clotting does not get out of hand.
6. Contrast anticoagulant medications with thrombolytic agents.
7. Contrast effects of blood type prevalence as you address these questions: (a) Is hemolytic disease of the newborn (HDN) more likely to occur among Asians and Native Americans or among white persons? Explain why. (b) Which cultural group has the greatest incidence of Type O blood, and the lowest incidence of the other three types?

MASTERY TEST: ■ CHAPTER 14

Questions 1 and 2: Arrange the answers in correct sequence.

_____ _____ _____ 1. Events in the coagulation process:
  A. Retraction or tightening of fibrin clot
  B. Fibrinolysis or clot dissolution by plasma
  C. Clot formation

_____ _____ _____ 2. Stages in the clotting process:
  A. Formation of prothrombinase
  B. Conversion of prothrombin to thrombin
  C. Conversion of fibrinogen to fibrin

3. Identify the value that is most likely to be outside the normal range for Ms. Marty, a 28-year old woman:
    A. RBCs—4.6 million/µL blood
    B. WBCs—9000/µL blood
    C. Platelets—1 million/µL blood
    D. Hemoglobin—13.5 g/100 mL blood
4. All of the following correctly match parts of blood with principal functions *except*:
    A. RBC: carry oxygen and CO_2
    B. Plasma: carries nutrients, wastes, hormones, enzymes
    C. WBCs: defense
    D. Platelets: determine blood type
5. Which of the following chemicals is an enzyme that converts fibrinogen to fibrin?
    A. Heparin   C. Prothrombin
    B. Thrombin   D. Tissue factor

Questions 6–10: Circle T (true) or F (false). If the statement is false, change the underlined word or phrase so that the statement is correct.

T F 6. Hemoglobin is made of <u>the protein named heme and the nonprotein globin.</u>

T F 7. Plasmin is an enzyme that facilitates clot <u>formation.</u>

T F 8. Hemolytic disease of the newborn (HDN) is most likely to occur with an Rh <u>positive mother and her Rh negative babies.</u>

T F 9. In both white and black populations in the United States, type O is <u>most</u> common and type AB is <u>least</u> common of the ABO blood groups.

T F 10. In 100 ml of blood there are usually about <u>15 mL</u> of formed elements, most of which are red blood cells, and this value is known as the <u>hematocrit.</u>

Questions 11–15: Fill-ins. Answer questions or complete sentences with the word or phrase that best fits.

_____ 11. Name five types of substances transported by blood.

_____ 12. Name three types of plasma proteins.

_____ 13. Name the type of leukocyte that is transformed into plasma cells, which then produce antibodies.

_____ 14. Write a value for a normal leukocyte count: _____/µL.

_____ 15. All blood cells and platelets are derived from ancestor cells called _____ .

ANSWERS TO MASTERY TEST: ■ CHAPTER 14

Arrange
1. C A B
2. A B C

Multiple Choice
3. C
4. D
5. B

True or False
6. F. The nonprotein named heme and the protein globin
7. F. Dissolution or fibrinolysis
8. F. Negative mother and her Rh positive babies
9. T
10. F. 42 (females) to 47 (males), hematocrit

Fill-ins
11. Oxygen, carbon dioxide, nutrients, wastes, water, hormones, enzymes
12. Albumin, globulins, and fibrinogen
13. B lymphocyte
14. 5000–10,000
15. Pluripotent stem cells

FRAMEWORK 15
The Cardiovascular System: Heart

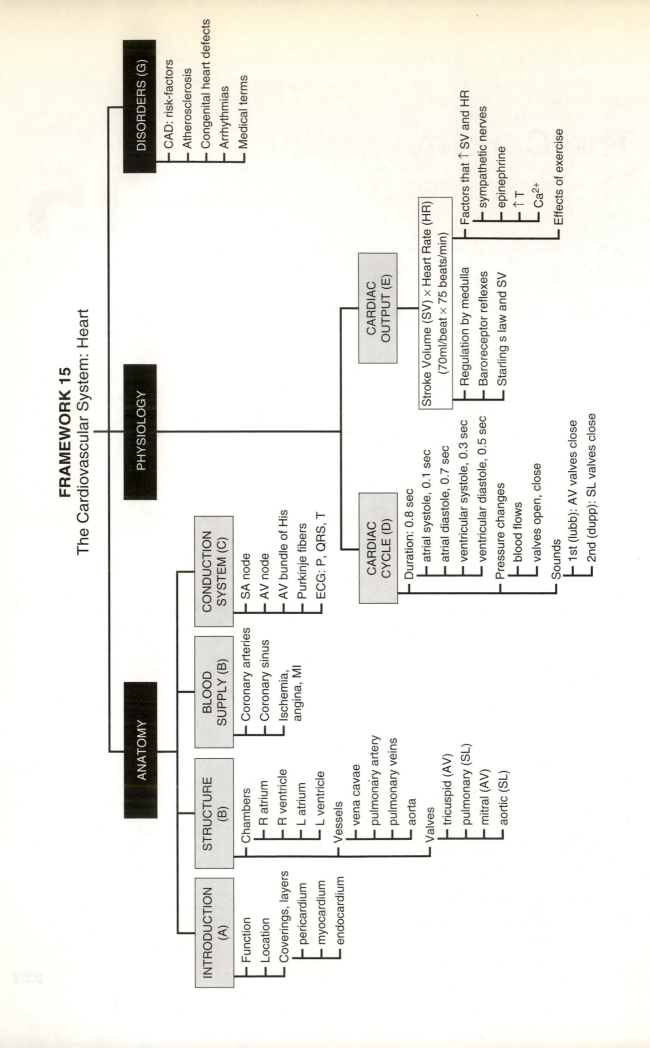

PHYSIOLOGY

ANATOMY

DISORDERS (G)
- CAD: risk-factors
- Atherosclerosis
- Congenital heart defects
- Arrhythmias
- Medical terms

INTRODUCTION (A)
- Function
- Location
- Coverings, layers
  - pericardium
  - myocardium
  - endocardium

STRUCTURE (B)
- Chambers
  - R atrium
  - R ventricle
  - L atrium
  - L ventricle
- Vessels
  - vena cavae
  - pulmonary artery
  - pulmonary veins
  - aorta
- Valves
  - tricuspid (AV)
  - pulmonary (SL)
  - mitral (AV)
  - aortic (SL)

BLOOD SUPPLY (B)
- Coronary arteries
- Coronary sinus
- Ischemia, angina, MI

CONDUCTION SYSTEM (C)
- SA node
- AV node
- AV bundle of His
- Purkinje fibers
- ECG: P, QRS, T

CARDIAC CYCLE (D)
- Duration: 0.8 sec
  - atrial systole, 0.1 sec
  - atrial diastole, 0.7 sec
  - ventricular systole, 0.3 sec
  - ventricular diastole, 0.5 sec
- Pressure changes
  - blood flows
  - valves open, close
- Sounds
  - 1st (lubb): AV valves close
  - 2nd (dupp): SL valves close

CARDIAC OUTPUT (E)
- Stroke Volume (SV) × Heart Rate (HR) (70ml/beat × 75 beats/min)
- Regulation by medulla
- Baroreceptor reflexes
- Starling s law and SV
- Factors that ↑ SV and HR
  - sympathetic nerves
  - epinephrine
  - ↑ T
  - Ca²⁺
- Effects of exercise

The Cardiovascular System: Heart

CHAPTER 15

Make a fist, then open and squeeze tightly again. Repeat this about once a second as you read the rest of this Overview. Envision your heart as a muscular pump about the size of your fist. Unfailingly, your heart exerts pressure on your blood, moving it onward through the vessels to reach all body parts. The heart has one job, it is simply a pump. But this function is critical. Without the force of the heart, blood would come to a standstill and tissue would be deprived of fluids, nutrients, and other vital chemicals. To serve as an effective pump, the heart requires a rich blood supply to maintain healthy muscular walls, a specialized nerve conduction system to synchronize actions of the heart, and intact valves to direct blood flow correctly. Heart sounds and ECG recordings as well as a variety of more complex diagnostic tools provide clues to the status of the heart.

Is your fist tired yet? The heart ordinarily pumps 24 hours a day without complaint and seldom reminds us of its presence. As you complete this chapter on the heart, keep in mind the value and indispensability of this organ. Start by studying the Chapter 15 Framework and the key terms for each section. Then carefully examine the Chapter 15 Topic Outline and check off each objective after you meet it.

TOPIC OUTLINE AND OBJECTIVES

A. Location and coverings of the heart

- ☐ 1. Describe the location of the heart and the structure and functions of the pericardium.

B. Heart structure and blood flow

- ☐ 2. Describe the layers of the heart wall and the chambers of the heart.
- ☐ 3. Identify the major blood vessels that enter and exit the heart.
- ☐ 4. Describe the structure and functions of the valves of the heart.
- ☐ 5. Explain how blood flows through the heart.
- ☐ 6. Describe the clinical importance of the blood supply of the heart.

C. Conduction system, ECG

- ☐ 7. Explain how each heartbeat is initiated and maintained.

- ☐ 8. Describe the meaning and diagnostic value of an electrocardiogram.

D. The cardiac cycle

- ☐ 9. Describe the phases of the cardiac cycle.

E. Cardiac output; exercise and the heart

- ☐ 10. Define cardiac output, explain how it is calculated, and describe how it is regulated.
- ☐ 11. Explain the relationship between exercise and the heart.

F. Common disorders, medical terminology

WORDBYTES

Study each wordbyte, its meaning, and an example of its use in a term. Check your understanding by jotting meanings of wordbytes in the margins. Identify other examples of terms that contain these wordbytes as you continue through the text and *Learning Guide*.

Wordbyte	Meaning	Example(s)	Wordbyte	Meaning	Example(s)
angio-	blood vessel	*angio*gram	myo-	muscle	*myo*cardial infarction
auscult-	listening	*auscult*ation	peri-	around	*peri*cardium
cardi-	heart	*cardi*ologist	phleb-	vein	*phleb*itis
coron-	crown	*coron*ary arteries	port-	carry	*port*al vein
-cuspid	point	tri*cuspid*	pulmon-	lung	*pulmon*ary artery
ectop-	displaced	*ectop*ic pacemaker	-sclerosis	hard	arterio*sclerosis*
endo-	within	*endo*carditis	tri-	three	*tri*cuspid
hepat-	liver	*hepat*ic artery	tunica	sheath	*tunica* intima
-lunar	moon	semi*lunar*	vaso-	vessel	*vaso*constriction
med-	middle	tunica *med*ia	ven-	vein	*ven*ipuncture

CHECKPOINTS

A. Location and coverings of the heart   (pages 365–366)

■ **A1.** Do this exercise about heart function.

a. What is the primary function of the heart? _____

b. Visualize a coffee cup; now imagine that sitting next to it is a gallon-size milk container. Keep in mind that a gallon holds about _____ quarts, or approximately 4 liters. Now take a moment to imagine a number of containers lined up and holding the amount of blood pumped out by the heart.

  1. With each heartbeat, the heart pumps a stroke volume equal to about $\frac{1}{3}$ cup (70 mL), since 1 cup = about

     _____ mL.

  2. Each minute, the heart pumps enough blood to fill up 1.3 gallons (_____ liters).

  3. Within the 1440 minutes in a day, the heart would fill up about 1980 gallons (over _____ liters).
     (Note: since the right and left sides of the heart *each* pump this volume, the amount pumped per day by the

     entire heart is actually double this amount—or _____ liters of blood each day.)

c. Reflecting on the work your heart is doing for you, what message would you give to your heart?

■ **A2.** Closely examine Figure 15.1, page 365 in your text. Consider the location, size, and shape of your heart as you do this exercise. Trace its outline on your body.

    a. Your heart lies in the _____ between your two _____ . About *(one-third? one-half? two-thirds?)* of the mass of your heart lies to the left of the midline of your body.

    b. Your heart is about the size and shape of your _____ .

    c. The *(apex? base?)* of the heart points downward, whereas the large blood vessels of the heart are attached to the "top" of the heart, the rather blunt *(apex? base?)*. The apex of the heart is formed mostly of the *(left? right?)* ventricle.

    d. The heart is located between two bony structures, the _____ and the _____ . This location

    makes possible the clinical procedure of CPR, which is _____ _____ .

■ **A3.** Arrange in order from most superficial to deepest.

_____ _____ _____ _____ _____ _____

E. Endocardium	PC. Pericardial cavity (site of pericardial fluid)
FP. Fibrous pericardium	PP. Parietal pericardium
M. Myocardium	VP. Visceral pericardium (epicardium)

A buildup of pericardial fluid or blood in the pericardial cavity is known as cardiac _____ . Explain why this condition is serious.

B. Heart structure and blood flow (pages 366–372)

B1. Almost all of the heart consists of *(peri? myo? endo?)*-cardium. Explain how intercalated discs and gap junctions contribute to effective myocardial function.

■ **B2.** _____-carditis is likely to damage heart valves because these valves are composed of _____-cardium.

■ **B3.** Refer to Figure LG 15.1 to do the following activity.

    a. Identify all structures with letters (A—I) by writing labels on lines A—I.

    b. Draw arrows on Figure LG 15.1 to indicate direction of the blood flow.

    c. Color red the chambers of the heart and vessels that contain highly oxygenated blood; color blue the regions in which blood is low in oxygen and high in carbon dioxide.

    d. On Figure LG 15.1, label the four valves that control blood flow through the heart.

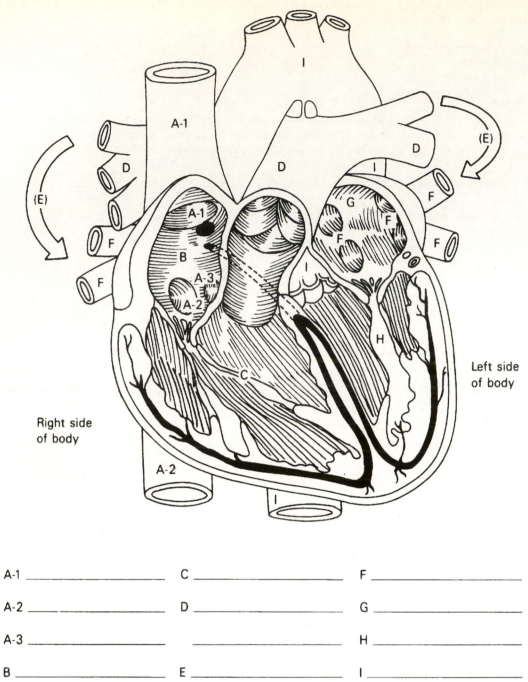

A-1

(E)

D

D

D

(E)

I

A-1

G

F

F

F

B

F

A-3

F

A-2

F

H

C

Left side
of body

Right side
of body

A-2

I

I

A-1 _____   C _____   F _____

A-2 _____   D _____   G _____

A-3 _____   _____   H _____

B _____   E _____   I _____

Figure LG 15.1   Diagram of a frontal section of the heart. Letters follow the path of blood through the heart. Label, color, and draw arrows as directed in Checkpoints B3 and C1.

■ **B4.**   Complete this Checkpoint about heart structure by choosing answers from the box.

LA. Left atrium	RA. Right atrium
LV. Left ventricle	RV. Right ventricle

_____ a. Thickest-walled chamber of the heart

_____ b. Chamber that receives blood from the superior and inferior vena cavae

_____ c. Chambers on either side of the interatrial septum (two answers)

■ **B5.** Check your understanding of heart structure by matching the structures in the box with correct descriptions below.

AOR. Aorta	PV. Pulmonary vein
AUR. Auricle	SV. Semilunar valves
PA. Pulmonary artery	SVC/IVC. Superior and inferior vena cavae
PC. Papillary muscles and chordae tendineae	

_____ a. Vessels that return "blue blood" (low in oxygen, high in wastes) to the heart

_____ b. Blood vessel that transports "blue blood" to lungs for oxygenation

_____ c. Blood vessels seen on the posterior of the heart (Figure 15.3b, page 368 of your text) that carry oxygenated blood from lungs to left atrium

_____ d. Structures that anchor tricuspid and mitral valves into walls of ventricles

_____ e. Structures located in the pulmonary artery and aorta

_____ f. An earlike flap that is part of each atrium

_____ g. A vessel that carries blood out of the left ventricle

■ **B6.** Before beginning this Checkpoint, be sure you can trace blood flow through the heart (follow arrows in Checkpoint B3b). Remember that *valves prevent flow of blood backward into the chamber from which blood has just come.* Now match valves with descriptions below. Some descriptions require more than one answer.

A. Aortic semilunar	B. Bicuspid	P. Pulmonary semilunar	T. Tricuspid

_____ a. Also called the mitral valve

_____ b. Prevents backflow of blood from the right ventricle to the right atrium

_____ c. Prevents backflow from the pulmonary trunk to the right ventricle

_____ d. Prevents backflow of blood into the left atrium

_____ e. Have half-moon-shaped leaflets or cusps (two answers)

_____ f. Also called atrioventricular (AV) valves (two answers)

■ **B7.** Narrowing of the mitral valve is a condition known as mitral valve (*incompetence? stenosis?*). Valvular

incompetence is also known as valvular _____ .

B8. Defend or dispute this statement: "The myocardium receives all the oxygen and nutrients it needs from blood that is passing through its four chambers."

■ **B9.** Refer to Figure 15.3(b-c), page 368 of your text and complete this exercise.

a. Identify the location of the right and left coronary arteries. They both arise from the (*aorta? pulmonary artery?*). This is logical because these vessels carry blood (*high? low?*) in oxygen to the heart wall.

b. Blood in coronary arteries passes into small vessels in the heart wall, and is then collected in veins that drain into

the coronary _____ located on the (*anterior? posterior?*) of the heart. (See Figure 15.3b, page 368 in the text).

c. The coronary sinus empties its blood into the (*left? right?*) atrium (Figure 15.3c, page 368 of the text).

229

B10. Define the term *anastomosis,* and explain how it relates to the heart.

B11. Defend or dispute this statement "Human myocardium lacks regenerative capabilities."

C. Conduction system, ECG   (pages 372–374)

■ **C1.** In this exercise describe how the heart beats regularly and continuously.

a. Label the parts of the conduction system on Figure LG 15.1.

b. The normal pacemaker of the heart is the *(SA? AV?)* node which is located in the *(left? right?)* atrium.

   The SA mode normally fires at _____ beats/min. However the vagus nerves which are *(sympathetic?*

   parasympathetic?) nerves, slow down this pace to about _____ beats per minute.

c. Lack of adequate coronary artery supply can injure the SA node. If so, the AV node has the ability to initiate

   action potentials at _____ beats per minute. Remaining structures of the conduction system can set up impulses

   at _____ beats per minute.

d. Which part of the conduction system is the only structure that permits transmission of action potentials between
   atria and ventricles? _____ Explain.

■ **C2.** What do the letters ECG (or EKG) stand for? _____ An ECG is a recording of

_____ . State two purposes of ECGs.

■ **C3.** Match the answers in the box with descriptions of parts of the ECG below.

| P. P wave | QRS. QRS complex | T. T wave |

_____ a. Related to depolarization and contraction   _____ c. Related to repolarization and relaxation
           of atria   of ventricles

_____ b. Related to depolarization and contraction
           of ventricles

C4. Describe an *artificial pacemaker* and explain how it stimulates the heart.

D. The cardiac cycle (pages 374–375)

■ **D1.** Refer to Figure 15.8, page 375 of your text, and complete this overview of the cardiac cycle.

a. If your heart beats at 60 seconds/minute, each heart beat (or cardiac cycle) takes _____ sec. If your pulse rate

increases to 120 beats/second, the duration of each cardiac cycle is about _____ second. In fact, extremely rapid heart rates (such as 200–400 beats/minute) provide *(too much? insufficient?)* time for events of the cardiac cycle to take place adequately.

b. If your heart beats at 75 seconds/minute, each heart beat (or cardiac cycle) requires slightly *(more? less?)* than

one minute, or about _____ sec. This is the length of a typical cardiac cycle.

c. In the relaxation phase, all four chambers of the heart are in *(systole? diastole?)*. During this time, blood *(fills? is*

ejected from?) the heart. This phase takes about _____ sec. (With very rapid heart rates, this phase is reduced so much that the heart cannot properly fill with blood.)

d. The start of atrial systole is marked by the *(P? QRS? T?)* wave. *Systole* refers to myocardial *(contraction? relax-*

ation) that pushes more blood from atria to ventricles. Atrial systole requires _____ sec.

e. During ventricular systole (_____ sec.), blood *(fills? is ejected from?)* the ventricles due to the *(high? low?)* pressure in those chambers. The *(P? QRS? T?)* wave occurs at the start of this phase. The *(P? QRS? T?)* wave signals the end of ventricular systole and the start of the relaxation phase of a new cardiac cycle.

f. *For extra review.* Refer to a figure of the heart, such as Figure LG 15.1, as you review this activity.

■ **D2.** Answer these additional questions about the cardiac cycle.

a. What factor normally causes 70–80 percent of blood in atria to move to ventricles? _____

b. Are atria or ventricles ever completely empty, with only air left in these chambers? *(Yes? No?)*

c. The average heart rate is 75 beats/min, so an average cardiac cycle requires _____ sec. During that 0.8 sec,

atria are in systole for _____ sec and in diastole for _____ sec, whereas ventricles are in systole for

_____ sec and in diastole for _____ sec.

d. Are atria and ventricles ever in systole at the same time? *(Yes? No?)*

e. Are atria and ventricles ever in diastole at the same time? *(Yes? No?)*

f. What causes the first heart sound, "lubb"? *(Opening? Closing?)* of *(AV? SL?)* valves. What causes the second

sound, "dupp"? _____ of _____ valves.

g. Valve disorders such as stenosis or regurgitation cause "heart _____ ."

E. Cardiac output; exercise and the heart (pages 375–378)

■ **E1.** Determine the average cardiac output in a resting adult. Cardiac output = stroke volume × heart rate

= _____ mL/stroke × _____ strokes/min

= _____ mL/min (_____ liter/min)

■ **E2.** The two major factors (shown in Checkpoint E1) that control cardiac output (CO) are

_____ and _____ . Do this activity about stroke volume (SV).

a. During exercise, muscles surrounding blood vessels, especially in the legs, squeeze more blood back toward the heart. This increased blood returning to the heart stretches the myocardium cells of the heart wall *(more? less?)*.

b. Within limits, a stretched muscle contracts with *(greater? less?)* force than a muscle that is only slightly

stretched. This is a statement of _____'s law of the heart. As a result, during exercise the ventricles of the normal heart contract *(more? less?)* forcefully, and stroke volume *(increases? decreases?)*.

c. State two or more reasons why stroke volume may be decreased abnormally.

■ **E3.** *A clinical challenge.* Do this exercise about Ms. S, who has congestive heart failure (CHF).

a. Her weakened heart becomes overstretched much like a balloon (except much thicker!) that has been expanded 5000 times. Now her heart myofibers are stretched beyond the optimum length according to the Frank-Starling's

law. As a result, the force of her heart is _____-creased, and its stroke volume _____-creases (opposite of Checkpoint E2b).

b. Ms. S has right-sided heart failure, so her blood is likely to back up, distending vessels in *(lungs? systemic regions, such as in neck and ankles?)*; thus _____ edema results, with signs such as swollen hands and feet.

c. Write a sign or symptom of left-sided heart failure.

■ **E4.** Summarize normal effects of factors below upon heart rate (HR), stroke volume (SV), cardiac output (CO), and blood pressure (BP) by completing arrows: ↑ (for increase) or ↓ (for decrease). The first one is done for you.

a. ↑ exercise → ↑ return of blood to the heart → ↑ SV

b. ↑ SV (if HR stays the same) → _____ CO

c. ↑ heart contractility → _____ SV and _____ CO

d. ↓ body temperature (hypothermia) → _____ HR and _____ strength of contraction

e. ↑ sympathetic nerve impulses → _____ SV, _____ HR, and _____ CO

f. Vagus (parasympathetic) nerve impulses to the heart → _____ HR, _____ strength of heart contraction,

_____ SV, and _____ CO

g. Epinephrine → _____ SV, _____ HR, and _____ CO

■ **E5.** Check your understanding of baroreceptors in this Checkpoint.

a. Define *baroreceptors*.

Indicate where these are located.

b. How do nerve impulses pass from these receptors to the medulla?

c. If these receptors detect that blood pressure (BP) is high, what response will help to lower BP?

■ **E6.** Describe effects of exercise on the entire body in this Checkpoint.

a. Athletic training normally has the effect of _____-creasing heart rate at rest to less than _____ beats/min.

This condition is known as _____-cardia. Maximal cardiac output is likely to increase to

_____ times that of sedentary persons as a result of a(n) _____-crease in stroke volume.

b. To meet increased needs for oxygen, athletic training leads to a(n) _____-crease in hemoglobin level and to

a(n) _____-crease in capillary beds in skeletal muscles.

c. List several other benefits of regular exercise.

F. Common disorders and medical terminology   (pages 379–381)

F1. List ten or more risk factors for coronary artery disease (CAD). Circle those that can be modified by a healthy life-style.

a. _____   f. _____

b. _____   g. _____

c. _____   h. _____

d. _____   i. _____

e. _____   j. _____

■ **F2.** Outline the sequence of changes in atherosclerosis in this exercise.

a. Atherosclerosis involves damage to walls of arteries. Which type of lipoproteins are more associated with this condition? *(High? Low?)* density lipoproteins, known by the acronym _____ . Which lipoproteins are the "good cholesterol" because they help to prevent atherosclerosis? _____

b. Name the two types of cells in arterial walls that are implicated in this process: _____ and _____ . These cells secrete substances that attract monocytes and convert them into _____ , cells that play roles in inflammatory responses.

c. Macrophages ingest oxidized LDLs and are then known as _____ cells, based on their appearance. Together with T lymphocytes, foam cells form a _____ streak which is the start of an atherosclerotic plaque.

d. _____ cells then form a cap over the plaque, walling it off from the blood stream. Explain how a heart attack result from this process.

■ **F3.** *A clinical challenge.* Do this exercise about "heart attacks."

a. "Heart attack" is a common name for a myocardial _____ (MI). What does the term *infarction* mean? _____

b. List one or more immediate cause(s) of a MI.

c. Define *ischemia* and explain how it is related to *angina pectoris*.

■ **F4.** Match each congenital heart defect in the box with the related description below.

ASD. Atrial septal defect	TF. Tetralogy of Fallot
PDA. Patent ductus arteriosus	VSD. Ventricular septic defect
PS. Pulmonary stenosis	

_____ a. Connection between aorta and pulmonary artery is retained after birth, allowing backflow of blood to right ventricle

_____ b. Septum between ventricles does not develop properly

_____ c. Narrowing of the valve reduces blood flow out of the right ventricle

_____ d. PS, IVSD, right ventricular hypertrophy, and aorta that emerges from both ventricles

_____ e. Incomplete closure of the foramen ovale

234

■ **F5.** Do this exercise on arrhythmias.

a. *(All? Not all?)* arrhythmias are serious.

b. Conduction failure across the AV node results in an arrhythmia known as _____.

c. Which is more serious? *(Atrial? Ventricular?)* fibrillation. Explain why.

■ **F6.** Fill in the name of each heart disorder or diagnostic procedure:

a. Valve disease that follows a streptococcal infection: _____

b. Enlarged heart: _____

c. Thickened right ventricle due to a chronic lung disorder such as emphysema and related

   pulmonary hypertension: _____

d. Insertion of a narrow tube into the heart or coronary arteries for diagnostic purposes: _____

ANSWERS TO SELECTED CHECKPOINTS: CHAPTER 15

A1. (a) It is a pump. (b) 4 (1 gallon = 4 quarts or 3.86 liters). (b1) 237. (b2) 5 (1.3 gallons = 5 liters). (b3) 7200 (1.3 gallons/min × 1440 min/day = 1800 gallons/day = over 7,000 liters; 14,000. (c) Perhaps: "Thanks!" "Great job!" "Amazing!" or "You deserve the best of care, and I intend to see that you get it."

A2. (a) Thorax, lungs; two-thirds. (b) Fist. (c) Apex, base; left. (d) Sternum, vertebrae; cardiopulmonary resuscitation.

A3. FP PP PC VP M E; tamponade (like a tampon, the fluid takes all the space available); the fluid compresses the heart, preventing its expansion (or filling with blood).

B2. Endo, endo.

B3. (a, d) See Figure LG 15.1A. (b) see text Figure 15.5. (c) Blue: A–D and first part of E; red: last part of E, F–I.

B4. (a) LV. (b) RA. (c) RA and LA.

B5. (a) SVC/IVC. (b) PA. (c) PV. (d) PC. (e) SV. (f) AUR. (g) AOR.

B6. (a) B. (b) T. (c) P. (d) B. (e) A, P. (f) B, T.

B7. Stenosis; insufficiency.

B9. (a) Aorta; high. (b) Sinus, posterior. (c) Right.

C1. (a) See Figure LG 15.1A. (b) SA, right; 100; parasympathetic, 75. (c) 40–60; 20–35. (d) AV node because it bridges connective tissue that otherwise separates atria from ventricles.

C2. Electrocardiogram; electrical changes (action potentials) associated with impulse conduction in the heart; diagnoses of abnormal cardiac rhythms and detection of fetal heartbeat.

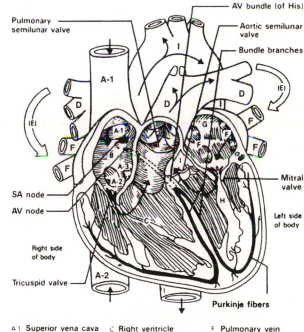

Figure LG 15.1A Diagram of a frontal section of the heart.

A 1	Superior vena cava	C	Right ventricle	F	Pulmonary vein
A 2	Inferior vena cava	D	Pulmonary trunk	G	Left atrium
A 3	Coronary sinus		and arteries	H	Left ventricle
B	Right atrium	E	Vessels in lungs	I	Aorta

C3. (a) P. (b) QRS. (c) T.

D1. (a) 1.0; 0.5; insufficient. (b) Less, 0.8. (c) Diastole; fills; 0.4. (d) P; contraction; 0.1. (e) 0.3, is ejected from, high; QRS; T.

D2. (a) Gravity. (b) No, there is always some blood in heart chambers, and normally no air is present. (c) 0.8 sec/cycle = 60 sec/75 cycles; 0.1, 0.7, 0.3, 0.5. (d) No. (e) Yes, the entire heart relaxes for about half of each cardiac cycle. (f) Closing (of) AV; closing (of) SL. (g) Murmurs.

E1. CO = 70 ml/stroke × 75 strokes (beats)/min = 5250 ml/min (5.25 liters/min).

E2. Stroke volume (SV) and heart rate (HR). (a) More. (b) Greater; Frank-Starling's; more, increases. (c) Damaged heart, for example, following a myocardial infarction, or excessive blood loss, so less blood returns to the heart.

E3. (a) De, de. (b) Systemic regions, such as in neck and ankles; peripheral or systemic. (c) Difficulty breathing or shortness of breath (dyspnea), which is a sign of pulmonary edema.

E4. (b) ↑ CO. (c) ↑ SV, ↑ CO. (d) ↓ HR, ↓ strength of contraction. (e) ↑ SV, ↑ HR, ↑ CO. (f) ↓ HR, ↓ strength of heart contraction, ↓ SV, ↓ CO, (g) ↑ SV, ↑ HR, ↑ CO.

E5. (a) Nerve cells sensitive to changes in blood pressure (BP); they are located in certain large arteries (such as aorta and carotids). (b) By cranial nerves, such as CN IX and X. (c) Decrease sympathetic and increase parasympathetic nerve impulses to the heart (so HR, SV, CO, and BP all decrease).

E6. (a) De, 60; brady, two, in. (b) In, in. (c) Reduced blood pressure, anxiety, and depression, weight control, and increased ability to control blood clotting.

F2. (a) Low, LDLs; HDLs. (b) Endothelial cells (in the tunica interna) and smooth muscle cells (in the tunica media); macrophages. (c) Foam; fatty. (d) Smooth muscle; the cap can break open in response to chemicals secreted by foam cells and lead to clot formation.

F3. Infarction; death of tissue because of lack of blood flow to that tissue. (b) Thrombus (clot) or embolus (such as mobile clot), or spasm of coronary artery. (c) Angina is a sign of damage to myocardial cells resulting from reduced blood and oxygen supply (ischemia). Angina is a squeezing pain in the chest that may radiate to the neck, chin, or left arm.

F4. (a) PDA. (b) IVSD. (c) PS. (d) TF. (e) ASD.

F5. (a) Not all. (b) Heart block. (c) Ventricular; atrial fibrillation reduces effectiveness of the heart by only about 20–30 percent, but ventricular fibrillation causes the heart to fail as a pump.

F6. (a) Rheumatic fever, (b) Cardiomegaly. (c) Cor pulmonale. (d) Cardiac catheterization.

CRITICAL THINKING: CHAPTER 15

1. Prepare a tour guide's description of a trip through the human heart and lungs. Name and describe the rooms (chambers), hallways (vessels), and structure and purpose of doors (valves) that you pass along the way.

2. Relate opening and closing of AV and semilunar heart valves to contraction and relaxation of atria and ventricles.

3. Contrast terms in the following pairs of terms: (a) stroke volume/cardiac output, (b) myocardial ischemia/myocardial infarction, (c) AV node/AV bundle of His, (d) effects of cardiac accelerator/vagus nerves on heart rate, (e) pulmonary edema/peripheral edema, (f) atrial fibrillation/ventricular fibrillation.

4. Describe these forms of pathology of the heart: (a) valvular regurgitation, (b) myocardial ischemia, (c) coronary artery disease (CAD), (d) heart palpitations.

5. Explain how the following chemicals are thought to contribute to coronory artery disease (CAD): (a) homocysteine; (b) C-reactive proteins; (c) lipoprotein (a).

MASTERY TEST: ■ CHAPTER 15

Questions 1–2: Arrange the answers in correct sequence.

_____ _____ _____ _____ 1. Pathway of the conduction system of the heart:
    A. AV node
    B. AV bundle and bundle branches
    C. SA node
    D. Purkinje fibers

_____ _____ _____ _____ _____ 2. Route of a red blood cell now in the right atrium:
    A. Left atrium
    B. Left ventricle
    C. Right ventricle
    D. Pulmonary artery
    E. Pulmonary vein

3. All of the following are correctly matched *except:*
   A. Myocardium—heart muscle
   B. Visceral pericardium—epicardium
   C. Endocardium—forms heart valves
   D. Pericardial cavity—space between fibrous pericardium and parietal layer of serous pericardium

4. Which of the following factors tends to decrease heart rate?
   A. Release of the transmitter substance norepinephrine in the heart
   B. Activation of neurons in the cardioaccelerator center
   C. Increase of vagal nerve impulses
   D. Increase of sympathetic nerve impulses

5. The average cardiac output for a resting adult is about _____ per minute.
   A. 1 quart   D. 0.5 liter
   B. 5 pints   E. 2000 mL
   C. 1.25 gallons (5 liters)

6. When ventricular pressure exceeds atrial pressure, what event occurs?
   A. AV valves open
   B. AV valves close
   C. Semilunar valves open
   D. Semilunar valves close

7. All of the following are defects involved in tetralogy of Fallot *except:*
   A. Ventricular septal defect
   B. Hypertrophied right ventricle
   C. Stenosed mitral valve
   D. Aorta emerging from both ventricles

Questions 8–10: Circle T (true) or F (false). If the statement is false, change the underlined word or phrase so that the statement is correct.

T   F   8. The blood in the left chambers of the heart contains <u>higher</u> oxygen content than blood in the right chambers.

T   F   9. The pulmonary <u>artery carries</u> blood from the lungs to the left atrium.

T   F   10. The normal cardiac cycle <u>does not</u> require direct stimulation by the autonomic nervous system.

Questions 11–15: Fill-ins. Complete each sentence with the word or phrase that best fits.

_____ 11. A defect in the interventricular septum would permit blood to flow directly between the _____ .

_____ 12. A myocardial infarction is commonly known as a _____ .

_____ 13. An ECG is a recording of _____ of the heart.

_____ 14. _____ is a term that means abnormality or irregularity of heart rhythm.

_____ 15. Are atrioventricular and semilunar valves ever open at the same time during the cardiac cycle? _____

ANSWERS TO MASTERY TEST: ■ CHAPTER 15

Arrange
1. C A B D
2. C D E A B

Multiple Choice
3. D
4. C
5. C
6. B
7. C

True or False
8. T
9. F. Veins carry
10. T

Fill-Ins
11. Left and right ventricles
12. "Heart attack" or a "coronary"
13. Electrical changes or currents that precede myocardial contractions
14. Arrhythmia
15. No

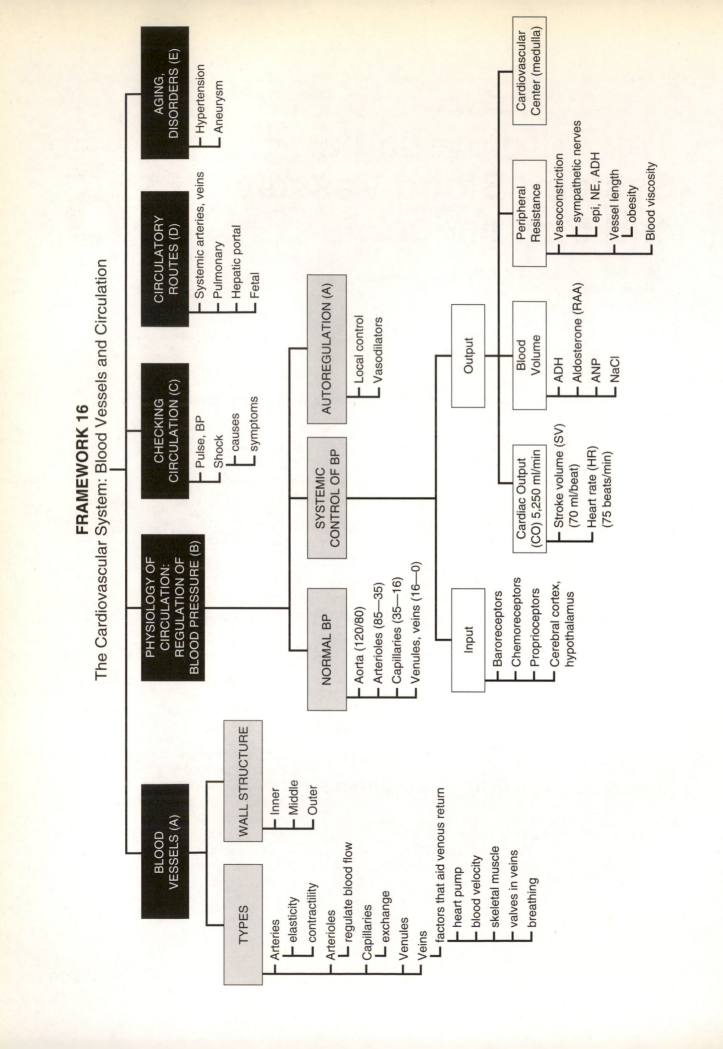

FRAMEWORK 16

The Cardiovascular System: Blood Vessels and Circulation

BLOOD VESSELS (A)

TYPES
- Arteries
  - elasticity
  - contractility
- Arterioles
  - regulate blood flow
- Capillaries
  - exchange
- Venules
- Veins
  - factors that aid venous return
    - heart pump
    - blood velocity
    - skeletal muscle
    - valves in veins
    - breathing

WALL STRUCTURE
- Inner
- Middle
- Outer

PHYSIOLOGY OF CIRCULATION: REGULATION OF BLOOD PRESSURE (B)

NORMAL BP
- Aorta (120/80)
- Arterioles (85—35)
- Capillaries (35—16)
- Venules, veins (16—0)

SYSTEMIC CONTROL OF BP

Input
- Baroreceptors
- Chemoreceptors
- Proprioceptors
- Cerebral cortex, hypothalamus

Output

Cardiac Output (CO) 5,250 ml/min
- Stroke volume (SV) (70 ml/beat)
- Heart rate (HR) (75 beats/min)

Blood Volume
- ADH
- Aldosterone (RAA)
- ANP
- NaCl

Peripheral Resistance
- Vasoconstriction
  - sympathetic nerves
  - epi, NE, ADH
- Vessel length
  - obesity
- Blood viscosity

Cardiovascular Center (medulla)

AUTOREGULATION (A)
- Local control
- Vasodilators

CHECKING CIRCULATION (C)
- Pulse, BP
- Shock
  - causes
  - symptoms

CIRCULATORY ROUTES (D)
- Systemic arteries, veins
- Pulmonary
- Hepatic portal
- Fetal

AGING, DISORDERS (E)
- Hypertension
- Aneurysm

The Cardiovascular System: Blood Vessels and Circulation

Blood is pumped by heart action to all regions of the body. The network of vessels is organized much like highways, major roads, side streets, and tiny alleys to provide interconnecting routes through all sections of the body. In city traffic, tension mounts when traffic is heavy or side streets are blocked. Similarly, pressure builds (arterial blood pressure) when volume of blood increases or when constriction of small vessels (arterioles) occurs. In a city, traffic may be backed up for hours, with vehicles unable to deliver passengers to destinations. A weakened heart may lead to similar and disastrous effects, backing up fluid into tissues (edema) and preventing transport of oxygen to tissues (ischemia, shock, and death of tissues).

Careful map study leads to trouble-free travel along the major arteries of a city. A similar examination of the major arteries and veins of the human body facilitates an understanding of the normal path of blood as well as clinical applications, such as the exact placement of a blood pressure cuff over the brachial artery or the expected route of a deep venous thrombus (pulmonary embolism) to the lungs.

Begin your route through circulatory pathways at the Chapter 16 Framework and familiarize yourself with the names of important routes. As you begin this chapter, carefully examine the Chapter 16 Topic Outline and check off each objective after you meet it.

TOPIC OUTLINE AND OBJECTIVES

A. Blood vessel structure and function

☐ 1. Compare the structure and function of the different types of blood vessels.

☐ 2. Describe how substances enter and leave the blood in capillaries.

☐ 3. Explain how venous blood returns to the heart.

B. Blood flow through blood vessels: blood pressure

☐ 4. Define blood pressure and describe how it varies throughout the systemic circulation.

☐ 5. Identify the factors that affect blood pressure and vascular resistance.

☐ 6. Describe how blood pressure and blood flow are regulated.

C. Checking circulation

☐ 7. Explain how pulse and blood pressure are measured.

D. Circulatory routes

☐ 8. Compare the major routes that blood takes through various regions of the body.

☐ 9. Identify the four principal parts of the aorta and locate the major arterial branches arising from each.

☐ 10. Identify the three arteries that branch from the arch of the aorta.

☐ 11. Identify the two major branches of the common iliac arteries.

☐ 12. Identify the three systemic veins that return deoxygenated blood back to the heart.

13. Identify the three major veins that drain blood from the head.
14. Identify the principal veins that drain the upper limbs.
15. Identify the principal veins that drain the lower limbs.

E. Aging, common disorders, and medical terminology

16. Describe the effects of aging on the cardiovascular system.

WORDBYTES

Study each wordbyte, its meaning, and an example of its use in a term. Check your understanding by jotting meanings of wordbytes in the margins. Identify other examples of terms that contain these wordbytes as you continue through the text and *Learning Guide*.

Wordbyte	Meaning	Example(s)	Wordbyte	Meaning	Example(s)
hepat-	liver	*hepat*ic artery	sphygmo-	pulse	*sphygmo*-manometer
med-	middle	tunica *med*ia	tunica	sheath	*tunica* intima
phleb-	vein	*phleb*itis	vaso-	vessel	*vaso*constriction
port-	carry	*port*al vein	ven-	vein	*ven*ipuncture
retro-	backward, behind	*retro*peritoneal			

CHECKPOINTS

A. Blood vessel structure and function (pages 386–390)

■ **A1.** Use the names of vessels in the box to answer the following description.

> Aorta and other large arteries
> Arteriole
> Capillary
> Inferior vena cava (large vein)
> Small artery
> Small vein
> Venule

_____ a. Site of gas, nutrient, and waste exchange with tissues

_____ b. Plays primary role in regulating moment-to-moment distribution of blood and in regulating blood pressure

_____ c. Reservoirs for about 60 percent of the volume of blood in the body (three answers)

_____ d. Thickest of all of these vessels, with thick middle layer of smooth muscle and elastic tissue

_____ e. Thinnest of all of these vessels; formed mostly of endothelium

■ **A2.** Describe properties of blood vessels in this Checkpoint.

a. *(Contractility? Elasticity?)* is the ability of a blood vessel to expand and then recoil. Which vessels have a higher proportion of elastic fibers relative to smooth muscle?

A. Large diameter arteries such as the aorta and carotid arteries

B. Medium-sized arteries such as the brachial and radial arteries

b. Contractility of a vessel is based primarily on the smooth muscle tissue in the *(inner? middle?)* layer of the vessel wall. Contraction of this muscle _____ -creases the size of the lumen. This process, called vaso-

_____ , is a response to *(sympathetic? parasympathetic?)* nerve stimulation to the vessel wall. Widening of a vessel, due to relaxation of smooth muscle, results in vaso-_____ .

240

■ **A3.** Check your understanding of capillaries in this Checkpoint.

    a. List four areas of the body that have rich capillary supplies.

    b. Now list four areas that have little or no capillary supply.

    c. Describe the function of *precapillary sphincters*.

       Whether these muscles contract or relax is the basis of _____ of a particular tissue.

    d. Sinusoids (which are found in the _____) have (*a tight endothelial layer? gaps between endothelial cells?*). Explain the significance of this structure.

    e. Capillaries of the brain have (*a tight endothelial layer? gaps between endothelial cells?*). Explain the significance of this structure.

■ **A4.** Do this exercise about capillary exchange.

    a. Exchanges of gases, nutrients, and wastes between blood and tissues occur across (*arterioles? capillaries? venules?*). One reason for this is that capillary walls are very (*thick? thin?*). Another is that velocity of blood flow is least in (*arterioles? capillaries? venules?*).

    b. Two major forces contribute to movement of substances across capillary membranes. One is a *pushing* pressure that is based on the amount of _____ (BP) in capillaries. The other, known as blood _____ (COP) *pulls* water and other chemicals back into blood plasma. The major factor that creates this pull is _____ .

    c. As shown in Figure 16.3 (page 389 of the text), at the arterial end of capillaries where capillary BP exceeds capillary COP, substances move out of plasma into interstitial fluid by a process called (*filtration? reabsorption?*). At the venous end of capillaries, where capillary BP is less than COP, most of that fluid returns to blood by the process called _____ . The small of fluid not returned to veins is normally "mopped up" by the _____ system.

■ **A5.** Complete this Checkpoint about structure of veins.

    a. Veins have (*thicker? thinner?*) walls than arteries, with a (*wider? more narrow?*) lumen than arteries. This structural feature relates to the fact that the pressure in veins is (*more? less?*) than in arteries. The pressure difference is demonstrated when a vein is cut; blood leaves a cut vein in (*rapid spurts? an even flow?*).

    b. Gravity exerts back pressure on blood in veins located inferior to the heart. To counteract this, veins contain _____ .

    c. When valves weaken, veins become enlarged and twisted. This condition is called _____ . This occurs more often in (*superficial? deep?*) veins. Why?

A6. Explain how these factors increase venous return during exercise:

a. Contractions of the heart

b. Skeletal muscle contractions and valves in veins (Figure 16.4, page 390 of the text)

c. Breathing

B. Blood flow through blood vessels; blood pressure   (pages 390–394)

■ **B1.** Refer to Figure 16.5 (page 391 of the text), and complete this exercise about blood pressure shown in that figure.

a. In the aorta and brachial artery, blood pressure (BP) is about 120 mm Hg immediately following ventricular con-

traction. This is called *(systolic? diastolic?)* BP. As ventricles relax (or go into _____), blood is no longer ejected into these arteries. However, the normally *(elastic? rigid?)* walls of these vessels recoil

against blood, pressing it onward with a diastolic BP of _____ mm Hg.

b. Blood tends to flow from an area of *(high? low?)* pressure to an area of _____ pressure.

c. Write the normal ranges of BP values for the following types of vessels: arterioles, _____ mm Hg; capillaries,

_____ mm Hg; venules, veins, and venae cavae, _____ mm Hg.

d. On Figure 16.5, the steepest decline in the pressure curve occurs as blood passes through *(aorta? arterioles? capillaries?),* indicating the greatest resistance to flow through these vessels.

■ **B2.** In Checkpoint E1 of Chapter 15 we discussed two factors that control cardiac output. Review these by completing the following equation:

$$\text{Cardiac output (CO)} = \underline{\hspace{4cm}} \times \underline{\hspace{4cm}}$$
$$= 70 \text{ mL/beat} \times 75 \text{ beats/min}$$

$$= \underline{\hspace{4cm}} \text{ liters/min}$$

■ **B3.** On Figure LG 16.1 fill in blanks and complete arrows to show details of all the nervous output, hormonal action, and other factors that increase blood pressure. *Hint:* It may help to refer to Checkpoints E2–E4 in Chapter 15 of the Learning Guide.

■ **B4.** *A clinical challenge.* Hypertension means *(high? low?)* blood pressure. It may be caused by an excess of any of the factors shown in Figure LG 16.1. If directions of arrows in Figure LG 16.1 are reversed, blood pressure can be *(increased? decreased?).*

■ **B5.** Describe mechanisms that signal homeostatic responses when BP is slightly low.
Hint: Again refer to Checkpoint E5 in LG Chapter 15.

a. The cardiovascular (CV) center located in the *(cerebrum? medulla?)* controls BP. The center responds to several types of stimuli. Name five such stimuli.

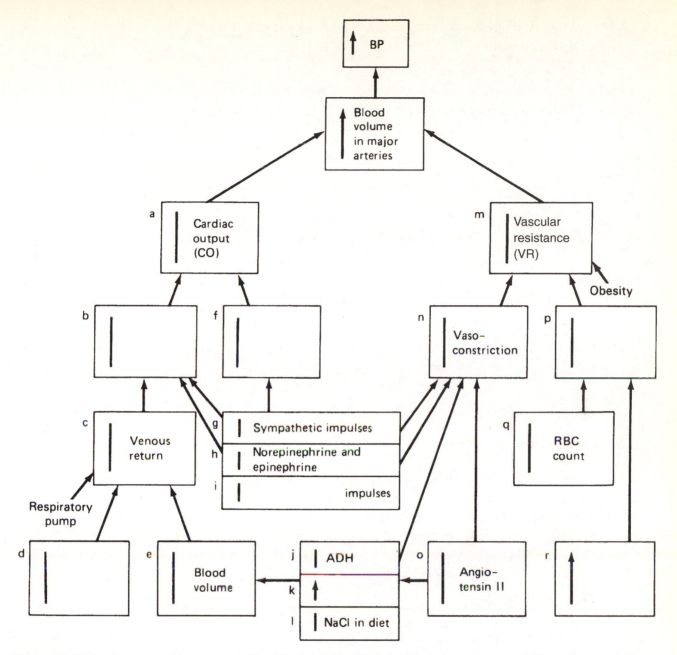

Figure LG 16.1 Summary of factors controlling blood pressure (BP). Arrows indicate increase (↑) or decrease (↓).
Complete as directed in Checkpoint B3.

b. Signals to increase BP include the following: baroreceptors respond to decrease in _____ ; chemoreceptors

respond to high blood levels of _____ or _____ ; they also react to low levels of _____ . Where
are these receptors located?

c. The CV center can then respond in two ways to raise BP: first by increasing heart rate (HR) and stroke volume
(SV). The CV center acts much like the controls that regulate a car. To make a car/heart go faster, you must press
your foot on the accelerator (gas pedal) which is comparable to *(sympathetic? parasympathetic?)* nerve impulses.
Simultaneously, you must take your foot off of the brake which is analogous to *(sympathetic? parasympathetic?)*

nerves. Visualize writing the word "vagus" on your brake pedal because this nerve (cranial nerve ____) serves as
the entire *(sympathetic? parasympathetic?)* nerve supply to the heart.

d. The second control mechanism for BP involves change in diameter of arterioles. To increase BP, the CV center

____-creases sympathetic impulses to arterioles to organs that are considered less vital during stressful times.

243

Name several sites of these arterioles: _____. As these vessels vaso-

_____, resistance in these vessels ____-creases. Such increase in vascular resistance in less vital arteries

results in ____-creased blood flow in major arteries and a(n) -crease in systemic BP.

e. These efforts to raise BP when it is too low are examples of _____ feedback mechanisms that attempt to regain homeostasis.

■ **B6.** Circle all factors that directly or indirectly *lower* BP:

ADH	Aldosterone	ANP	Dehydration
Epinephrine	Exercise	Parasympathetic (vagus) nerves	
Polycythemia	Sympathetic nerves	Vasodilator medications	

C. Checking circulation   (page 394)

C1. Define *pulse*.

C2. List five places where pulse can readily be palpated. Try to locate pulses at these points on yourself or on a friend.

_____   _____   _____

_____   _____

■ **C3.** Normal pulse rate is about _____ beats per minute. *Tachycardia* means *(rapid? slow?)* pulse rate;

_____ means slow pulse rate.

■ **C4.** Describe the method commonly used for taking blood pressure by answering these questions about the procedure.

a. Name the instrument used to check BP. _____

b. The cuff is usually placed over the _____ artery and inflated.

c. As the cuff is deflated, the first sound heard indicates _____ pressure, since reduction of pressure from the cuff below systolic pressure permits blood to begin spurting through the artery during diastole.

d. Diastolic pressure is indicated by the point at which the sounds _____ . At this time blood can once again flow freely through the artery.

e. Is a blood pressure reading of 120/80 considered normal for an adult? *(Yes? No?)* Which value is the diastolic pressure? *(120? 80?).*

■ **C5.** Describe causes of shock and compensation for shock in this activity.

a. Shock is said to occur when tissues receive inadequate _____ supply to meet their demands for oxygen, nutrients, and waste removal.

b. Write two or more causes of shock.

c. Compensatory mechanisms activate *(sympathetic? parasympathetic?)* nerves that _____ -crease both heart rate and vasoconstriction. One common indicator of this compensatory mechanism is *(warm, pink, and dry? cool,*

pale, and clammy?) skin related to vaso- _____ , and _____ -creased sweating.

d. Which hormones cause water retention, which helps to increase BP in shock?

e. Write one other mechanism that helps to compensate for fluid loss associated with shock.

D. Circulatory routes   (pages 394–413)

■ **D1.** Use Figure LG 16.2 to do this learning activity about systemic arteries.

a. The major artery from which all systemic arteries branch is the _____ . It exits from the chamber of the heart known as the *(right? left?)* ventricle. The aorta can be divided into three portions. They are

the _____ (N1 on the figure), the _____ (N2), and the

_____ (N3).

b. The first arteries to branch off from the aorta are the _____ arteries. Label these (at O) on

the figure. These vessels supply blood to the _____ .

c. Three major arteries branch from the aortic arch. Write the names of these (at K, L, and M) on the figure, beginning with the vessel closest to the heart.

d. Locate the two branches of the brachiocephalic artery. The right subclavian artery (letter _____) supplies

blood to _____

_____ .

Vessel E is named the *(right? left?)* _____ artery. It supplies the right side of the head and neck.

e. As the subclavian artery continues into the axilla and arm, it bears different names (much as a road does when it passes into new towns). Label G and H. Note that H is the vessel you studied as a common site for measurement

of _____ .

f. Label vessels I and J, and then continue the drawing of these vessels into the forearm and hand. Note that they anastomose in the hand. Vessel *(I? J?)* is often used for checking pulse in the wrist.

g. The right and left vertebral arteries (letter _____) supply the back of the _____ arterial

circle (circle of _____).

■ **D2.** Identify the abdominopelvic organs supplied with blood by the following arteries. Label each of the arteries at letters on Figure LG 16.2.

a. Branches from the celiac trunk (P): gastric to the _____ , splenic to the

_____ , and hepatic to the _____ .

b. Superior mesenteric (Q) and inferior mesenteric (U) to _____ .

c. Renal (S) to the _____ .

d. Ovarian or testicular (T) to _____ or _____ .

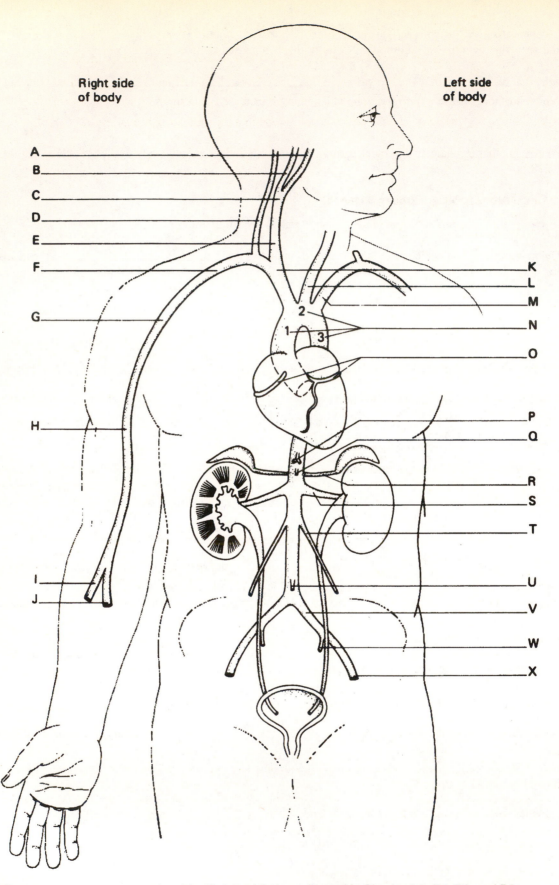

Right side
of body

Left side
of body

A

B

C

D

E

F

G

H

I

J

K

L

M

N

O

P

Q

R

S

T

U

V

W

X

1 2 3

Figure LG 16.2 Major systemic arteries. Identify letter labels as directed in Checkpoints D1, D2, and D3.

■ **D3.** The aorta ends at about the level of the umbilicus ("belly button") by dividing into the right and left

_____ arteries. Label these at V on Figure LG 16.2. Branches of these vessels supply blood

to internal pelvic organs (W) and to _____ (X).

■ **D4.** Name the three main vessels that empty venous blood into the right atrium of the heart.

_____ _____ _____ (Refer to Fig. LG 15.1,
page 228.)

■ **D5.** Do this exercise about venous return from the head.

a. Blood from all the dural vascular sinuses eventually drains into the _____ veins, which

descend in the neck. These veins are positioned close to the _____ arteries.

b. Where are the jugular arteries and carotid veins located?

■ **D6.** Name the visceral veins that empty into the abdominal portion of the inferior vena cava. Then color them blue on Figure LG 21.1 (page LG 325). Contrast accompanying systemic arteries by coloring them red on that figure. Note on Figure LG 21.1 that the inferior vena cava normally lies on the *(right? left?)* side of the aorta. One way to remember this is that the inferior vena cava is returning blood to the *(right? left?)* side of the heart.

■ **D7.** Describe differences between arterial and venous divisions of the cardiovascular system.

a. All arteries carry blood *(away from? towards?)* the heart. Virtually all arteries carry blood that appears more *(red? blue?)* based on its low CO_2 and high O_2 content. List two exceptions, that is, arteries that appear more blue

based on high CO_2 and low O_2 content: _____ and _____ .

b. The human body has more *(arteries? veins?)*. State a rationale for this difference.

c. Which veins lie parallel to arteries and have similar names? *(Deep? Superficial?)* veins. Circle the names of all veins that are superficial:

Anterior tibial	Axillary	Basilic	Brachial	Cephalic
Femoral	Great saphenous	Radial	Small saphenous	

■ **D8.** Check your understanding of blood flow through the liver in this Checkpoint.

a. Blood enters the liver via two vessels. One is the _____ artery, a branch off of the celiac

trunk. The other is the _____ vein, which drains blood from

_____ organs and from the spleen. Blood mixes as it passes through capillary-like vessels known as *(sinuses? sinusoids?)* in the liver.

b. What functions are served by this special hepatic portal circulation?

c. Through what vessels does blood exit from the liver? _____ veins, which empty into the *(superior? inferior?)* vena cava.

■ **D9.** Match the correct term in the box next to each description of a part of the circulatory system listed below.

| Cer. Cerebral | Cor. Coronary | Hep. Hepatic portal | Pulm. Pulmonary | Umb. Umbilical |

_____ a. Blood supply to the brain

_____ b. Blood supply to the heart

_____ c. Blood supply to the developing fetus

_____ d. Blood passing from digestive organs to the liver

_____ e. Blood pathways from heart to and through lungs, and back to the heart

D10. What aspects of fetal life require the fetus to possess special cardiovascular structures not needed after birth?

■ **D11.** Complete the table below about fetal structures. (Note that the structures are arranged in order of blood flow from fetal aorta back to fetal aorta.)

Fetal Structure	Structures Connected	Function	Fate of Structure After Birth
a.		Carries fetal blood low in oxygen and nutrients and high in wastes	
b. Placenta	(Omit)		
c.	Placenta to liver and ductus venosus		
d.		Branch of umbilical vein, bypasses liver	
e. Foramen ovale			
f.			Becomes ligamentum arteriosum

E.   Aging, common disorders, and medical terminology   (pages 412–416)

■ **E1.** Complete this Checkpoint on effects of aging on the cardiovascular system.

a. One normal aging change is increased *(stiffness? flexibility?)* of the aorta.

b. With normal aging, cardiac output is likely to _____ -crease, whereas the maximum heart rate is likely to _____ -crease.

c. *(Systolic? Diastolic?)* blood pressure is likely to increase with aging.

d. Aging kidneys and brain typically receive a(n) *(increased? diminished?)* blood flow.

e. A healthy lifestyle *(can? cannot?)* minimize such aging changes.

248

■ **E2.** In this exercise, fill in each body system that is aided by the cardiovascular system.

a. About 20% of resting cardiac output passes to organs in this system where blood is cleaned by filtration:

_____

b. Wound healing is facilitated as blood delivers clotting factors; movement of blood to this system aids in cooling

the body temperature: _____

c. Tissue in this system can grow and heal as calcium and phosphate are delivered here:

_____

d. Passage of blood to this system allows blood to receive oxygen and give up carbon dioxide:

_____

e. Blood removes lactic acid from these hard-working tissues: _____

f. Passage of blood into organs of this system allows blood to pick up hormones and also to vasocongest organs to

improve function of this system: _____

g. The heart pumps blood containing antibodies, lymphocytes, and macrophages to virtually all body parts:

_____

h. Chemicals from foods are delivered from this system by the heart and blood vessels:

_____

i. This system delivers its releasing hormones to the anterior pituitary gland and tropic hormones to target hormones

via blood vessels: _____

■ **E3.** Complete this exercise about blood pressure (BP).

a. Normal blood pressure refers to systolic blood pressure (BP) of _____ mm Hg or less and diastolic BP of _____ mm Hg or less.

b. A consistent BP of 190/112 indicates *(high-normal BP? hypertension?)*.

c. Hypertension increases risk of many serious disorders. Name three.

E4. List seven health practices that can help in management of hypertension.

E5. List four possible causes of *aneurysm.*

■

E6. Explain how a *deep venous thrombosis* (DVT) of the left femoral vein can lead to a pulmonary embolism.

■ **E7.** Match the disorders in the box with correct descriptions below.

An. Aneurysm	P. Phlebitis
Occ. Occlusion	S. Syncope
OH. Orthostatic hypotension	

_____ a. Fainting due to inadequate blood flow to the brain

_____ b. Balloonlike sac in an artery

_____ c. Obstruction of a blood vessel, usually by atherosclerotic plaque

_____ d. Inflammation of a vein

_____ e. Low blood pressure (often accompanied by faintness) upon sitting or standing up

ANSWERS TO SELECTED CHECKPOINTS: CHAPTER 16

A1. (a) Capillaries. (b) Arterioles. (c) Venules, small veins, and large veins (such as inferior vena cava). (d) Aorta and other large arteries. (e) Capillaries.

A2. (a) Elasticity; A. (b) Middle; de; constriction, sympathetic; dilation.

A3. (a) Muscles, liver, kidney, and nervous system. (b) Tendons, ligaments, covering and lining epithelia, cartilage, cornea, and lens. (c) Their contraction or relaxation determines amount of blood flow through that capillary bed at any given moment; autoregulation. (d) Liver, gaps between endothelial layer; plasma proteins (made in the liver) can pass into the blood stream, despite their large size. (e) A tight endothelial layer; creates the blood-brain barrier that protects the brain from harmful substances.

A4. (a) Capillaries; thin; capillaries. (b) Blood pressure; colloid osmotic pressure; plasma proteins. (c) Filtration; reabsorption; lymphatic.

A5. (a) Thinner, wider; less; an even flow. (b) Valves. (c) Varicose veins; superficial; skeletal muscles around deep veins limit overstretching.

B1. (a) Systolic; diastole; elastic, 80–70. (b) High, low. (c) 85–35; 35–16; 16–0. (d) Arterioles.

B2. CO = stroke volume (SV) × heart rate (HR) = 5.2 liters/min.

B3. (a) ↑ CO. (b) ↑ Stroke volume. (c) ↑ Venous return. (d) ↑ Exercise. (e) ↑ Blood volume. (f) ↑ Heart rate. (g) ↑ Sympathetic impulses. (h) ↑ Norepinephrine and epinephrine. (i) ↓ Vagal impulses. (j) ↑ ADH. (k) ↑ Aldosterone. (l) ↑ NaCl in diet. (m) ↑ Vascular resistance (PR). (n) ↑ Vasoconstriction. (o) ↑ Angiotensin II. (p) ↑ Viscosity of blood. (q) ↑ RBC count. (r) ↑ Plasma proteins.

B4. High; decreased.

B5. (a) Medulla; emotions, change in body temperature, movements of muscles and joints (sensed by proprioceptors), and changes in BP and chemistry of blood. (b) BP; CO_2 or H^+, O_2; in walls of aorta and carotid arteries. (c) Sympathetic, parasympathetic; X, parasympathetic. (d) In; skin, GI tract, kidneys, and spleen; constrict, in; in; in. (e) Negative.

B6. ANP, dehydration, parasympathetic (vagus) nerves; vasodilator medications.

C3. 70–80; rapid; bradycardia.

C4. (a) Sphygmomanometer. (b) Brachial. (c) Systolic. (d) Suddenly become faint or stop. (e) Yes; 80.

C5. (a) Blood (or O_2 and nutrients). (b) Loss of blood or other body fluids (as in sweating, dehydration, vomiting, diarrhea, or burns). (c) Sympathetic, in; cool, pale, and clammy; constriction, in. (d) Aldosterone and ADH. (e) Thirst, which increases desire for fluid intake.

D1. (a) Aorta; left; ascending aorta, arch of the aorta, descending aorta. (b) Coronary; heart. (c) K, brachiocephalic; L, left common carotid; M, left subclavian. (d) F; the right extremity and right side of thorax, neck, and head; right common carotid. (e) G, axillary; H, brachial; blood pressure. (f) I, radial; J, ulnar; I. (g) D; cerebral, Willis.

D2. (a) Stomach, spleen, liver. (b) Intestines (small and large). (c) Kidneys. (d) Ovaries, testes.

D3. Common iliac; legs.

D4. Superior vena cava, inferior vena cava, coronary sinus.

D5. (a) Internal jugular; common carotid. (b) There are no such vessels.

D6. See key, page 325 of the *Learning Guide*. Right, right.

D7. (a) Away from; red; pulmonary artery (postnatally) and umbilical arteries (in fetal circulation): See Checkpoint D11. (b) Veins; blood in veins is under lower pressure than blood in arteries so is more sluggish in returning to the heart. (c) Deep; basilic, cephalic, great and small saphenous.

D8. (a) Hepatic; hepatic portal, digestive (or GI); sinu-soids. (b) While in the liver, blood is cleaned and detoxified, and ingested nutrients and other chem-icals are metabolized and stored there. (c) Hepatic, inferior.

D9. (a) Cer. (b) Cor. (c) Umb. (d) Hep. (e) Pulm.

D11.

Fetal Structure	Structures Connected	Function	Fate of Structure After Birth
a. Umbilical arteries (2)	Fetal internal iliac arteries to placenta	Carries fetal blood low in oxygen and nutrients and high in wastes	Medial umbilical ligaments
b. Placenta	(Omit)	Site where maternal and fetal blood exchange gases, nutrients, and wastes	Delivered as "afterbirth"
c. Umbilical vein (1)	Placenta to liver and ductus venosus	Carries blood high in oxygen and nutrients, low in wastes	Round ligament of the liver (ligamen-tum teres)
d. Ductus venosus	Umbilical vein to inferior vena cava	Branch of umbilical vein, bypasses liver	Ligamentum venosum
e. Foramen ovale	Right and left atria	Bypasses lungs	Fossa ovalis
f. Ductus arteriosus	Pulmorary artery to aorta	Bypasses lungs	Becomes ligamentum arteriosum

E1. (a) Stiffness. (b) De, de. (c) Systolic. (d) Dimin-ished. (e) Can.

E2. (a) Urinary. (b) Integumentary. (c) Skeletal. (d) Respiratory. (e) Muscular. (f) Reproductive. (g) Lymphatic and immune. (h) Digestive. (i) Endocrine.

E3. (a) 140, 90. (b) Hypertension (Stage 3). (c) Heart failure, kidney disease, stroke.

E6. Embolus (a "clot-on-the-run") travels through femoral vein to iliac veins to inferior vena cava to right side of heart and pulmonary vessels.

E7. (a) S. (b) An. (c) Occ. (d) P. (e) OH.

CRITICAL THINKING: CHAPTER 16

1. Explain how the structure of the following types of blood vessels is admirably suited to their functions: arteries, arterioles, capillaries, and veins.

2. The typical human body contains about 5–6 liters of blood. Describe mechanisms that regulate where this blood goes at any given moment, for example, at rest versus during a period of vigorous exercise. Be sure to include the following terms in your dis-cussion: precapillary sphincter, arteriole, sympa-thetic, and vasoconstriction.

3. Write a summary of factors and mechanisms that may lower blood pressure.

4. In most cases, arteries and veins are parallel in both name and locations, such as the brachial artery and brachial vein lying close together in the arm. Iden-tify major arteries and veins that differ in names and locations.

5. How can you identify whether it is an artery or a vein that is bleeding from a wound?

6. Mr. Rebold, age 82, stands up suddenly after taking a nap and feels dizzy as a result. Explain. (*Hint*: refer to page 416)

MASTERY TEST: ■ CHAPTER 16

Questions 1–5: Circle T (true) or F (false). If the statement is false, change the underlined word or phrase so that the statement is correct.

T F 1. The wall of the femoral artery is <u>thicker</u> than the wall of the femoral vein.

T F 2. Most of the smooth muscle in arteries is in the <u>inner layer</u>.

T F 3. Decrease in the size of the lumen of a blood vessel by contraction of smooth muscle is called <u>vaso-constriction</u>.

T F 4. Cool, clammy skin is a sign of shock that results from <u>sympathetic stimulation of blood vessels and sweat glands</u>.

T F 5. In its passage from an artery to a vein a red blood cell must ordinarily travel through a <u>capillary</u>.

6. In order for a RBC in the left brachial vein to flow to an artery in the right arm, which structures must the RBC pass through?
   A. Right side of the heart
   B. Left side of the heart
   C. A lung
   D. Capillaries in the left arm

Questions 7–10: Arrange the answers in correct sequence.

_____ _____ _____ _____ 7. Route of a drop of blood from the right side of the heart to the left side of the heart:
   A. Pulmonary artery
   B. Arterioles in lungs
   C. Capillaries in lungs
   D. Venules and veins in lungs

_____ _____ _____ _____ _____ 8. Route of a drop of blood from the stomach to heart:
   A. Gastric vein
   B. Hepatic portal vein
   C. Small vessels in the liver
   D. Hepatic vein
   E. Inferior vena cava

_____ _____ _____ _____ _____ _____ 9. Route of a drop of blood from the gastrocnemius muscle to the right atrium:
   A. Inferior vena cava
   B. Popliteal vein
   C. Common iliac vein
   D. Posterior tibial vein
   E. External iliac vein
   F. Femoral vein

_____ _____ _____ _____ 10. Blood pressure in vessels, from highest to lowest:
   A. Aorta and other arteries
   B. Arterioles
   C. Capillaries
   D. Venules and veins

Questions 11–15: Fill-ins. Complete each sentence or answer the question with the word(s) or phrase that best fits.

_____ 11. A weakened section of a blood vessel forming a balloonlike sac is known as a(n) _____ .

_____ 12. _____ refers to a sudden, dramatic drop in blood pressure upon standing or sitting up straight.

_____ 13. Name the two blood vessels that are locations of baroreceptors and chemoreceptors for regulation of blood pressure.

_____ 14. In taking a blood pressure, the cuff is first inflated over an artery. As the cuff is then slowly deflated, the first sounds heard indicate the level of _____ blood pressure.

_____ 15. Most arteries are paired (one on the right side of the body, one on the left). Name three or more major arteries that are unpaired.

ANSWERS TO MASTERY TEST: ■ CHAPTER 16

True or False

1. T
2. F. Middle layer
3. T
4. T
5. T

Multiple Answers

6. A B C

Arrange

7. A B C D
8. A B C D E
9. D B F E C A
10. A B C D

Fill-ins

11. Aneursym
12. Orthostatic hypotension
13. Aorta and carotids
14. Systolic
15. Aorta, anterior communicating cerebral, basilar, brachiocephalic, celiac, gastric, hepatic, splenic, superior and inferior mesenteric

FRAMEWORK 17
The Lymphatic and Immune System and Resistance to Disease

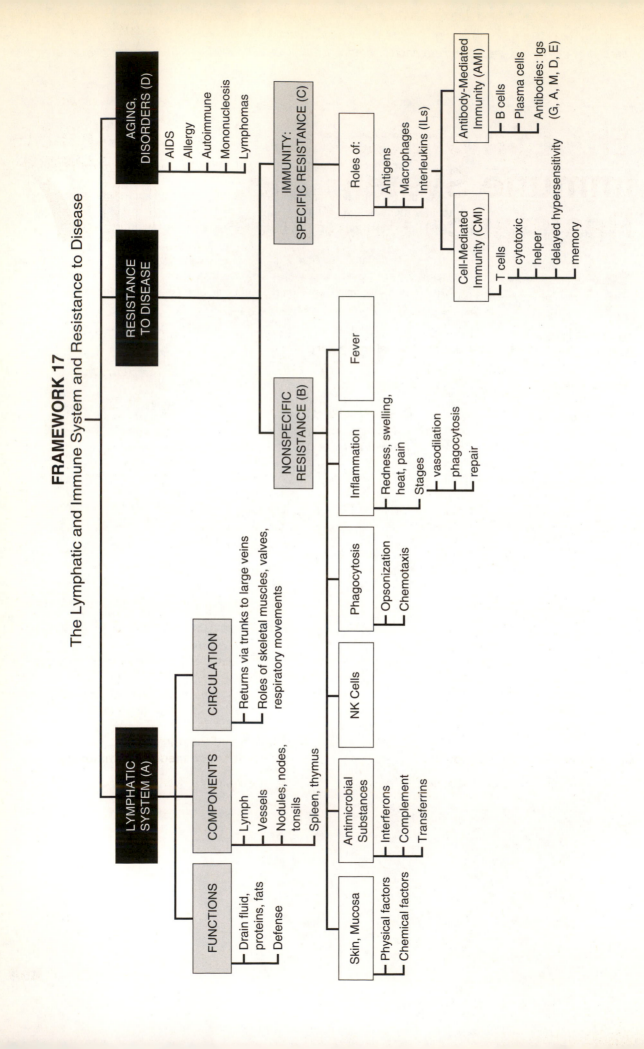

The Lymphatic and Immune System and Resistance to Disease

The human body is continually exposed to foreign (nonself) entities: bacteria or viruses, pollens, chemicals from a mosquito bite, or foods or medications that provoke allergic responses. Resistance to specific invaders is normally provided by a healthy immune system made up of antibodies and battalions of T lymphocytes.

Nonspecific resistance is offered by a variety of structures and mechanisms, such as skin, mucus, white blood cells assisting in inflammation, and even by a rise in temperature (fever) that limits the survival of invading microbes. The lymphatic system provides sentinels at points of entry (as in tonsils) and at specific sites (lymph nodes) near entry to body cavities (inguinal and axillary regions). The lymphatic system also wards off the detrimental effects of edema by returning proteins and fluids to blood vessels.

Start your study of the physical defense mechanisms of the body by previewing the Chapter 17 Framework and the key terms it contains. As you begin this chapter, carefully examine the Chapter 17 Topic Outline and check off each objective after you meet it.

TOPIC OUTLINE AND OBJECTIVES

A. Lymphatic system

☐ 1. Describe the components and major functions of the lymphatic system.

☐ 2. Describe the organization of lymphatic vessels and the circulation of lymph.

☐ 3. Compare the structure and functions of the primary and secondary lymphatic organs and tissues.

B. Innate immunity (nonspecific resistance)

☐ 4. Describe the various components of innate immunity.

C. Adaptive immunity (specific resistance)

☐ 5. Define adaptive immunity and compare it with innate immunity.

☐ 6. Explain the relationship between an antigen and an antibody.

☐ 7. Compare the functions of cell-mediated immunity and antibody-mediated immunity.

D. Aging and the immune system; common disorders and medical terminology

☐ 8. Describe the effects of aging on immunity.

WORDBYTES

Study each wordbyte, its meaning, and an example of its use in a term. Check your understanding by jotting meanings of wordbytes in margins. Identify other examples of terms that contain these wordbytes as you continue through the text and *Learning Guide*.

Wordbyte	Meaning	Example(s)	Wordbyte	Meaning	Example(s)
anti-	against	*anti*body	hetero-	other	*hetero*graft
auto-	self	*auto*immune	iso-	same	*iso*graft
axilla-	armpit	*axilla*ry nodes	lact-	milk	*lact*eals
chyl-	juice	cisterna *chyl*i	meta-	beyond	*meta*stasis
gen-	to produce	anti*gen*, *gen*etic	stasis	to stand	meta*stasis*

CHECKPOINTS

A. Lymphatic system (pages 421–425)

■ **A1.** List five types of harmful agents to which your body can be exposed on a daily basis.

A2. Summarize the body's defense mechanisms against invading microorganisms and other potentially harmful substances. Be sure to use these terms in your paragraph: *pathogens, resistance, susceptibility*.

■ **A3.** Write *A* next to adaptive immune responses and *I* next to innate responses.

       _____ a. Barriers to infection, such as skin and mucous membranes

       _____ b. Chemicals such as gastric and vaginal secretions, which are both acidic

       _____ c. Destruction of a cancer cell or microbe by a T lymphocyte

       _____ d. Known as nonspecific defenses, they are present from birth, and provide immediate protection

       _____ e. Responses are slower but have a "memory" component

■ **A4.** Identify components of the lymphatic system in this Checkpoint.

a. _____ , which is the fluid portion of the lymphatic system

b. _____ , which transport lymph

c. Lymph nodes, tonsils, and spleen which contain _____ .

d. _____ where blood cells (including lymphocytes) first develop.

■ **A5.** Describe the functions of the lymphatic system in this exercise.

a. Lymphatic vessels drain fluid from _____ . They also transport _____ and fat-soluble vitamins

(_____ , _____ , _____ , and _____) from the gastrointestinal (GI) tract to the bloodstream.

b. The defensive functions of the lymphatic system are directed against harmful cells and other "invaders" such as

_____ and _____ .

■ **A6.** Check your understanding of lymphatic vessels in this Checkpoint.

a. Endothelial cells that form lymphatic capillaries are arranged *(tightly? so they overlap?)*. Describe the effect of this arrangement.

b. Larger lymph vessels most resemble *(arteries? veins?)* but are *(thicker? thinner?)* and have *(more? fewer?)* valves than veins.

c. About three-fourths of all lymph finally drains into the vessel known as the *(right lymphatic duct? thoracic duct?),* whereas about one-fourth of lymph drains into the _____

_____ duct. Both of these ducts empty into *(arteries? veins?)* in the lower part of the

neck. Specifically, the site is at the junction of the internal _____ and the subclavian veins.

d. In summary, lymph fluid starts out originally in plasma or cells and circulates as _____ fluid. Then as lymph it undergoes extensive cleaning as it passes through nodes. Finally, fluid and other sub-

stances in lymph are returned to _____ in veins of the neck.

■ **A7.** The lymphatic system does not have a separate heart for pumping lymph. Describe two principal factors that are responsible for return of lymph from the entire body to major blood vessels in the neck. (Note that these same factors also facilitate venous return.)

A8. Define *edema* and list reasons why this may occur.

■ **A9.** Contrast primary and secondary lymphatic organs by writing P (primary) or S (secondary) next to each phrase.

_____ a. Sites where immune responses take place

_____ b. Sites where stem cells differentiate into B cells or T cells

_____ c. Red bone marrow and thymus

_____ d. Lymph nodes, lymphatic nodules, and spleen

■ **A10.** Describe the thymus gland in this exercise.

a. This gland is located in the *(neck? thorax? abdomen?)*. Its position is just posterior to the *(heart? sternum?)*.

b. The thymus gland is the site where *(B? T?)* lymphocytes that arrive from the red marrow can mature. About *(98% 2%)* of the T cells here do become mature T lymphocytes.

■ **A11.** Refer to Figure 17.1(page 422) in the text and then circle regions of the body that contain groups of lymph nodes:

Abdomen	Ankles	Axillae (armpits)
Cervical (neck) region		Inguinal (groin) region
Knees	Neck	Wrists

A12. Describe how lymph is "processed" as it circulates through lymph nodes. How does this function of lymph nodes explain the fact that nodes ("glands") become enlarged and tender during infection?

■ **A13.** Spread, or _____ , of cancer can occur via the lymphatic system. All (*benign? malignant?*) tumors eventually metastasize. How can cancerous lymph nodes be differentiated from infected lymph nodes?

■ **A14.** The spleen is located in the (*upper right? upper left? lower right? lower left?*) quadrant of the abdomen,

immediately inferior to the _____ . B and T cells carry out immune responses within (*red? white?*) pulp of the spleen.

List three functions of the red pulp: _____ _____

_____

A15. Contrast *lymph nodes* with *lymphatic nodules*.

List three sites in the body where lymphatic nodules are located.

■ **A16.** State an advantage of the locations of tonsils.

Which tonsils are most commonly removed during a "tonsillectomy"? (*Lingual? Pharyngeal? Palatine?*)

Which tonsils are commonly called "adenoids"? _____

B. Innate immunity (nonspecific resistance) (pages 425–428)

B1. Review types of innate and adaptive resistance to disease in Checkpoint A3.

■ **B2.** *A clinical challenge.* Match the following conditions or factors with the "first line of defense" barriers that provide innate immunity.

Innate (nonspecific) Resistance Lost	Condition/Factor
_____ a. Cleansing of oral mucosa	A. Long-term smoking
_____ b. Cleansing of vaginal mucosa	B. Excessive dryness of skin
_____ c. Flushing of microbes from urinary tract	C. Medications that have side effects of decreasing salivation
_____ d. Lacrimal fluid with lysozyme	D. Thinning of the vaginal wall related to normal aging
_____ e. Respiratory mucosa with cilia	E. Dry eyes related to decreased production of tears, as in aging
_____ f. Closely packed layers of keratinized cells	F. Blockage of urine pathways by enlargement of the prostate gland
_____ g. HCl production; pH 1 or 2	G. Skin wound
_____ h. Sebum production	H. Partial gastrectomy (removal of part of stomach)

■ **B3.** Describe other innate defenses described as the "second line of defense," in this Checkpoint.

a. Interferons (IFNs) are proteins produced by cells such as _____ , _____ , and _____ that are

infected with viruses. IFNs then stimulate *(infected? uninfected?)* cells to produce _____ proteins.

b. State two locations of complement proteins. _____ and _____ .
Explain why these proteins are called *complement*.

Describe these processes that they facilitate: *cytolysis* and *opsonization*.

c. _____ are proteins that bind to iron (Fe) in body fluids such as blood, saliva, tears, and milk. Explain how these proteins inhibit bacterial growth.

d. NK cells, or _____ _____ , make up about 5–10% of *(lymphocytes? neutrophils?)* that attack

microbes or tumor cells. NK cells are one type of cell that is deficient or defective in patients with _____ .

e. The primary type of white blood cells involved in phagocytosis is the _____ .

Other leukocytes, known as _____ , travel to infection sites and develop into highly phagocytic

_____ cells. Name five organs that contain fixed macrophages.

■ **B4.** Complete this exercise about another innate defense, *inflammation*.

a. Write three main functions of inflammation.

b. Defend or dispute this statement: "Inflammation is a process that can be both helpful and harmful."

c. List the four cardinal signs or symptoms of inflammation.

_____   _____

_____   _____

d. Explain how leakage of clotting proteins into the injured tissue enhances the inflammatory response.

■ **B5.** List in correct sequence the events in a typical inflammation: _____ _____ _____ _____

A. Monocytes arrive at the scene and transform into wandering macrophages.

B. Neutrophils emigrate through vessel walls; antibodies and clotting chemicals also move into interstitial areas.

C. Chemicals cause blood vessels in the injured area to become dilated and more permeable.

D. Phagocytes are attracted to the microbes by chemotaxis (enhanced by opsonization).

E. Pus forms from damaged tissue and dead phagocytes; abscess and ulcer possibly result.

B6. Explain the mechanism by which bacterial infection can cause fever.

How can fever help to combat the infection?

C. Adaptive immunity (specific resistance)   (pages 428–437)

■ **C1.** Adaptive (specific) immunity involves production of (*antigens? antibodies?*) as well as certain cells that destroy (*antigens? antibodies?*).

■ **C2.** Describe development of lymphocytes in this exercise.

a. Two categories of lymphocytes are involved with immunity: _____ cells and _____ cells. Both types of cells initially develop from stem cells within the (*bone marrow? heart? thymus?*).

b. T cells are called T cells because many of them travel to the (*bone marrow? heart? thymus?*) for processing.

Later they move to _____ tissue. B cells leave bone marrow and migrate to

_____ tissue.

c. By the time they leave the thymus, many of the T cells have specific proteins inserted in their plasma membranes.

Name two types of these T cells: _____ _____ .

■ **C3.** Contrast two types of immunity by writing AMI before descriptions of *antibody-mediated immunity* and CMI before those describing *cell-mediated immunity*.

_____ a. Involves plasma cells (derived from B cells) that produce antibodies (like bullets) that destroy the antigen.

_____ b. Uses T lymphocytes (much like an army of soldiers) that directly attack the antigen

■ **C4.** Do this exercise on antigens.

a. An antigen is defined as "any chemical substance that, when introduced into the body,

_____ ." An antigen also has the ability to stimulate the

body to _____ .

b. Can an entire microbe serve as an antigen *(Yes? No?)*? List parts of microbes that may be antigenic.

c. If you are allergic to pollen in spring or fall or to certain foods, the pollen or foods act as *(antigens? antibodies?)* to you.

d. In general, antigens are *(parts of the human body? foreign substances?)*. Usually the body's own chemicals *(are? are not?)* recognized as self and *(are? are not?)* normally antigenic to us. This characteristic of humans is known as *(self-tolerance? autoimmune disease?)*.

■ **C5.** Describe roles of "self antigens" or major histocompatibility complex (MHC) antigens in this exercise.

a. Describe the "good news" (how they help you) and the "bad news" (how they may be harmful) about MHC antigens.

b. For what purpose is histocompatibility testing performed?

■ **C6.** Refer to Figure 17.6a (page 429) in the text and discuss structure of antibodies in this activity.

a. Chemically, all antibodies are composed of _____ . These consist of *(2? 4? 8?)* polypeptide chains.

Because of the flexibility of the antibody, it can bind with at least _____ identical antigens at a time.

b. *(Constant? Variable?)* portions of antibodies are diverse for different antibodies because these serve as binding sites for different antigens. *(Constant? Variable?)* portions are *(different? are identical?)* for all antibodies in the same class.

■ **C7.** Because antibodies are classified as globulin types of proteins that are involved in immunity, they are

called immunoglobulins, abbreviated _____ . Describe classes of Ig's by selecting answers in the box that fit descriptions below.

A. IgA	D. IgD	E. IgE	G. IgG	M. IgM

_____ a. First antibody to be secreted after initial exposure to antigen; make up 5–10% of all antibodies in blood

_____ b. Make up 80% of all antibodies in blood; only type of antibody to cross placenta from mother to fetus

_____ c. Make up about 15% of all antibodies in blood; also offer protection because present in sweat, tears, saliva, mucus, and mother's milk

_____ d. Activate B cells (two answers)

C8. Complete this Checkpoint as nonstressfully as possible!

a. In stressful situations such as college exams, adrenal cortical hormones are likely to _____ -crease. Name the

major glucocorticoid: _____ . This hormone *(stimulates? inhibits?)* immune responses. In fact, it is said to be "anti-inflammatory."

b. What message does the field of PNI offer about healthy responses to stress?

■ **C9.** Describe the processing and presenting of antigens in this activity.

a. For an immune response to occur, either _____ or _____ cells must recognize the presence of a foreign antigen. *(B? T?)* cells can recognize antigens located in extracellular fluid (ECF) and not attached to any cells. However, for *(B? T?)* cells to recognize antigens, the antigens must have been processed by a cell, and then they must

be presented in association with MHC (or _____ _____ _____) proteins.

b. Cells that present these antigenic proteins are called APCs or a _____ p _____

c _____ . These cells are generally found in *(lymph nodes? at sites where antigens are likely to enter the body, such as skin or mucous membranes?)*. Name three examples of APCs.

c. Place in correct sequence the steps in processing and presenting an antigen? _____ _____ _____ _____

1. An APC digests the antigen into fragments and combines a fragment with an MHC protein

2. The APC enters a lymph vessel and migrates to a lymphatic tissue

3. The APC ingests an antigen

4. The APC meets T cells with the correct antigen receptors to fit the processed antigen

C10. Write a paragraph comparing the activation of T cells to the process of starting the car. Be sure to include these terms: *antigen, T cell receptor, antigen recognition, co-stimulator (IL-2)*.

■ **C11.** Match types of cells involved in immunity to descriptions below.

APC. Antigen-presenting cells	H. Helper T cells
C. Cytotoxic T cells	M. Memory T cells

_____ a. Secrete perforin, lymphotoxin, and gran-
ulysin

_____ b. Programmed to recognize the original invader; can initiate dramatic responses to
reappearance of the intruder

_____ c. These cells recognize and bind to cells
infected with microbes, and directly kill
antigens

_____ d. Secrete gamma-infection which enhances
phagocytosis

_____ e. Help in many aspects of immunity by stimulating growth of T cells, enhancing action
of macrophages, and costimulating B cells
to produce Ab's

_____ f. Process and present antigens; include
macrophages, B cells, and dendritic cells

■ **C12.** Contrast activities of B cells with those of T cells by circling correct answers in this Checkpoint.

a. Which cells pass out of lymphatic tissue to the site of the invader? *(B? T? both B and T?)* cells

b. Which cells stay in lymph nodes, spleen, or lymph tissue in the GI tract and "do their work from there"? *(B? T? both B and T?)* cells

c. Which cells do not require that the antigen be presented by APCs? *(B? T?)* cells

d. Which cells secrete IL-2 that helps to activate B cells and macrophages? *(B? T?)* cells

e. Which cells are activated to enlarge, divide, and differentiate into plasma cells that secrete antibodies? *(B? T? both B and T?)* cells

■ **C13.** List five or more mechanisms by which antibodies can attack antigens.

C14. Define the term monoclonal antibody (MAb) and discuss the clinical use of MAbs.

■ **C15.** Do this exercise on immunizations.

a. Once a specific antigen has initiated an immune response, either by infection or by a(n) *(initial? booster dose?)*

immunization, the person produces some long-lived B cells called _____ .

b. Upon subsequent exposure to the same antigen, such as during another infection or by a

_____ dose of vaccine, memory cells provide a *(more? less?)* intense response.

Memory cells can live for *(hours? months? decades?)*. They can then form _____ cells or _____
cells. The antibodies formed from plasma cells are *(more? less?)* than those formed during a primary response.

■ **C16.** *A clinical challenge*. Contrast types of immunity in each case. Choose from answers in the box.

AAAI.	Artificially acquired active immunity
AAPI.	Artificially acquired passive immunity
NAAI.	Naturally acquired active immunity
NAPI.	Naturally acquired passive immunity

_____ a. Kelly gives her baby Crystal antibodies in breast milk and also provided antibodies across the placenta during pregnancy

_____ b. Jenny is exposed to hepatitis B virus through her work in the emergency room; she takes intravenous gamma globulin to prevent infection

_____ c. Elena takes her 6-week-old baby Olivia to the clinic for her first immunizations

D. Aging and the immune system; common disorders and medical terminology (pages 437–441)

D1. List several normal aging changes in the immune system that make older adults more susceptible to infections.

D2. List specific examples of the impact of lymphatic and immune system activities on the following systems:

a. Digestive

b. Reproductive

c. Respiratory

■ **D3.** Complete this Checkpoint about AIDS.

a. AIDS is an acronym for _____ _____ _____ . This condition was first identified in the

United States in 19 _____ .

b. The causative agent of AIDS is the _____ _____ virus. What are the major effects and likely causes of death resulting from this viral infection?

c. Circle the four fluids through which the HIV virus has been found to be transmitted in sufficient quantities to be infective.

Blood   Breast milk   Mosquito venom   Saliva

Semen   Tears   Vaginal fluids

d. List the groups of people who have been most infected in the United States.

e. As many as _____ million people worldwide have been infected with AIDS.

f. The HIV virus is considered *(fragile? hardy?)*. List several methods of reducing HIV transmission.

g. The HIV virus contains its genetic material in (DNA? RNA?). It *(does? does not?)* depend on a living cell for

replication. Which cells are most infected? _____

h. Explain how "HIV-positive" status is usually determined.

Describe how the diagnosis of "AIDS" established.

i. HAART therapy refers to _____ _____ _____ therapy. Name the categories of drugs that are included in

this protocol: _____ inhibitors (RTIs) and _____ inhibitors PIs). In which category are the drugs ZDV

(formerly called AZT) and ddI? _____ . List three barriers to use of this drug regimen.

■ **D4.** Mr. Ruskin who is allergic to egg whites eats baked goods that contain small amounts of egg whites. Complete this exercise about him.

a. The egg whites are allergens which are *(antigens? antibodies?)*. The antigenic components of the egg are likely to

attach to Ig ____ antibodies already present in his body, and trigger reactions within *(minutes? days?)*. Name two chemicals likely to be released from Mr. Ruskin's mast cells and basophils in this allergic reaction.

b. These chemicals are likely to vaso-*(constrict? dilate?)* blood vessels and _____-crease their permeability,

_____-creasing mucus secretion which can clog airways and possibly _____-crease blood pressure. Also,

airways are likely to broncho-*(constrict? dilate?)*. If severe, these manifestations are signs of _____ shock.

This type of reaction is known as a Type ____ allergic reaction.

■ **D5.** Maureen breaks out with poison ivy 18 hours after exposure to it. This is an example of a(n) *(allergic? delayed hypersensitivity?)* or Type _____ allergic reaction. The poison ivy toxin is presented by _____ cells to *(B? T?)* cells that initiate an inflammatory response.

■ **D6.** Name one example of a type II (cytotoxic) reaction.

■ **D7.** Select the disorder in the box that best fits each description below.

CFS.	Chronic fatigue syndrome
HL.	Hodgkin disease (or Hodgkin lymphoma)
IM.	Infectious mononucleosis
NHL.	Non-Hodgkin lymphoma
SLE.	Systemic lupus erythematosus

_____ a. A chronic, systemic, autoimmune disease that occurs almost entirely in young women; named after the skin lesions that were thought to resemble a wolf bite; its most serious effects are on kidneys

_____ b. A contagious disease often transmitted by kissing; caused by the Epstein-Barr virus, it affects B cells and leads to fever, sore throat, and enlarged and tender lymph nodes

_____ c. May be first recognized by an enlarged but painless lymph node; with early diagnosis, survival rate may approach 95%

_____ d. A disorder primarily of young adult females; involves extreme fatigue for months in the absence of other known diseases

■ **D8.** A valve from a pig that is transplanted into a human heart is an example of a(n) _____-graft. Skin transplanted from one part of a person to burned area of the same person is a(n) _____-graft.

ANSWERS TO SELECTED CHECKPOINTS: CHAPTER 17

A1. Pathogenic microbes, cuts, sunlight, burns, chemical toxins.

A3. (a-b) I. (c) A. (d) I. (e) A.

A4. (a) Lymph. (b) Lymph vessels such as lymph capillaries and lymphatics (c) Lymphatic tissue. (d) Bone marrow.

A5. (a) Interstitial spaces; fats, A, D, E, and K. (b) Cancer cells and microbes.

A6. (a) So they overlap; these vessels act like one-way swinging doors by allowing lymph to enter but not to flow back out of these vessels. (b) Veins, thinner, more. (c) Thoracic duct, right lymphatic; veins; jugular. (d) Interstitial; plasma.

A7. Skeletal muscle contraction squeezing lymphatics, which contain valves that direct flow of lymph. Respiratory movements.

A9. (a) S. (b) P. (c) P. (d) S.

A10. (a) Thorax; sternum. (b) T; 2%.

A11. Abdomen, axillae (armpits), cervical (neck) region, inguinal (groin) region.

A13. Metastasis; malignant; cancerous lymph nodes enlarge, but feel firm and fixed to underlying tissues; they are also likely to be painless.

A14. Upper left, diaphragm; white; removal of red cells and platelets from blood, storage of platelets, production of blood cells in the fetus.

A16. Their strategic location surrounding entrances to major airways provides defense against invading microorganisms; palatine; pharyngeal.

B2. (a) C. (b) D. (c) F. (d) E. (e) A. (f) G. (g) H. (h) B.

B3. (a) Lymphocytes, macrophages, and fibroblasts; uninfected, antiviral. (b) In blood plasma and on plasma membranes; when activated, they "complement" or enhance other defense mechanisms; they punch holes in microbes (cytolysis) and coat bacteria so they are "tastier" to phagocytes. (c) Transferrins; they prevent bacterial access to iron that microbes need for growth. (d) Natural killer, lymphocytes; AIDS. (e) Neutrophil; monocytes, macrophage; skin, liver, lungs, brain, and spleen.

B4. (a) Removal of microbes, toxins, and debris at the site of injury; prevent spread; prepare the site for tissue repair. (b) It can limit infection and help repair, but does involve swelling and pain. (c) Redness, heat, swelling (edema) and pain. (d) The resulting clot isolates the invading microbes and their toxins.

B5. C D B A E.
C1. Antibodies, antigens.
C2. (a) B, T; bone marrow. (b) Thymus; lymphoid; lymphoid. (c) CD4, CD8.
C3. (a) B. (b) T.
C4. (a) Is recognized as foreign; produce specific antibodies or T cells that react with the antigen. (b) Yes; flagella, capsules, cell walls, and toxins made by bacteria, and viral proteins. (c) Antigens. (d) Foreign substances; are, are not; self-tolerance.
C5. (a) Good news: they help T cells to recognize foreign invaders because antigenic proteins must be processed and presented in association with MHC antigens before T cells can recognize the antigenic proteins; bad news: MHC antigens in transplanted tissue present that tissue as "foreign" to the body, causing it to be rejected. (b) Organ transplant such as kidney or liver transplant or tissue transplant (such as bone marrow).
C6. Proteins, 4; 2. (b) Variable; constant, are identical;
C7. (a) M. (b) G. (c) A. (d) G, M.
C9. (a) B, T; B, T; major histocompatibility complex. (b) Antigen-presenting cells; at sites where antigens are likely to enter the body, such as skin or mucous membranes; B cells, macrophages, and dendritic cells. (c) 3 1 2 4.
C11. (a) C. (b) M. (c) C. (d) C. (e) H. (f) APC.
C12. (a) T. (b) B. (c) B (although they respond more intensely to presented antigens). (d) T (helper T cells). (e) B cells.
C13. Neutralize or agglutinate antigen; immobilize bacteria; activate complement; or enhance phagocytosis
C15. (a) Initial, memory. (b) Booster, more; decades; plasma or cytotoxic; more.

C16. (a) NAPI. (b) AAPI. (c) AAAI.
D3. (a) Acquired immunodeficiency syndrome; 81. (b) Human immunodeficiency; by destroying the immune system, AIDS places the patient at high risk for opportunistic infections (normally harmless) that can affect skin, respiratory and GI tracts, and nervous system, as well as cancers. (c) Blood, semen, vaginal fluids, breast milk. (d) Persons who have practiced unprotected anal, vaginal, or oral sex with an HIV-infected individual, IV drug users who share needles, and hemophiliacs who received contaminated blood products before blood testing 1985. (e) 40. (f) Fragile; sexual abstinence, use of latex condoms, avoidance of sharing needles among IV drug-users, anti-HIV medications to HIV-infected pregnant women, cleaning or heating items that could transmit the virus. (g) RNA; does; helper T cells. (h) Test for presence of anti-HIV antibodies in blood; helper T cell count drops (from a normal of about 800–1000/uL) to a level less than 200/uL or a major opportunistic infection develops. (i) Highly active antiretroviral; reverse transcriptase, protease; RTIs; cost, side effects, and complex dosing schedule.
D4. (a) Antigens; E, minutes; histamine, prostaglandins. (b) Dilate, in, in, de; constrict; anaphylactic.
D5. Delayed hypersensitivity; APCs, T.
D6. Reaction to an incompatible blood type, for example, reaction to a person with type A blood to a type B transfusion.
D7. (a) SLE. (b) IM. (c) NHL. (d) CFS.
D8. Xeno; auto.

CRITICAL THINKING: CHAPTER 17

1. Explain how the lymphatic system contributes to your body's defenses.
2. Contrast the structure of lymphatic vessels with the structure of veins. Explain how that structure is related to lymphatic function.
3. Contrast specific innate immunity with adaptive immunity. Give examples of each type of resistance.
4. Contrast primary and secondary lines of defense against pathogens.
5. Explain what physiological changes during inflammation lead to the four classic signs or symptoms of inflammation.
6. Contrast four types of lymphocytes according to their functions.
7. Describe stages of HIV infection and AIDS, including major infections and related signs and symptoms that commonly occur. Also outline several ways to avoid contracting AIDS.
8. Defend or dispute this statement: "AIDS is now curable, so precautions against HIV infection are no longer needed."

Questions 1–7: Circle the letter preceding the one best answer to each question. Take particular note of underlined terms.

1. Both the thoracic duct and the right lymphatic duct empty directly into:
    A. Axillary lymph nodes
    B. Superior vena cava
    C. Cisterna chyli
    D. Subclavian arteries
    E. Junction of internal jugular and subclavian veins

2. All of these are examples of nonspecific defenses *except*:
    A. Antibodies   D. Interferon
    B. Saliva   E. Skin
    C. Complement   F. Phagocytes

3. Choose the *false* statement about lymphatic vessels.
    A. Lymph capillaries are <u>more</u> permeable than blood capillaries.
    B. Lymphatics have <u>thinner</u> walls than veins.
    C. Like arteries, lymphatics contain <u>no</u> valves.
    D. Skeletal muscle contraction <u>aids</u> lymph flow.

4. Choose the *false* statement about T cells.
    A. Some are called <u>memory cells</u>.
    B. They are called T cells because many of them are processed in the <u>thymus</u>.
    C. They are involved primarily in <u>antibody-mediated immunity (AMI)</u>.
    D. Like B cells, they originate from <u>cells in bone marrow</u>.

5. Choose the *false* statement about lymphatic organs and nodules.
    A. The <u>palatine tonsils</u> are the ones most often removed in a tonsillectomy.
    B. The <u>spleen</u> is the largest mass of lymphatic tissue in the body.
    C. The thymus reaches its maximum size at age <u>70</u>.
    D. The spleen is located in the <u>upper left quadrant of the abdomen</u>.

6. Choose the one *true* statement about AIDS.
    A. It appears to be caused by a <u>fungus</u>.
    B. AIDS kills primarily by <u>destroying muscle cells</u>.
    C. People with AIDS experience a decline in <u>helper T cells</u>.
    D. AIDS is transmitted primarily through <u>semen and saliva</u>.

7. Ability to "recognize one's own tissues as self," known as immunological _____ , is deficient in people with autoimmune disease.
    A. Tolerance
    B. Escape
    C. Surveillance

Questions 8–10: Circle T (true) or F (false). If the statement is false, change the underlined word or phrase so that the statement is correct.

T   F   8. Interferon is a chemical made by <u>cells infected by viruses, causing these cells to produce antiviral proteins</u>.

T   F   9. <u>Homeostasis</u> is a term that means "spread of cancer."

T   F 10. Histamine <u>increases</u> permeability of capillaries so leukocytes can more readily reach the infection site.

Questions 11–15: Fill-ins. Complete each sentence with the word or phrase that best fits.

_____ 11. _____ lymphocytes provide antibody-mediated immunity (AMI), and _____ lymphocytes and macrophages offer cell-mediated immunity (CMI).

_____ 12. AIDS is an acronym for the disease named _____ .

_____ 13. List the four fundamental signs or symptoms of inflammation.

_____ 14. The term *immunoglobulins* refers to _____ .

_____ 15. List examples of mechanical factors that provide innate (nonspecific) immunity.

ANSWERS TO MASTERY TEST: ■ CHAPTER 17

Multiple Choice

1. E
2. A
3. C
4. C
5. C
6. C
7. A

True or False

8. F. Cells near cells infected by viruses, causing these cells to produce antiviral proteins
9. F. Metastasis
10. T

Fill-ins

11. B, T
12. Acquired immunodeficiency syndrome
13. Redness, warmth, pain, and swelling
14. Antibodies
15. Skin, mucosa, cilia, also flushing of microbes by tears, saliva, urine, vaginal secretions, vomiting, and defecation.

FRAMEWORK 18
Respiratory System

ANATOMY (A)
- Nose
- Pharynx
- Larynx
- Trachea
- Bronchi
- Bronchioles
- Alveoli of lungs
- Respiratory membrane

PHYSIOLOGY (B)

PULMONARY VENTILATION
(atmosphere—alveoli)
- Pressure changes
  - inhalation active
  - exhalation passive
- Surfactant and compliance
- Lung volumes
  - TV, IRV, ERV, RV
- Lung capacities
  - TLC, VC, IC, FRC

EXTERNAL RESPIRATION
(alveolar air—pulmonary blood)
- pO_2, pCO_2
- Thin membrane, large surface area, pulmonary blood flow

INTERNAL RESPIRATION
(systemic blood—tissues)
- Transport of O_2: on Hb, dissolved
- Transport of CO_2: as HCO_3^-, on Hb, dissolved

TRANSPORT OF GASES (C)
- O_2
  - hemoglobin (98.5%)
  - dissolved (1.5%)
- CO_2
  - as HCO_3^- (78%)
  - on hemoglobin (13%)
  - dissolved (9%)

CONTROL OF RESPIRATION (D)
- Control centers
  - medullary rhythmicity center
- Regulation
  - primary stimulus: ↑ H^+ or ↑ CO_2
  - secondary stimulus: ↓ O_2
  - other factors: BP, pain, temperature, irritation, proprioceptors
  - cortical influence
  - inflation reflex

EXERCISE, AGING, AND DISORDERS (E)
- Lung cancer
- Asthma, bronchitis, emphysema
- Pneumonia, TB
- RDS, SIDS
- Colds, flu
- Pulmonary edema
- CF

The Respiratory System

Oxygen is available all around us. But to get an oxygen molecule to a muscle cell in the stomach or a neuron of the brain requires coordinated efforts of the respiratory system with its transport partner, the cardiovascular system. Oxygen traverses a system of ever-narrowing and diverging airways to reach millions of air sacs (alveoli). Each alveolus is surrounded by a meshwork of pulmonary capillaries, much like a balloon encased in nylon stocking. Here oxygen changes places with the carbon dioxide wastes in blood, and oxygen travels through the blood to reach the distant stomach or brain cells. Interference via weakened respiratory muscles (as in normal aging) or extremely narrowed airways (as in asthma or bronchitis) can limit oxygen delivery to lungs. Decreased diffusion of gases from alveoli to blood (as in emphysema, pneumonia, or pulmonary edema) can profoundly affect total body function and potentially lead to death.

Take a couple of deep breaths, and look at the Chapter 18 Framework and the key terms for each section. As you begin this chapter, carefully examine the Chapter 18 Topic Outline and check off each objective after you meet it.

TOPIC OUTLINE AND OBJECTIVES

A. Organs of the respiratory system

☐ 1. Describe the structure and functions of the nose, pharynx, larynx, trachea, bronchi, bronchioles, and lungs.

B. Respiratory physiology: ventilation, gas exchange

☐ 2. Explain how inhalation and exhalation take place.
☐ 3. Define the various lung volumes and capacities.
☐ 4. Describe the exchange of oxygen and carbon dioxide between alveolar air and blood (external respiration) and between blood and body cells (internal respiration).

C. Transport of respiratory gases

☐ 5. Describe how the blood transports oxygen and carbon dioxide.

D. Control of respiration

☐ 6. Explain how the nervous system controls breathing and list the factors that can alter the rate and depth of breathing.

E. Exercise, aging, common disorders, and medical terminology

☐ 7. Describe the effects of exercise on the respiratory system.
☐ 8. Describe the effects of aging on the respiratory system.

WORDBYTES

Study each wordbyte, its meaning, and an example of its use in a term. Check your understanding by jotting meanings of wordbytes in margins. Identify other examples of terms that contain these wordbytes as you continue through the text and *Learining Guide*.

Wordbyte	Meaning	Example(s)	Wordbyte	Meaning	Example(s)
atele-	incomplete	*atele*ctasis	intra-	within	*intra*-alveolar
-baros	pressure	hyper*baric*	-pnea	breathe	*a*pnea
dia-	through	*dia*phragm	pneumo-	air	*pneumo*thorax
dys-	bad, difficult	*dys*pnea	pulmo-	lung	*pulmo*nary
-ectasis	dilation	atele*ctasis*	rhin-	nose	*rhin*orrhea
ex-	out	*ex*piration	spir-	breathe	in*spir*ation
in-	in	*in*spiration			

CHECKPOINTS

A. Organs of the respiratory system (pages 446–452)

A1. List five functions of the respiratory system.

State the reason why the body requires a continuous supply of oxygen (O_2).

A2. Describe the medical specialty of *otorhinolaryngology*.

Which structure is surgically altered in a *rhinoplasty*?

■ **A3.** Refer to Figure LG 18.1 and visualize the process of respiration in yourself as you do this exercise.

a. Label the first phase of respiration at (1) on the figure. This process involves transfer of air from your environment to *(alveoli? pulmonary capillaries?)* in your lungs. Inflow of air rich in *(O_2? CO_2?)* is known as _____-halation (or _____-spiration), whereas outflow of air concentrated in *(O_2? CO_2)* is known as _____-halation (or _____-spiration).

b. Phase 2 of respiration involves exchange of gases between _____ in alveoli and

_____ in pulmonary capillaries. This process is know as *(external? internal?)* respiration. Label phase 2 on the figure.

c. Phase 3 of respiration starts with transport of O_2 and CO_2 through your _____ . After oxygenated blood returns to your heart via your pulmonary *(artery? veins?)*, blood is pumped to your tissues where gas

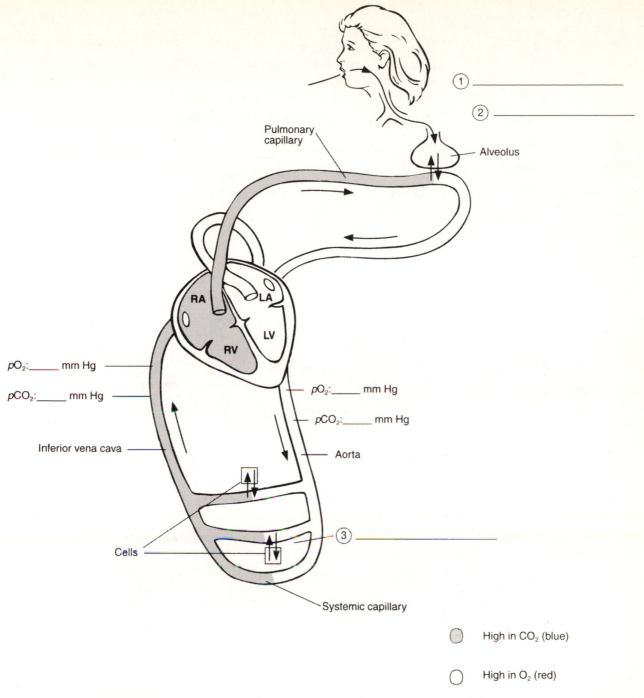

Pulmonary capillary

① _____

② _____

Alveolus

RA LA

LV

RV

pO_2: _____ mm Hg

pCO_2: _____ mm Hg

pO_2: _____ mm Hg

pCO_2: _____ mm Hg

Inferior vena cava

Aorta

③ _____

Cells

Systemic capillary

High in CO_2 (blue)

High in O_2 (red)

Figure LG18.1   Phases of respiration and circulatory routes. Numbers refer to Checkpoint A3. Fill in short lines according to Checkpoint B9h.

exchange occurs between _____ and _____ . This process is known as *(external? internal?)* respiration. Label phase 3 on the figure.

d. Blood high in CO_2 then returns to your heart where it passes directly into the pulmonary *(artery? vein?)* to reach the lungs. Here diffusion of CO_2 from pulmonary capillaries to alveoli occurs; this is another aspect of *(external? internal?)* respiration (phase _____). Finally, you exhale this air, a process that is part of *(external respiration? pulmonary ventilation?)* (phase _____).

e. Which process takes place within tissue cells, utilizing O_2 and producing CO_2 as a waste product? *(Cellular? Internal?)* respiration.

f. *For extra review,* color all blood vessels on Figure LG 18.1 according to color code ovals on the figure.

■ **A4.** Identify parts of the respiratory system in this activity.

a. Arrange respiratory structures listed in the box in correct sequence from first to last in the pathway of inspiration.

ADA. Alveolar ducts and alveoli	MN. Mouth and nose
BB. Bronchi and bronchioles	P. Pharynx
L. Larynx	T. Trachea

_____ → _____ → _____ → _____ → _____ → _____

b. Write asterisks (*) next to structures forming the upper respiratory system.

c. Most of the structures listed in the box above are parts of the *(conducting? respiratory?)* zone of the respiratory system.

■ **A5.** Name the structures of the nose that are designed to carry out each of the following functions.

a. Warm, moisten, and filter air

b. Sense smell

c. Assist in speech

A6. Identify the location of the following structures on yourself:

External nares	Nasal septum	Nasal conchae
Olfactory epithelium	Thyroid cartilage	Cricoid cartilage

■ **A7.** Write LP (laryngopharynx), NP (nasopharynx), or OP (oropharynx) to indicate locations of each of the following structures.

_____ a. Adenoids

_____ b. Palatine and lingual tonsils

_____ c. Openings from auditory (Eustachian) tubes

_____ d. Opening from oral cavity

_____ e. Opening into larynx and esophagus

■ **A8.** Describe parts of the larynx in this Checkpoint.

a. The *(arytenoid? cricoid? epiglottis? thyroid?)* cartilage is known as the "Adam's apple" and forms the *(anterior? posterior?)* of the larynx.

b. The _____ is a space between the true vocal cords. The _____ is

a cartilage that covers this space to prevent food from entering the lower airways. _____ cartilages are pyramid-shaped cartilages attached to vocal cords.

c. The _____ cartilage forms a complete ring that marks the *(superior? inferior?)* border of the larynx.

274

A9. Tell how the larynx produces sound. Explain how pitch is controlled and what causes male pitch usually to be lower than female pitch.

■ **A10.** Complete this Checkpoint on the trachea.

a. The trachea is part of the *(upper? lower?)* respiratory tract. Commonly know as the

_____ , the trachea terminates at a Y-shaped intersection where the two primary

_____ arise.

b. Arrange these structures from most anterior to most posterior: _____ _____ _____

      C. Cervical vertebrae   E. Esophagus   T. Trachea

c. The tracheal wall is lined with a _____ membrane that is *(ciliated? nonciliated?)*. The

tracheal wall is strengthened by 16–20 C-shaped rings composed of _____ . State one function of these rings.

A11. Identify tracheobronchial tree structures J–N on Figure LG 18.2 and color each of those structures indicated with color code ovals.

■ **A12.** Bronchioles *(do? do not?)* have cartilage rings. How is this fact significant during an asthma attack?

A13. On Figure LG 18.2 color the two layers of the pleura and the diaphragm according to color code ovals. Also cover the key and identify all other parts of the diagram from A to I. The right lung has *(2? 3?)* lobes. Label the

superior, middle, and *inferior lobes.* Lobes of the lungs are divided further into _____ . The *(right? left?)* lung has a cardiac notch. What is the function of this notch?

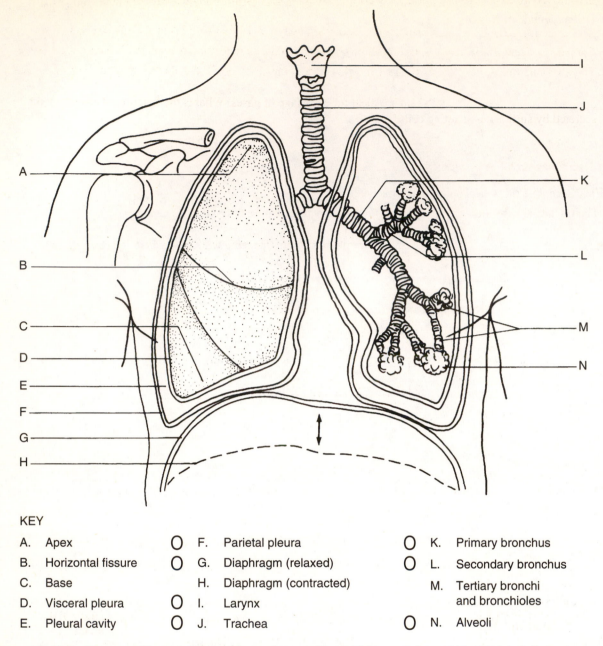

KEY

A.	Apex	◯	F.	Parietal pleura	◯	K. Primary bronchus
B.	Horizontal fissure	◯	G.	Diaphragm (relaxed)	◯	L. Secondary bronchus
C.	Base		H.	Diaphragm (contracted)		M. Tertiary bronchi
D.	Visceral pleura	◯	I.	Larynx		and bronchioles
E.	Pleural cavity	◯	J.	Trachea	◯	N. Alveoli

Figure LG 18.2 Diagram of lungs with pleural coverings and bronchial tree. Color and identify labeled structures as directed in Checkpoints A11 and A13. Also refer to Checkpoint B1.

■ **A14.** Describe the structures of a lobule in this exercise.

a. Arrange in order the structures through which air passes as it enters a lobule en route to alveoli.

___ ___ ___

A. Alveolar ducts   R. Respiratory bronchiole   T. Tertiary bronchiole

b. In order for air to pass from alveoli to blood in pulmonary capillaries, it must pass through the

_____ membrane. Identify structures in this pathway: A to E on Figure LG 18.3.

c. *For extra review*. Now label layers 1–4 of the alveolar-capillary membrane in the insert in Figure LG 18.3.

d. The alveolar wall layer of the respiratory membrane contains three types of cells. Forming most of the layer are

_____ cells that are ideal for exchange of gases. What is the function of alveolar fluid secreted by surfactant-secreting cells?

Macrophage cells help to remove _____ from lungs.

KEY

A. Alveolus
B. Alveolar wall
C. Interstitial space
D. Capillary wall
E. Blood plasma and blood cells

Figure LG 18.3 Diagram of alveoli and pulmonary capillary. Insert enlarges respiratory membrane. Numbers and letters refer to Checkpoint A14. Complete partial pressures (in mm Hg) as directed in Checkpoint B9.

B. Respiratory physiology: ventilation, internal and external respiration (pages 453–459)

■ **B1.** Refer to Figure LG 18.2 and describe the process of inhalation (inspiration) in this Checkpoint.

a. Most of the work of inhalation is accomplished by the *(diaphragm? external intercostals? internal intercostals?)*.

   This large muscle receives its innervation from the _____ nerves which are branches from *(cervical? thoracic?)* spinal nerves 3, 4, and 5.

b. Contraction of the diaphragm causes it to become *(flatter? more dome-shaped?)* which increases the distance from superior to inferior within the thoracic cavity. List several reasons why the diaphragm may not be able to descend.

c. In addition, the size of the thorax increases from anterior to posterior by contraction of the *(external? internal?)* intercostals muscles. List three other muscles that can assist during labored inhalations.

■ **B2.** Refer to Figure 18.8, page 455 in the text, and explain the pressure changes that accompany ventilation.

a. In diagram (a), just before the start of inspiration, pressure within the lungs (called _____

   _____) is _____ mm Hg. This *(more than? less than? the same as?)* atmospheric pressure.

b. The first step in inspiration occurs as muscles of the diaphragm and thoracic wall contract, _____-creasing the volume of the thorax (as discussed in Checkpoint B1). Because this increase in volume of a closed space (lungs)

   simultaneously _____ pressures there, alveolar pressure changes to _____ mm Hg.

c. A pressure gradient is now established. Air flows from high-pressure area *(alveoli? atmosphere?)* to low-pressure are *(alveoli? atmosphere?)*. So air flows *(into? out of?)* lungs. Thus *(inspiration? expiration?)* occurs. By the end of inspiration, sufficient air will have moved into the lungs to make pressure there equal to atmospheric

   pressure, that is, _____ mm Hg.

d. Inspiration is a(n) *(active? passive?)* process, whereas expiration is normally a(n) *(active? passive?)* process. In certain respiratory disorders, accessory muscles help to force air out, Name two of these:

e. List two factors that normally prevent lungs from excessive collapse.

f. A clinical challenge. Mr. Stratton has a collapsed lung following an auto accident. This condition is known

   as _____ . State two reasons why his lung may have collapsed.

■ **B3.** Maureen is breathing at the rate of 15 breaths per minute. She has a tidal volume of 480 mL per breath. Her

minute volume (MV) is _____ .

■ **B4.** Of the total amount of air that enters the lungs with each breath, about *(99 percent? 70 percent? 30 percent? 5 percent?)* actually enters the alveoli. The remaining amount of air is much like the last portion of a crowd trying to rush into a store: It does not succeed in entering the alveoli during an inspiration, but just reaches airways (entrance areas in the store) and then is quickly ushered out during the next expiration. Such air is known as anatomic

_____ and constitutes about _____ mL of a typical breath.

■ **B5.** Match the lung volumes and capacities with the descriptions given. You many find it helpful to refer to Figure 18.9 (page 456 in the text).

ERV. Expiratory reserve volume	TLC. Total lung capacity
FRC. Functional residual capacity	TV. Tidal volume
RV. Residual volume	VC. Vital capacity

_____ a. The amount of air taken in with each inhalation during normal breathing is

called _____ .

_____ b. At the end of a normal exhalation the volume of air left in the lungs is called

_____ .

_____ c. Forced exhalation can remove some of the air in FRC. The maximum volume of air that can be expired beyond normal exhalation is called

_____ .

_____ d. Even after the most strenuous expiratory effort, some air still remains in the lungs; this amount, which cannot be removed voluntarily, is called

_____ .

_____ e. The volume of air that represents a person's maxium breathing ability is called

_____ . This is the sum of ERV, TV, and IRV.

_____ f. Adding RV to VC gives

_____ .

■ **B6.** Indicate normal volumes for each of the following.

a. TV = _____ mL (_____ liter)

b. TLC = _____ mL (_____ liters) in males and _____ mL (_____ liters) in females.

c. VC = _____ mL in males and _____ mL

B7. Contrast terms in the following pairs.

a. Diaphragmatic breathing – costal breathing

b. Coughing – sighing

c. Crying – sobbing

d. Sneezing – hiccupping

■ **B8.** Do this exercise about respiratory gases.

a. The atmosphere contains enough gaseous molecules to exert pressure on a column of mercury to make it rise

about _____ mm. The atmosphere is said to have a pressure of 760 mm Hg.

b. Name four types of gases present in atmospheric air.

c. Of the total 760 mm Hg of atmospheric pressure, a certain amount is due to each type of gas. Determine the portion of the total pressure due to oxygen molecules:

$$21\% \times 760 \text{ mm Hg} = _____ \text{ mm Hg}$$

This is the partial pressure of oxygen, or _____ .

■ **B9.** Answer these questions about external and internal respiration.

a. A primary factor in the diffusion of gas across a membrane is the difference in concentration of the gas (reflected

by _____ pressures) on the two sides of the membrane. On the left side of Figure LG 18.3 write values for pO_2 (in mm Hg) in each of the following areas. (Refer to text Figure 18.10, page 458 for help.)

Atmospheric air

Alveolar air (Note that this value is lower than that for atmospheric pO_2 because some alveolar O_2 enters blood.)

Blood entering lungs

b. Calculate the pO_2 difference (gradient) between alveolar air and blood entering lungs.

$$_____ \text{ mm Hg} - _____ \text{ mm Hg} = _____ \text{ mm Hg}$$

c. Three other factors that increase exchange of gases between alveoli and blood are *(large? small?)* surface area of lungs, *(thick? thin?)* respiratory membrane, and *(increased? decreased?)* blood flow through lungs (as in exercise).

d. By the time blood leaves lungs to return to heart and systemic artieries, its pO_2 is normally *(greater than? about the same as? less than?)* pO_2 of alveoli. Write the correct value on Figure LG 18.3.

e. Now fill in all three pCO_2 values on Figure LG 18.3.

f. A *clinical challenge.* Jenny, age 50 has had blood drawn from her radial artery to determine her arterial blood

gases. Her pO_2 is 56 and her pCO_2 is 48. Are these typical values for a healthy adult? _____ .

g. At high altitude, air is "thinner"; that is, gas molecules are farther apart, so atmospheric pressure is *(higher? lower?)* than 760 mm Hg. Yet oxygen still makes up 21% of the air at high altitudes, so the pO_2 of the air you breathe on mountains is *(higher? lower?)* than 160 mm Hg.

h. Fill in blood gas values (pO_2 and pCO_2) on the short lines on Figure LG 18.1, page 273 of the *Learning Guide*.

280

C. Transport of respiratory gases (pages 459–461)

■ **C1.** Answer these questions about oxygen transport. Refer to Figure 18.11 (page 460 in your text).

a. About 98.5 percent of oxygen is carried in blood as _____ . Only a small amount

(_____ percent) of oxygen is carried dissolved in blood plasma.

b. Oxygen is attached to the _____ atoms in hemoglobin. The chemical formula for oxy-

hemoglobin is _____ . When hemoglobin carries all of the oxygen it can hold, it is said to

be fully _____ . High pO_2 in alveoli tends to *(increase? decrease?)* oxygen saturation of
hemoglobin.

c. List three factors that enhance the dissociation of oxygen from hemoglobin so that oxygen can enter tissues.
(*Hint:* Think of conditions within active muscle tissue.)

■ **C2.** Carbon monoxide has about _____ times the affinity that oxygen has for hemoglobin. State the significance
of this fact.

■ **C3.** Complete this Checkpoint about transport of carbon dioxide.

a. Write the percentage of CO_2 normally carried in each of these forms: _____ percent is present in bicarbonate

ion (HCO_3^-); _____ percent is bound to hemoglobin; _____ percent is dissolved in plasma.

b. Carbon dioxide (CO_2) produced by cells of your body diffuses into red blood cells (RBCs) and combines with

water to form _____ . Name the enzyme within red blood cells that greatly enhances this

process. _____

c. Carbonic acid tends to dissociate into two products. One is H^+, which binds to _____ .

The other product is _____ (bicarbonate), which is carried in *(RBCs? plasma?)*.

d. Now write the entire sequence of reactions described in (b) and (c). Note that the reactions show that increase in
CO_2 in the body (as when respiratory rate is slow) tends to cause a buildup of acid (H^+) in the body.

$$CO_2 + \text{_____} \rightarrow \text{_____} \rightarrow H^+ + \text{_____}$$

$$\qquad\qquad\qquad\qquad\qquad\qquad \downarrow \qquad\qquad\quad \downarrow$$

$$\qquad\qquad\qquad\qquad\qquad\quad \text{binds to} \quad\quad \text{shifts to plasma}$$

$$\qquad\qquad\qquad\qquad\qquad\qquad \text{Hb}$$

C4. List the major steps that occur in the lungs so that CO_2 can be exhaled. Note that as the red blood cells reach
lung capillaries, the same reactions you just studied in Checkpoint C3 occur, but in reverse.

C5. Write two paragraphs discussing the effects of a cigarettes on (a) the respiratory system; (b) other systems of the body.

■ **C6.** Match the terms in the box to related descriptions below. Use each answer once.

CAR.	Carcinogens	E. Emphysema
CB.	Chronic bronchitis	N. Nicotine
CO.	Carbon monoxide	

_____ a. Chemical that decreases airflow into/out of lungs by constricting terminal bronchioles.

_____ b. Chemical that binds to hemoglobin and reduces oxygen-carrying capacity of hemoglobin.

_____ c. Condition in which alveolar walls are progressively destroyed.

_____ d. Condition in which walls of airways are inflamed and secrete excessive amounts of mucus, leading to infections and "smokers cough."

_____ e. Cancer-causing agents such as benzopyrene, N-nitrosamines, radon, and polonium.

D. Control of respiration (pages 461–465)

■ **D1.** Describe events regulating the basic rhythm of breathing in this Checkpoint.

a. The *(pons? medulla? cerebrum?)* controls the basic pattern of breathing. Nerves in this part of the brainstem

stimulate the _____ nerves to the diaphragm with impulses lasting for about _____

seconds. When the diaphragm contracts, _____ -creasing the volume of the lungs, air *(enters? exits from?)* the lungs.

b. Once impulses from the inspiratory center cease, the diaphragm *(contracts? relaxes?)* for about _____ seconds. Lung walls then *(actively contract? passively recoil?)* as exhalation occurs.

c. What causes lungs to exhale more forcefully? Impulses from the _____ area of the

medullary rhythmicity area which stimulate muscles named _____ and

_____ muscles to increase intensity of exhalation.

■ **D2.** Answer these questions about respiratory control.

a. The main chemical change that stimulates respiration is increase in blood level of _____ , which is directly related to *(decrease in pO$_2$? increase in pCO$_2$?)* of blood.

b. Cells most sensitive to changes in blood CO$_2$ are located in the *(medulla? pons? aorta and carotid arteries?)*.

c. An increase in arterial blood pCO$_2$ is called _____ . Write an arterial pCO$_2$ value that is

hypercapnic. _____ mm Hg *(Even slight? Only severe?)* hypercapnia stimulates the respiratory system, leading to *(hyper? hypo?)*-ventilation.

d. State two locations of chemoreceptors sensitive to changes in pO_2.

(Even slight? Only large?) decreases in pO_2 level of blood stimulate these chemoreceptors and lead to hyper-

ventilation. Give an example of a pO_2 low enough to evoke such a response. _____ mm Hg.

e. Increase in body temperature (as in fever), as well as stretching of the anal sphincter, causes _____ -crease in the respiratory rate. Think of several other factors that have altered your rate of respiration.

f. Take a deep breath. Imagine the _____ receptors in your airways being stimulated. These cause (excitation? inhibition?) of the inspiratory and apneustic area, resulting in expiration. This reflex, known

as the _____ reflex, prevents overinflation of lungs.

E. Exercise, aging, common disorders, and medical terminology (pages 465–468)

■ **E1.** Describe effects of exercise on the respiratory system in this Checkpoint.

a. During exercise, cardiac output typically _____ -creases. As a result, blood flow through lungs _____ -

creases. As more blood and air enters lungs, surface areas of alveolar-capillary membranes _____ -crease to

permit a(n)_____ -crease in diffusion surface.

b. With moderate exercise, ventilation (depth? rate?) increases more. At the start of a vigorous exercise session, ventilation increases abruptly due to activation of (baroreceptors? proprioceptors?).

■ **E2.** Describe effects of aging on the respiratory system in this exercise.

a. Airways, alveoli, and chest wall typically become (stiffer? more elastic?)

b. Total lung capacity and vital capacity typically _____ -crease with aging.

c. Older adults are generally at (higher? lower?) risk for infections. List two or more factors that cause this effect.

■ **E3.** Describe impact of the respiratory system on different body systems in this exercise. Identify the affected system(s) in each case.

a. Lungs produce angiotensin converting enzyme (ACE): _____

b. Olfactory receptors are located in the nose: _____

c. Hairs in the nose and cilia in airways help to prevent infections: _____

d. Deep breathing by respiratory muscles assists with expulsion of abdominopelvic contents:

_____

e. Increased oxygen intake supplies active muscles: _____

■ **E4.** Match the condition with the correct description.

Ast. Asthma	D. Dyspnea	Ra. Rales
Asp. Aspiration	Em. Emphysema	Rhin. Rhinitis
CB. Chronic bronchitis	Epi. Epistaxis	TB. Tuberculosis
CF. Cystic fibrosis	P. Pneumonia	W. Wheeze

_____ a. Permanent overinflation of lungs due to loss of elasticity; rupture and merging of alveoli; characterized by "barrel chest"

_____ b. Inflammation of bronchi with excessive mucus production

_____ c. Acute infection or inflammation of alveoli that fill with fluid

_____ d. Spasms of small passageways with wheezing and dyspnea

_____ e. Caused by a species of *Mycobacterium;* lung tissue is destroyed and replaced with inelastic connective tissue

_____ f. Difficult, painful breathing; shortness of breath

_____ g. Nosebleed

_____ h. A hereditary disease in which thick mucus obstructs airways and certain digestive organs, such as pancreas and liver. Death typically occurs from lung disease.

_____ i. Inflammation of nasal mucosa which may cause "runny nose"

_____ j. Abnormal lung sounds related to excessive mucus or other fluid in lungs

_____ k. Inhalation of a substance other than air into airways

_____ l. A high-pitched sound during breathing; a result of partial airway obstruction, as in asthma

ANSWERS TO SELECTED CHECKPOINTS: CHAPTER 18

A3. (a) Label (1): pulmonary ventilation; alveoli; O_2, in (in), CO_2, ex (ex). (b) Air, blood; label (2): external respiration. (c) Blood; veins, blood in systemic capillaries, tissue cells; internal; label (3): internal respiration. (d) Artery, external, 2; pulmonary ventilation, 1. (e) Cellular. (f) Refer to Figure 18.10, page 458 in you text.

A4. (a–b) MN* → P * → L → T → BB → ADA. (c) Conducting.

A5. (a) Mucosa lining nose, septum, conchae, and sinuses, lacrimal drainage, and coarse hairs. (b) Olfactory region lies superior to superior nasal conchae. (c) Sounds resonate in nose and paranasal sinuses.

A7. (a) NP. (b) OP. (c) NP. (d) OP. (e) LP.

A8. (a) Thyroid; anterior. (b) Glottis; epiglottis; arytenoid. (c) Cricoid; inferior.

A10. (a) Lower; windpipe, bronchi. (b) T E C. (c) Mucous, ciliated; cartilage; prevent collapse of the airway during exhalation.

A12. Do not; muscle spasms can collapse airways that lack cartilaginous support.

A14. (a) T R A. (b) Respiratory. See KEY to Figure LG 18.3A. (c) 1, alveolar cells; 2, epithelial basement membrane; 3, capillary basement membrane; 4, capillary endothelium. (d) Simple squamous epithelial; reduces tendency of alveoli to collapse; debris.

B1. (a) Diaphragm; phrenic, cervical. (b) Flatter; injury to the cervical spinal cord or phrenic nerves; pregnancy, obesity, a full stomach. (c) External; sternocleidomastoid, scalenes, and pectoralis minor.

B2. (a) Alveolar pressure, 760; the same as. (b) In; de, 758. (c) Atmosphere, alveoli; into; inhalation; 760. (d) Active, passive; abdominal, internal intercostals. (e) Surfactant and negative intrapleural pressure (756 mm Hg). (f) Atelectasis; chest wounds or surgery following the accident may cause pneumothorax or hemothorax (blood in the pleural cavity). Respiratory infections can also block airways and lead to atelectasis.

B3. 7200mL/min (= 15 breaths/min × 480 mL/breath).

B4. 70 percent (350 mL/500 mL); dead space, 150.

B5. (a) TV. (b) FRC. (c) ERV. (d) RV. (e) VC. (f) TLC.

B6. (a) 500 (0.5). (b) 6000, 6, 4200 (4.2). (c) 4800, 3100.

B8. (a) 760. (b) N_2, O_2, CO_2 and H_2O vapor. (c) About

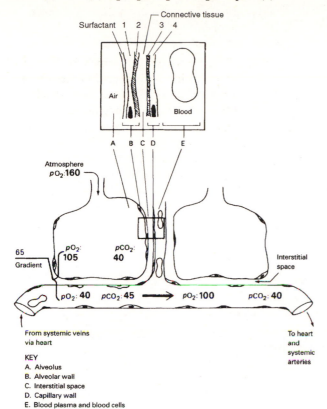

Figure LG 18.3A Diagram of alveoli and pulmonary capillary. Insert enlarges alveolar–capillary membrane.

160 mm Hg = pO_2.

B9. (a) Partial; see Figure LG 18.3A. (b) 105 − 40 = 65. (c) Large; thin; increased. (d) About the same as; 100 to 105. (e) See Figure LG 18.3A. (f) No. Typical radial arterial values are same as for alveoli or blood leaving lungs: $pO_2 = 100$, $pCO_2 = 40$. Jenny's values indicate inadequate gas exchange. (g) Lower; lower. (h) Inferior vena cava: $pO_2 = 40$; $pCO_2 = 45$; aorta: $pO_2 = 100$; $pCO_2 = 40$.

C1. (a) Oxyhemoglobin; 1.5. (b) Iron; HbO_2; saturated; increase. (c) Increase in temperature, pCO_2, and acidity.

C2. 200; oxygen-carrying capacity is drastically reduced in carbon monoxide poisoning.

C3. (a) 78, 13, 9. (b) H_2CO_3; carbonic anhydrase. (c) Hemoglobin (as H·Hb); HCO_3^-, plasma. (d) $CO_2 + H_2O \rightarrow H_2CO_3$ (carbonic acid) $\rightarrow H^+ + HCO_3^-$ (bicarbonate).

C6. (a) N. (b) CO. (c) E. (d) CB. (e) CAR.

D1. (a) Medulla; phrenic, 2; in, enters. (b) Relaxes, 3; passively recoil. (c) Expiratory, internal intercostals and abdominal.

D2. (a) H^+, increase in pCO_2. (b) Medulla. (c) Hypercapnia; any value higher than 40; even slight, hyper. (d) Aortic and carotid bodies; only large; usually below 50. (e) In; stimulation from limbic system (such as gasping to emotional responses of surprise, horror), proprioceptor stimulation or pain (as with sudden movements), and irritation of airways as by intense fumes. (f) Stretch; inhibition.

E1. (a) In; in; in, in. (b) Depth; proprioceptors.

E2. (a) Stiffer. (b) De. (c) Higher; decreased action of cilia and macrophages, and decreased thoracic cage movements that may cause more mucus (warm moist environment for microbes) to be retained in lungs.

E3. (a) Endocrine and cardiovascular, since ACE increases blood pressure by stimulating production of the vasoconstrictor hormone, angiotensin II. (b) Nervous/sensory. (c) Immune. (d) Digestive (defecation) and reproductive (labor and delivery). (e) Muscular.

E4. (a) Em. (b) CB. (c) P. (d) Ast (e) TB. (f) D. (g) Epi. (h) CF. (i) Rhin. (j) Ra. (k) Asp (l) W.

CRITICAL THINKING: CHAPTER 18

1. Contrast location, structure, and function of each of these parts of the respiratory system: nasopharynx/oropharynx, epiglottis/thyroid cartilage, primary bronchi/bronchioles, parietal pleura/visceral pleura.
2. After a larynx is removed (for example, due to cancer), what other structures would help the laryngectomee to speak?
3. Define and state one or more examples of causes of each of the following clinical conditions: asphyxia, aspiration, dyspnea, pulmonary edema, rales, and wheezing.
4. Contrast primary and secondary lines of defense against pathogens.
5. Discuss why smokers are at higher risk than nonsmokers for cancers in a variety of organs (mouth, cheek, throat, esophagus, stomach, pancreas, colon, kidneys, urinary bladder, cervical, and breast, as well as lungs).

MASTERY TEST: ■ CHAPTER 18

Questions 1–2: Arrange the answers in correct sequence.

_____ _____ _____ _____ _____   1. From superior to inferior:
- A. Bronchioles
- B. Bronchi
- C. Larynx
- D. Pharynx
- E. Trachea

_____ _____ _____ _____ _____ _____   2. Pathway of inspired air:
- A. Alveolar ducts
- B. Bronchioles
- C. Secondary bronchi
- D. Primary bronchi
- E. Tertiary bronchi
- F. Alveoli

Questions 3–4: Circle the letter preceding the one best answer to each question.

3. Which of these values (in mm Hg) would be normal for pO_2 of blood in the femoral artery?
   - A. 40
   - B. 45
   - C. 100
   - D. 160
   - E. 760
   - F. 0

4. Choose the correct formula for carbonic acid:
   - A. HCO_3^-
   - B. H_3CO_2
   - C. H_2CO_3
   - D. HO_3C_2
   - E. H_2C_3O

Questions 5–10: Circle T (true) or F (false). If the statement is false, change the underlined word or phrase so that the statement is correct.

T   F   5. When chemoreceptors sense <u>increase in pCO_2 or increase in acidity of blood (H^+)</u>, respiratory rate normally is stimulated.

T   F   6. The cricoid cartilage is located <u>inferior</u> to the thyroid cartilage.

T   F   7. The pneumotaxic and apneustic areas controlling respiration are located in the <u>pons</u>.

T   F   8. Most CO_2 is carried in the blood in the form of <u>bicarbonate</u>.

T   F   9. Vital capacity is normally <u>larger than</u> expiratory reserve volume.

T   F   10. The pO_2 and pCO_2 of blood leaving the lungs <u>are normally about the same</u> as pO_2 and pCO_2 of alveolar air.

Questions 11–15: Fill-ins. Complete each sentence with the word or phrase that best fits.

_____ 11. The process of exchange of gases between alveolar air and blood in pulmonary capillaries is known as _____ .

_____ 12. Take a normal breath and then let it out. The amount of air left in your lungs is the capacity called _____ and it usually measures about _____ mL.

_____ 13. _____ is a term that means "difficult or labored breathing."

_____ 14. The epiglottis, thyroid, and cricoid cartilages are all parts of the _____ .

_____ 15. _____ is a chemical that increases inflatability (compliance) of lungs.

ANSWERS TO MASTERY TEST: ■ CHAPTER 18

Arrange
1. D C E B A
2. D C E B A F

Multiple Choice
3. C
4. C

True or False
5. T
6. T
7. T
8. T
9. T
10. T

Fill-ins
11. External respiration (or pulmonary respiration, or diffusion)
12. Functional residual capacity (FRC), 1800 in females and 2400 in males
13. Dyspnea
14. Larynx
15. Surfactant

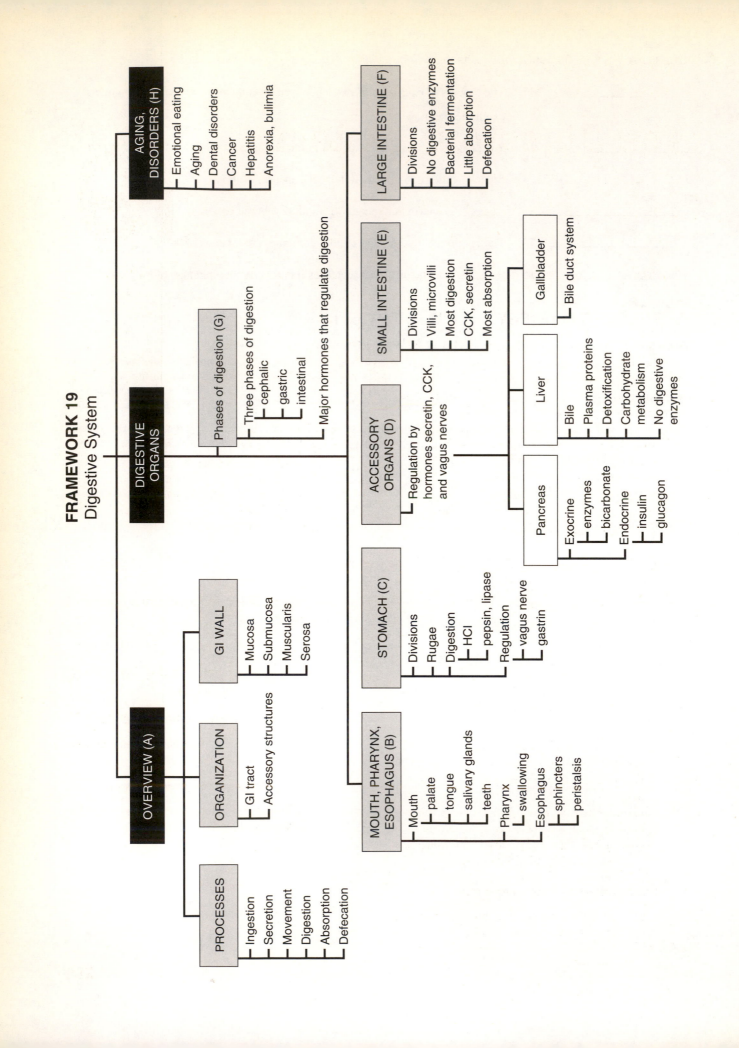

FRAMEWORK 19
Digestive System

DIGESTIVE ORGANS

AGING, DISORDERS (H)
- Emotional eating
- Aging
- Dental disorders
- Cancer
- Hepatitis
- Anorexia, bulimia

Phases of digestion (G)
- Three phases of digestion
  - cephalic
  - gastric
  - intestinal
- Major hormones that regulate digestion

LARGE INTESTINE (F)
- Divisions
- No digestive enzymes
- Bacterial fermentation
- Little absorption
- Defecation

SMALL INTESTINE (E)
- Divisions
- Villi, microvilli
- Most digestion
- CCK, secretin
- Most absorption

ACCESSORY ORGANS (D)
- Regulation by hormones secretin, CCK, and vagus nerves

Pancreas
- Exocrine
  - enzymes
  - bicarbonate
- Endocrine
  - insulin
  - glucagon

Liver
- Bile
- Plasma proteins
- Detoxification
- Carbohydrate metabolism
- No digestive enzymes

Gallbladder
- Bile duct system

STOMACH (C)
- Divisions
- Rugae
- Digestion
  - HCl
  - pepsin, lipase
- Regulation
  - vagus nerve
  - gastrin

OVERVIEW (A)

PROCESSES
- Ingestion
- Secretion
- Movement
- Digestion
- Absorption
- Defecation

ORGANIZATION
- GI tract
- Accessory structures

GI WALL
- Mucosa
- Submucosa
- Muscularis
- Serosa

MOUTH, PHARYNX, ESOPHAGUS (B)
- Mouth
  - palate
  - tongue
  - salivary glands
  - teeth
- Pharynx
  - swallowing
- Esophagus
  - sphincters
  - peristalsis

The Digestive System

CHAPTER 19

Food comes in very large pieces (like whole oranges, stalks of broccoli, and slices of bread) that must fit into very small spaces in the human body (like liver cells or brain cells). The digestive system makes such change possible. Foods are minced and enzymatically degraded so that absorption into blood en route to cells becomes a reality. The gastrointestinal (GI) tract provides a passageway complete with mucous glands to help food slide along, muscles that propel food forward and muscles (sphincters and valves) that regulate flow, and secretions that break apart foods or modify pH to suit local enzymes. Accessory structures such as teeth, tongue, pancreas, and liver lie outside the GI tract, yet each contributes to the mechanical and chemical dismantling of food from bite-sized to right-sized pieces. The ultimate fate of the absorbed products of digestion is the story line of metabolism (in Chapter 20).

You will find food for thought in the Chapter 19 Framework and its terminology. As you begin this chapter, carefully examine the Chapter 19 Topic Outline and check off each objective after you meet it.

TOPIC OUTLINE AND OBJECTIVES

A. Overview of the digestive system

☐ 1. Identify the organs of the digestive system and their basic functions.

☐ 2. Describe the four layers that form the wall of the gastrointestinal tract.

B. Gastrointestinal tract: mouth, pharynx, and esophagus

☐ 3. Identify the locations of the salivary glands, and describe the functions of their secretions.

☐ 4. Describe the structure and functions of the tongue.

☐ 5. Identify the parts of a typical tooth, and compare deciduous and permanent dentitions.

☐ 6. Describe the location, structure, and functions of the pharynx and the esophagus.

C. Gastrointestinal tract: stomach

☐ 7. Describe the location, structure, and functions of the stomach.

D. Accessory structures: pancreas, liver, and gallbladder

☐ 8. Describe the location, structure, and functions of the pancreas.

☐ 9. Describe the location, structure, and functions of the liver and gallbladder.

E. Gastrointestinal tract: small intestine

☐ 10. Describe the location, structure, and functions of the small intestine.

F. Gastrointestinal tract: large intestine

☐ 11. Describe the location, structure, and functions of the large intestine.

G. Phases of digestion

☐ 12. Describe three phases of digestion.
☐ 13. Describe the major hormones that regulate digestive activities

H. Aging and the digestive system, common disorders, and medical terminology

☐ 14. Describe the effects of aging on the digestive system.

WORDBYTES

Study each wordbyte, its meaning, and an example of its use in a term. Check your understanding by jotting meanings of wordbytes in margins. Identify other examples of terms that contain these wordbytes as you continue throught the text and *Learning Guide*.

Wordbyte	Meaning	Example(s)	Wordbyte	Meaning	Example(s)
amyl-	starch	*amyl*ase	gingiv-	gums	*gingiv*itis
-ase	enzyme	malt*ase*	ileo-	ileum	*ileo*cecal valve
caec-, cec-	blind	*cec*um	jejun-	empty	*jejun*ileostomy
chole-, cholecyst-	gallbladder	*cholecyst*itis	or-	mouth	*or*al
chym-	juice	*chym*otrypsin	-ose	sugar	lact*ose*
dent-	tooth	*dent*ures	-rhea	to flow	diar*rhea*
-ectomy	removal of	append*ectomy*	-stomy	mouth, opening	colo*stomy*
entero-	intestine	*entero*kinase	taen-	ribbon	*taen*ia coli
gastr-	stomach	*gastr*in			

CHECKPOINTS

A. Overview of the digestive system (pages 473–476)

A1. Explain why food is vital to life. Give two specific examples of uses of foods in the body.

■ **A2.** Cover the key and identify all digestive organs on Figure LG 19.1. Then color the structures with color code ovals; these five structures are all *(parts of the gastrointestinal tract? accessory structures?)*.

Food actually passes through *(organs of the GI tract? accessory organs?)*.

■ **A3.** List the six basic activities of the digestive system.

_____ _____ _____

_____ _____ _____

■ **A4.** _____ digestion occurs by action of enzymes (such as those in saliva) and intestinal

secretion, whereas _____ digestion involves action of the teeth and muscles of the stomach and intestinal wall.

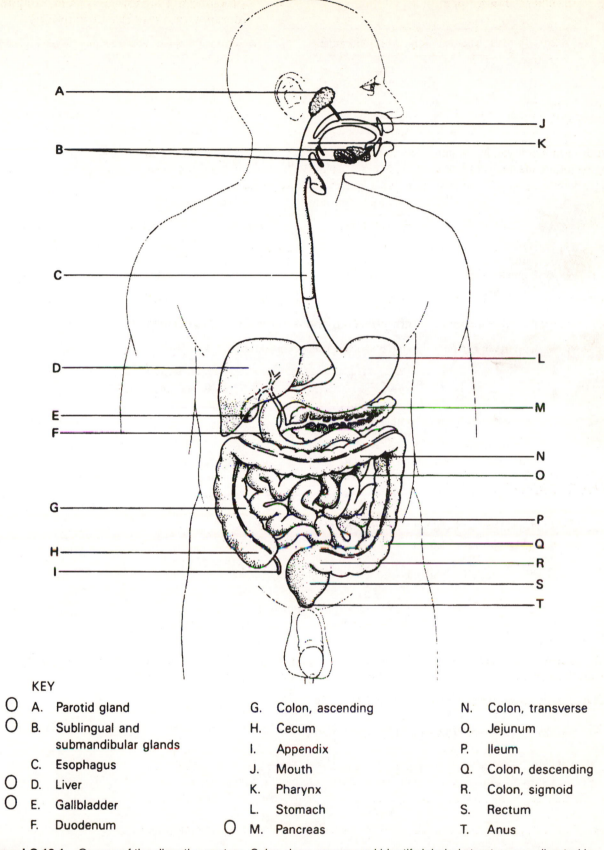

Figure LG 19.1 Organs of the digestive system. Color, draw arrows, and identify labeled structures as directed in Checkpoints A2, B2, C1, and F1.

■ **A5.** Match the names of layers of the GI wall with the correct description.

Muc. Mucosa	Mus. Muscularis	Ser. Serosa	Sub. Submucosa

_____ a. Also known as the peritoneum, it has parietal and visceral layers

_____ b. Innermost lining adjacent to lumen with epithelium in direct contact with food

_____ c. Connective tissue containing glands, nerves, blood, and lymph vessels

_____ d. Consists of an inner circular layer and an outer longitudinal layer

■ **A6.** Match the names of these peritoneal extensions with the correct descriptions.

G. Greater omentum	M. Mesentery

_____ a. Binds small intestine to posterior abdominal wall; provides route for blood and lymph vessels and nerves to reach small intestine

_____ b. "Fatty apron"; covers and helps prevent infection in small intestine

_____ c. Considerable fat in this structure contributes to a "beer belly"

B. Gastrointestinal tract: mouth, pharynx, and esophagus   (pages 476–480)

■ **B1.** Match the parts of the mouth in the box with descriptions below.

G. Gingiva	PL. Periodontal ligament
LF. Lingual frenulum	U. Uvula
Pap. Papillae	

_____ a. U-shaped projection of the soft palate

_____ b. Anchors tooth to bone

_____ c. Projections on the tongue covered with taste buds

_____ d. Attachment of the tongue to the floor of the mouth

_____ e. Gums

B2. Identify the three salivary glands on Figure LG 19.1 and visualize their locations on yourself.

■ **B3.** Complete this exercise about salivary glands.

a. Which glands are largest? *(Parotid? Sublingual? Submandibular?)*

b. State three functions of saliva.

c. Salivation is stimulated by *(sympathetic? parasympathetic?)* nerves. These nerve fibers are contained within

cranial nerves _____ and _____ .

d. Recall a stressful situation when you experienced "dry mouth." Explain what accounts for this condition.

■ **B4.** Arrange the following in correct sequence:

_____ _____ _____ a. From most superficial to deepest:
A. Root   B. Neck   C. Crown

_____ _____ _____ b. From most superficial to deepest within the crown of a tooth:
A. Pulp cavity   B. Dentin   C. Enamel

_____ _____ _____ c. From most superficial to deepest within a root:
A. Pulp cavity   B. Dentin   C. Cementum

_____ _____ _____ _____ d. From most anterior to most posterior in location:
A. Molars   B. Premolars   C. Cuspids   D. Incisors

_____ _____ _____ e. From greatest to least in number in one set of permanent teeth:
A. Molars   B. Incisors   C. Cuspids

■ **B5.** Salivary amylase is an enzyme that digests _____ . *(All? Only some?)* of the starch ingested is broken down by the time food leaves the mouth.

■ **B6.** Match each stage of deglutition with the correct description.

E. Esophageal   P. Pharyngeal   V. Voluntary

_____ a. Soft palate and epiglottis close off respiratory passageways.
_____ b. Tongue pushes food back into oropharynx.

_____ c. Peristaltic contractions push bolus from
pharynx to stomach.

■ **B7.** Failure of the lower esophageal sphincter to close results in the sensation of _____ . Consequently, the esophageal lining may be irritated by *(acidic? basic?)* contents of the stomach that enter the

esophagus. Resulting pain may be confused with _____ pain.

B8. Summarize digestion in the mouth, pharynx, and esophagus by completing parts *a* and *b* of Table LG 19.1.

C. Gastrointestinal tract: stomach   (pages 480–483)

■ **C1.** Refer to Figure LG 19.1 and arrange these regions of the stomach according to the pathway of food from first to last:

Body   Cardia   Fundus   Pylorus

_____ → _____ → _____ →

_____

Table LG 19.1 Summary of digestive organs and processes

Digestive Organs	Carbohydrate	Protein	Lipid	Other Functions
a. Mouth, salivary glands	Salivary amylase: digests starch to maltose			
b. Pharynx, esophagus				Deglutition, peristalsis
c. Stomach				1. Secretes intrinsic factor and HCl 2. Produces hormone gastrin 3. Secretes mucus
d. Pancreas			Pancreatic lipase: digests about 80% of fats	
e. Intestinal juices				
f. Liver	No enzymes for digestion of carbohydrates			
g. Large intestine	No enzymes for digestion of carbohydrates	No enzymes for digestion of proteins	No enzymes for digestion of fats	

■ **C2.** Complete this table about gastric secretions.

Name of Cell	Type of Secretion	Function of Secretion
a. Chief		
b. Mucous neck		
c.	HCl	
d.	Gastrin	Omit

■ **C3.** How does the structure of the stomach wall differ from that in other parts of the GI tract?

a. Mucosa

b. Muscularis

■ **C4.** Describe the following effects of the process of "gastric emptying."

a. Explain how gastric emptying prevents excessive amounts of chyme from overloading the duodenum.

b. Gastric emptying is fastest after a meal rich in (*carbohydrates? lipids? proteins?*), and slowest after a meal rich in _____.

■ **C5.** Answer these questions about vomiting.

a. Identify the two strongest stimuli for vomiting.

b. Explain how prolonged vomiting can lead to serious disturbances in homeostasis.

■ **C6.** Answer these questions about chemical digestion in the stomach.

a. Pepsin is most active at very (*acid? alkaline?*) pH.

b. State two factors that enable the stomach to digest protein without digesting its own cells (which are composed largely of protein).

c. If mucus fails to protect the gastric lining, the condition known as _____ may result. (*Hint:* See page 497 of the text.)

295

■ **C7.** The stomach is responsible for *(much? little?)* absorption of foods. What types of substances are absorbed by the stomach?

C8. Complete part *c* of Table LG 19.1, describing the role of the stomach in digestion.

D. Accessory structures: pancreas, liver, and gallbladder (pages 483–486)

D1. Refer to Figure LG 19.2 and color all of the structures with color code ovals according to colors indicated.

■ **D2.** Complete these statements about the pancreas.

a. The pancreas lies posterior to the _____ . Shaped roughly like a fish, its "head" lies in the

curve of the _____ , with its "tail" nudged up next to the _____ .

b. The pancreas produces hormones from its *(endo? exo?)*-crine glands and pancreatic juice from

_____ -crine glands. Pancreatic juice is *(acidic? alkaline?)* in nature because it contains

the chemical sodium _____ . Pancreatic juice neutralizes the chyme, which is *(acidic? alkaline?)* due to gastric HC1.

c. Trypsin is one enzyme that digests *(carbohydrates? fats? proteins?)*. Name two other pancreatic enzymes that

digest proteins. (Hint: refer to Table 19.1, page 489 in the text.) _____

_____ _____ All of these enzymes are produced in the *(active? inactive?)* state. Where are they activated?

d. Name the two nucleic acid–digestive enzymes produced by the pancreas.

e. All exocrine secretions of the pancreas empty into ducts (_____ and _____ on Figure LG 19.2). These

empty into the _____ .

f. The endocrine portions of the pancreas are know as the _____

_____ . List the two major hormones they secrete: _____

_____ Typical of all hormones, these pass into *(ducts? blood vessels?)*, specifically into

vessels that empty into the _____ vein.

g. Pancreatic cancer is *(rarely? usually?)* fatal. List five factors linked to development of this type of cancer.

D3. Fill in part *d* of Table LG 19.1, describing the role of the pancreas in digestion.

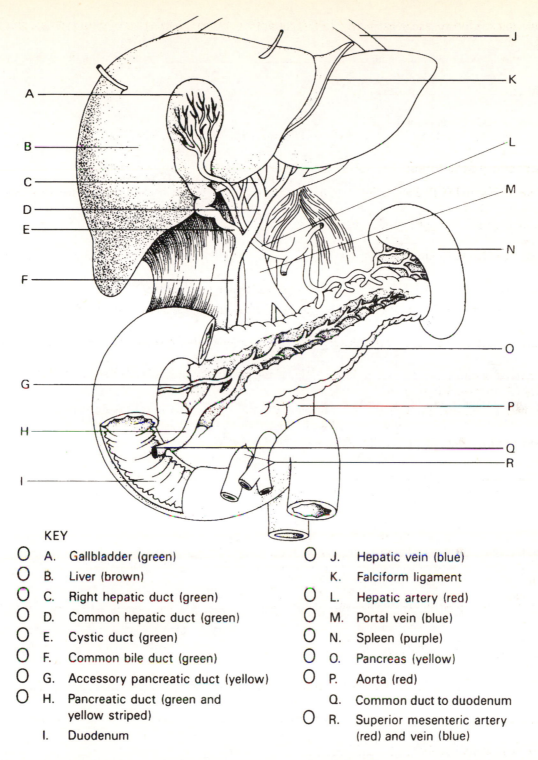

KEY

○ A.	Gallbladder (green)
○ B.	Liver (brown)
○ C.	Right hepatic duct (green)
○ D.	Common hepatic duct (green)
○ E.	Cystic duct (green)
○ F.	Common bile duct (green)
○ G.	Accessory pancreatic duct (yellow)
○ H.	Pancreatic duct (green and yellow striped)
I.	Duodenum

○ J.	Hepatic vein (blue)
K.	Falciform ligament
○ L.	Hepatic artery (red)
○ M.	Portal vein (blue)
○ N.	Spleen (purple)
○ O.	Pancreas (yellow)
P.	Aorta (red)
Q.	Common duct to duodenum
○ R.	Superior mesenteric artery (red) and vein (blue)

Figure LG 19.2 Liver, gallbladder, pancreas, and duodenum, with associated blood vessels and ducts. (Stomach has been removed.) Refer to Checkpoints D1, D2, and D6; color and draw arrows as directed.

■ **D4.** Complete this exercise about the liver.

a. This organ weighs about _____ kg (_____ lb.). It lies in the upper *(right? left?)* portion of the abdomen. The

falciform _____ separates the right and left lobes of the liver.

b. Blood enters the liver via vessels named _____ and _____ . In

liver lobules, blood mixes in channels called _____ before leaving the liver via vessels

named _____ .

■ **D5.** Complete this Checkpoint about bile.

a. Name the primary bile pigment in bile: _____ , which is a breakdown product of hemo-

globin in _____ cells. After bilirubin passes into the intestine, this chemical contributes

to the normal color of _____ .

b. Two functions of bile are _____ of fats and _____ of products of
fat digestion (and fat-soluble vitamins).

c. Bile is stored in the organ named the _____ . If bile crystallizes here, it can form

_____ , that can obstruct bile ducts. Name two clinical procedures used to treat this

condition: shock waves (or _____) to break apart stones so they can pass through ducts,

and _____ , which is surgical removal of the gallbladder.

■ **D6.** Complete Figure LG 19.2 as directed.

a. Identify the pathway of bile from liver to intestine by following the structures you colored green in that figure:
C D F H. The general direction of bile is from *(superior to inferior? inferior to superior?)*.

b. Now draw arrows to show the direction of blood through the liver. In order to reach the inferior vena cava, blood
must flow *(superiorly? inferiorly?)* through the liver.

■ **D7.** Describe other functions of the liver in this Checkpoint.

a. The liver can lower blood glucose levels by converting glucose to the polysaccharide

_____ or to triglycerides. Conversely, the liver can raise blood levels of glucose by

converting _____ , _____ , and _____ into glucose.

b. List three chemicals categorized as lipids that are produced by the liver.

c. Identify two or more types of chemicals that are detoxified by the liver.

d. Name several plasma proteins synthesized by the liver.

e. The liver stores the four fat-soluble vitamins named _____ , _____ , _____ , and _____ . It also stores a

vitamin necessary for erythropoiesis: vitamin _____ . And the liver cooperates with kidneys and skin to acti-

vate vitamin _____ .

f. Note that the liver *(does? does not?)* secrete digestive enzymes.

■ **D8.** Complete part *f* of Table LG 19.1.

298

E. Gastrointestinal tract: small intestine   (pages 486–491)

■ **E1.** Describe the structure of the intestine in this Checkpoint.

a. The average diameter of the small intestine is:
   A. 2.5 cm (1 inch)   B. 5.0 cm (2 inches)

b. In a living person, its average length is about:
   A. 3 m (10 feet)   B. 6 m (20 feet)

c. Name its three main parts (in sequence from first to last):

   _____ → _____ → _____

■ **E2.** Complete the table about types of cells within the wall of the small intestine and secretions of each. Cell types *a-e* are found in the intestinal mucosa.

Type of cell	Type of secretion
a. Goblet cells	
b.	Intestinal juice (enzymes) (See Table 19.1, page 489, for specifics)
c. S cells	
d.	Cholecystokinin (CCK)
e. K cells	
f. Duodenal glands of the submucosa	

■ **E3.** Name three modifications of the intestinal wall that increase surface area for digestion and absorption.

Which of these structural features is best described as "fingerlike projections up to 1 mm high" that give the

intestinal lining a velvetlike appearance? _____ Epithelial cells on the surface of villi *(do? do not?)* synthesize intestinal enzymes.

What is a lacteal?

E4. Contrast these two types of movements: *segmentation* and *peristaltic waves*.

E5. Complete the description of functions of intestinal juices by filling in part *e* of Table LG 19.1. Then review digestion of each of the three major food groups by reading your columns vertically. (Hint: Refer to text Table 19.1, page 489.)

■ **E6.** Describe the absorption of end products of digestion in this exercise.

a. Almost all absorption takes place in the *(large? small?)* intestine. Glucose, short-chain fatty acids, and some

amino acids move into epithelial cells lining the intestine by _____ transport. Glucose

and other monosaccharides are then transported into capillaries by the process of

_____ .

b. Long-chain fatty acids and monoglycerides first combine with _____ salts in the form of
(micelles? chylomicrons?). This enables fatty acids and monoglycerides to enter epithelial cells in the intestinal

lining via exocytosis and then enter vessels named _____ .

c. Aggregates of fats called _____ are coated with proteins. These pass from epithelial cells

lining villi and into tiny lymph vessels known as _____ and ultimately into larger lymph

vessels, the _____ duct, and into blood in the _____ vein.

Ultimately, chylomicrons are removed from the blood and stored in the _____ .

d. About _____ liters (or quarts) of fluids are ingested or secreted into the GI tract each day. Most of this fluid is
reabsorbed into blood capillaries within the walls of the *(small? large?)* intestine.

e. A *clinical challenge.* Which vitamins are absorbed with the help of bile? *(Water-soluble? Fat-soluble?)* Circle the
fat-soluble vitamins: A B_{12} C D E K.

E7. List several manifestations of *lactose intolerance.*

Explain the cause of this disorder.

F. Gastrointestinal tract: large intestine   (pages 491–493)

F1. Identify the regions of the large intestine in Figure LG 19.1. Draw arrows to indicate direction of movement of
intestinal contents.

■ **F2.** The total length of the large intestine is about _____ m (_____ feet). More than 90 percent of its length

consists of the part known as the _____ .

Arrange these parts of the large intestine in correct order in the pathway of wastes:

_____ → _____ → _____ → _____ → _____ → _____ → _____

AnC. Anal canal	R. Rectum
AsC. Ascending colon	SC. Sigmoid colon
C. Cecum	TC. Transverse colon
DC. Descending colon	

■ **F3.** Fill in the blank with the correct term related to the large intestine.

   a. Valve between the small intestine and the large intestine: _____

   b. Blind-ended tube attached to the cecum: _____

   c. Benign growth of the large intestinal mucosa; some have the potential of becoming malignant, and may be re-moved, for example, during a colonoscopy: _____

■ **F4.** Contrast types of movements of the large intestine in this exercise.

   a. Peristalsis in the large intestine normally occurs at a *(faster? slower?)* pace than in other parts of the gastrointestinal tract.

   b. Mass peristalsis occurs about _____ times a day, and moves wastes from the colon into the _____ .

   F5. Does the large intestine produce enzymes that help with digestion? *(Yes? No?)*
   Explain how bacteria in the colon contribute to the following processes:

   a. Fermentation of carbohydrates

   b. Synthesis of vitamins

   c. Breakdown of bilirubin

■ **F6.** Explain how the large intestine produces gases.

   Define the term *flatulence.*

■ **F7.** List the chemical components of feces.

   Which chemicals contribute to the odor of feces?

F8. Explain the roles of the following in the defecation reflex: *stretch receptors, parasympathetic nerves, internal anal sphincter, diaphragm* and *abdominal muscles, external anal sphincter.*

F9. Contrast these two conditions; include causes and consequences of each: *diarrhea* and *constipation.*

F10. Complete part *g* of Table LG 19.1, describing the roles of the large intestine in digestion.

G. Phases of digestion (pages 494-495)

G1. Match the correct phase of digestion in the box with the related description below.

C. Cephalic	G. Gastric	I. Intestinal

_____ a. Cranial nerves VII and IX activate salivary glands, and cranial nerve X stimulates glands in the wall of the stomach.

_____ b. Gastrin stimulates gastric glands to secrete gastric juice, increases gastric motility, strengthens contraction of the lower esophageal sphincter to prevent acid reflux, and relaxes the pyloric sphincter to promote gastric emptying.

_____ c. Hormones CCK and secretin stimulate secretion of pancreatic juice rich in enzymes and bicarbonate, release of bile from the gallbladder, and slow gastric emptying.

■ **G2.** *For extra review* of roles of GI organs in digestion, identify which chemicals in the list in the box are made by each of the following organs.

Alb. Albumin	Ins. Insulin
Amy. Amylase	L. Lipase
B. Bile	M. Mucus
CCK. Cholecystokinin	MLS. Maltase, lactase, sucrase
FP. Fibrinogen and prothrombin	N. Nucleases
Gas. Gastrin	Ps. Pepsin
GIP. Gastric inhibitory peptide	Pt. Peptidase
Glu. Glucagon	S. Secretin
Gly. Glycogen	TCC. Trypsin, chymotrypsin,
HCl. Hydrochloric acid	carboxypeptidase
IF. Intrinsic factor	

_____ a. Salivary glands and tongue (two answers)

_____ b. Pharynx and esophagus (one answer)

_____ c. Stomach (five answers)

_____ d. Pancreas (six answers)

_____ e. Small intestine (six answers)

_____ f. Liver (four answers)

■ **G3.** *For extra review* of secretions involved with digestion, match names of secretions listed for Checkpoint G2 with descriptions below. Blank lines follow some descriptions. On these indicate whether the secretion is classified as an enzyme (E), hormone (H), or neither of these (N). The first one is done for you.

CCK a. Causes contraction of gallbladder (__H__)

_____ b. Stimulates production of alkaline pancreatic fluids (_____)

_____ c. Stimulates pancreas to produce secretions rich in enzymes such as lipase and amylase (_____)

_____ d. Increases gastric activity (secretion and motility) (_____)

_____ e. Protein-digesting enzymes (three answers)

_____ f. Activates pepsinogen to pepsin (_____)

_____ g. A storage form of carbohydrate

_____ h. Starch-digesting enzymes secreted by salivary glands and pancreas

_____ i. Intestinal enzymes that complete carbohydrate breakdown, resulting in simple sugars

_____ j. Digest DNA and RNA (_____)

H. Emotional eating, aging, common disorders, and medical terminology of the digestive system   (pages 494-498)

H1. Describe how food can provide a biochemical "fix" for negative emotions and describe possible consequences.

H2. Contrast these conditions: *binge-eating* and *bulimia*.

H3. *A clinical challenge.* List physical changes with aging that may lead to a decreased desire to eat among the elderly popluation.

a. Related to the upper GI tract (above the stomach)

b. Related to the lower GI tract (stomach and beyond)

H4. Write a paragraph relating digestive activities to normal functioning of each of these systems: cardiovascular, urinary, skeletal, endocrine.

■ **H5.** Which of the following is an inflammation? *(Diverticulitis? Diverticulosis?)*

■ **H6.** Contrast causes of two forms of hepatitis:

a. Hepatitis A

b. Hepatitis B

■ **H7.** Choose the disorders or procedures listed in the box with related descriptions below.

AN. Anorexia nervosa	Fla. Flatulence
Cho. Cholecystitis	IBD. Irritable bowel disorder
Colit. Colitis	IBS. Irritable bowel syndrome
Colos. Colostomy	PD. Periodontal disease
Con. Constipation	PUD. Peptic ulcer disease

_____ a. Incision of the colon, creating artificial anus

_____ b. Inflammation and degeneration of gum and tooth structures

_____ c. Inflammation of the colon

_____ d. Chronic disorder with body-image disturbance and self-induced weight loss

_____ e. Crohn's disease and ulcerative colitis, involving inflammation of the small or large intestine; may lead to rectal bleeding

_____ f. Disease of the entire gastrointestinal tract with cramping and alternating diarrhea and constipation.

_____ g. Infrequent or difficult defecation

_____ h. Inflammation of the gallbladder

A2. Accessory structures; organs of the GI tract.
A3. Ingestion, secretion, motility, digestion, absorption, defecation.
A4. Chemical, mechanical.
A5. (a) Ser. (b) Muc. (c) Sub. (d) Mus.
A6. (a) M. (b) G. (c) G.
B1. (a) U. (b) PL. (c) Pap. (d) LF. (e) G.
B3. (a) Parotid; (b) Dissolving medium for foods, lubrication, source of lysozyme and amylase. (c) Parasympathetic; VII, IX. (d) Sympathetic nerves decrease salivary secretions.
B4. (a) C B A. (b) C B A. (c) C B A. (d) D C B A. (e) A B C.
B5. Starch; only some.
B6. Swallowing. (a) P. (b) V. (c) E.
B7. Heartburn; acidic; heart.
C1. Cardia → Fundus → Body → Pylorus.
C2.

Name of Cell	Type of Secretion	Function of Secretion
a. Chief	Pepsionogen	Precursor of pepsin
b. Mucous neck	Mucus	Protects gastric lining from acid and pepsin
c. Parietal	HCl Intrinsic factor	Activates pepsinogen to pepsin. Facilitates absorption of vitamin B$_{12}$
d. G cells	Gastrin	Omit

C3. (a) Arranged in rugae, and contains gastric pits lined with gastric glands. (b) It has three, rather than two, layers of smooth muscle; the extra one is an oblique layer located inside the circular layer.
C5. (a) Irritation and distension of the stomach. (b) Loss of electrolytes (such as HCl) as well as fluids can lead to fluid, electrolyte, and acid-base imbalances.
C6. (a) Acid. (b) Pepsin is released in the inactive state (pepsinogen); mucus protects the stomach lining from pepsin. (c) Ulcer.
C7. Little; some water, electrolytes, alcohol, some fatty acids, and certain drugs such as aspirin.
D2. (a) Stomach; duodenum, spleen. (b) Endo, exo; alkaline, bicarbonate; acidic. (c) Proteins; chymotrypsin, carboxypeptidase; inactive (so they don't digest the pancreas itself); duodenum. (d) RNA-ase and DNA-ase. (e) G, H; duodenum. (f) Pancreatic islets (of Langerhans); insulin and glucagon; blood vessels, portal. (g) Usually; fatty foods, high alcohol consumption, genetic factors, smoking, and chronic pancreatitis.
D4. (a) 1.4 (3.0); right; ligament. (b) Hepatic artery, portal vein; sinusoids, hepatic veins.
D5. (a) Bilirubin, red blood; feces. (b) Emulsification, absorption. (c) Gallbladder; gallstones;

lithotripsy, cholecystectomy.
D6. (a) Superior to inferior. (b) Draw arrows upward from L and M toward B and then J; superiorly.
D7. (a) Glycogen; amino acids, lactic acid, and other simple sugars. (b) Lipoproteins, triglycerides, cholesterol. (c) Alcohol, antibiotics, and certain hormones. (d) Albumin, globulins (used to make antibodies), and clotting proteins (prothrombin and fibrinogen). (e) A, D, E, K; B$_{12}$; D. (f) Does not.
E1. (a) A. (b) A. (c) Duodenum —> jejunum —> ileum.
E2.

Type of Cell	Type of Secretion
a. Goblet cells	Mucos
b. Intestinal glands	Intestinal juice: maltase, sucrase, lactase, peptidases; ribonuclease and deoxyribonuclease
c. S cells	Secretin
d. CCK cells	Cholecystokinin (CCK)
e. K cells	Glucose-dependent insulinotropic peptide (GIP)
f. Duodenal glands of the submucosa	Alkaline mucus

E3. Villi, microvilli, circular folds; villi; do; lymphatic capillary inside a villus.
E6. (a) Small; active; facilitated diffusion. (b) Bile, micelles; lacteals. (c) Chylomicrons; lacteals, thoracic, subclavian liver and adipose tissue. (d) 9.3 liters; small. (e) Fat-soluble; ADEK.
F2. 1.5 (5.0); colon; C → AsC → TC → DC → SC → R → AnC.
F3. (a) Ileocecal. (b) Appendix. (c) Polyps.
F4. (a) Slower, (b) Three to four; rectum.
F6. By bacterial fermentation of carbohydrates; the process releases gases (flatus); flatulence refers to the excessive amounts of these gases.
F7. Bacteria and their products, salts, mucosal cells, undigested food, and water; gases such as methane.
G1. (a) C. (b) G. (c) I.
G2. (a) Amy, M. (b) M. (c) Gas, HCl, IF, M, Ps. (d) Amy, Glu, Ins, L, N, TCC. (e) CCK, M, MLS, N, PtS. (f) Alb, B, FP, Gly.
G3. (b) S (H). (c) CCK (H). (d) Gas (H). (e) Ps, Pt, and TCC. (f) HCl (N). (g) Gly. (h) Amy. (i) MLS. (j) N (E) .
H5. Diverticulitis. (*Hint:* Remember that "itis" means inflammation.)
H6. (a) Fecal contamination. (b) Blood. (*Hint:* Remember A for anus (feces) and B for blood (transfusions)).
H7. (a) Colos. (b) PD. (c). Colit. (d) AN. (e) IBD. (f) IBS. (g) Con. (h) Cho.

CRITICAL THINKING: CHAPTER 19

1. Place in correct sequence each of these structures in the pathway of food through the GI tract. Then briefly describe each structure: anal canal, ascending colon, cecum, duodenum, esophagus, ileocecal sphincter, oropharynx, pyloric sphincter, sigmoid colon.
2. Describe nine functions of the liver. Identify which of these are essential to survival.
3. Describe the effects of the following hormones on digestion: secretin, CCK and gastrin.
4. Describe healthful effects of dietary fiber, both insoluble and soluble fiber.
5. Identify with which digestive organs the following structures are associated: dentin, lacteal, frenulum linguae, uvula, pyloric region, ileum, villi, microvilli, cecum, and sinusoids.
6. Identify with which digestive organs the following conditions are associated: cirrhosis, colorectal cancer, hepatitis C-E, hernia, Crohn's disease, nausea, traveler's diarrhea.

MASTERY TEST: ■ CHAPTER 19

Questions 1–4: Circle the letter preceding the one best answer to each question.

1. Which of these organs is not part of the GI tract, but is an accessory structure?
   A. Mouth   D. Small intestine
   B. Pancreas   E. Esophagus
   C. Stomach
2. All of the following are chemicals produced by the pancreas *except*:
   A. Amylase   C. Trypsinogen
   B. Bicarbonate   D. Pepsinogen

3. Which of the following is a hormone, not an enzyme?
   A. Lipase   C. Pepsin
   B. Gastrin   D. Trypsin
4. All of the following are enzymes involved in protein digestion *except*:
   A. Amylase   D. Pepsin
   B. Trypsin   E. Chymotrypsin
   C. Carboxypeptidase

Questions 5–8: Arrange the answers in correct sequence.

_____ _____ _____ 5. Arrange from greatest to least in number in a complete set of adult dentition:
   A. Molars
   B. Premolars
   C. Cuspids (canines)

_____ _____ _____ _____ 6. GI tract wall, from inside to outside of the wall:
   A. Mucosa   C. Serosa
   B. Muscularis   D. Submucosa

_____ _____ _____ _____ _____ 7. Pathway of chyme:
   A. Ileum   D. Duodenum
   B. Jejunum   E Pylorus
   C. Cecum

_____ _____ _____ _____ _____ 8. Pathway of wastes:
   A. Ascending colon   D. Descending colon
   B. Transverse colon   E. Rectum
   C. Sigmoid colon

Questions 9–10: Circle T (true) or F (false). If the statement is false, change the underlined word or phrase so that the statement is correct.

T   F   9.   In general, the parasympathetic nervous system <u>stimulates</u> secretions of the salivary glands.

T   F   10.   The principal chemical activity of the stomach is to begin digestion of <u>protein</u>.

Questions 11–15: Fill-ins. Write the word or phrase that best fits the description.

_____ 11.   Chief cells, parietal cells, mucous cells, and G cells are all found in the wall of the

_____ .

_____ 12.   Villi, circular folds, and the jejunum are all parts of the _____ .

_____ 13.   The three phases of digestion are _____, _____ and _____.

_____ 14.   The clinical specialty that deals with diseases of the stomach and intestines is

_____ .

_____ 15.   The products of digestion of each of the three major food types are _____ .

ANSWERS TO MASTERY TEST: ■ CHAPTER 19

Multiple Choice
1. B
2. D
3. B
4. A

Arrange
5. A B C
6. A D B C
7. E D B A C
8. A B D C E

True or False
9. T. Inhibits
10. T

Fill-ins
11. Stomach
12. Small intestine
13. Cephalic, gastric, intestinal
14. Gastroenterology
15. Carbohydrates: monosaccharides; proteins: amino acids; lipids: fatty acids and monoglycerides

FRAMEWORK 20
Nutrition and Metabolism

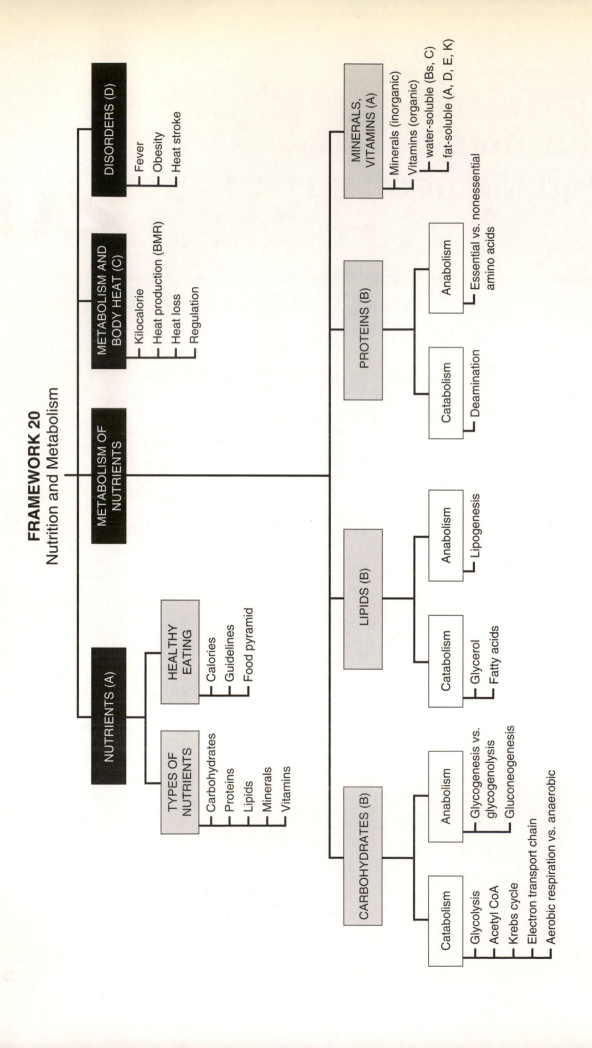

Nutrition and Metabolism

CHAPTER 20

Ingested foods, once digested, absorbed, and delivered, are used by body cells. This array of biochemical reactions within cells is known as metabolism. Products of digestion may be reassembled, as in formation of human protein from the amino acids of meat or beans, or in synthesis of fats from ingested oils. These reactions are examples of anabolism. The flip side of metabolic currency is catabolism, as in the breakdown of complex carbohydrates stored in liver or muscle to provide simple sugars for quick energy. Catabolism releases energy that maintains body heat and provides ATP to fuel activity. Vitamins and minerals play key roles in metabolism—for example, in enzyme synthesis and function.

The Chapter 20 Framework provides an organizational preview of the metabolism of the major food groups. Study the key terms and concepts it presents. As you begin this chapter, carefully examine the Chapter 20 Topic Outline and check off each objective after you meet it.

TOPIC OUTLINE AND OBJECTIVES

A. Nutrients

☐ 1. Define a nutrient and identify the six main types of nutrients.

☐ 2. List the guidelines for healthy eating.

B. Metabolism

☐ 3. Define metabolism and describe its importance in homeostasis.

☐ 4. Explain how the body uses carbohydrates, lipids, and proteins.

C. Metabolism and body heat

☐ 5. Explain how body heat is produced and lost.

☐ 6. Describe how body temperature is regulated.

D. Common disorders, and medical terminology

WORDBYTES

Study each wordbyte, its meaning, and an example of its use in a term. Check your understanding by jotting meanings of wordbytes in margins. Identify other examples of terms that contain these wordbytes as you continue through the text and *Learning Guide*.

Wordbyte	Meaning	Example(s)	Wordbyte	Meaning	Example(s)
ana-	up	*ana*bolism	-gen	to form	gluconeo*gen*esis
calor-	heat	*Calor*ie	-lysis	destruction	hydro*lysis*
cata-	down	*cata*bolism	neo-	new	gluco*neo*genesis
de-	remove, from	*de*amination			

CHECKPOINTS

A. Nutrients (pages 504–507)

A1. List three primary functions of nutrients.

■ **A2.** List the six principal types of nutrients.

■ **A3.** Contrast the caloric values of the three major food types.

a. Carbohydrates and proteins each produce _____ Cal/g or (because a pound contains 454 g) _____ Cal/lb.

b. Fats produce _____ Cal/g or _____ Cal/lb.

■ **A4.** Complete this Checkpoint about daily requirements of nutrients.

a. Fill in the number of Calories likely to be needed each day for each person.

   1. Emily Jensen, age 73: _____ Calories.

   2. Joel Jensen, age 38, who runs five miles a day four times a week: _____ Calories.

   3. Devon Jensen, age 15, a soccer player: _____ Calories.

   4. Dora Jensen, age 15, a dancer: _____ Calories.

   5. Kevin Jenson, age 8: _____ Calories

b. Complete the table about requirements for different types of nutrients based on a 2,000 Calorie daily intake:

Type of nutrients	% of 2,000 Calories needed each day	Calories/day	Calories/gram (See Checkpoint A3)	Grams/day
1. Carbohydrate		1000–1200		250–300
2. Fat			9	
3. Protein	12–15%			

A5. Write seven guidelines for healthy eating.

■ **A6.** Using the Food Pyramid as a guide (Figure 20.1, page 505 of the text), and identify the number of servings suggested per day for each food group listed.

_____ a. Bread, cereal, rice, pasta   _____ d. Meat, fish, poultry, dry beans, eggs, nuts

_____ b. Vegetables   _____ e. Milk, yogurt, and cheese

_____ c. Fruits   _____ f. Fats, oils, sweets

■ **A7.** Define *minerals*.

Minerals make up about _____ percent of body weight and are concentrated in _____ .

■ **A8.** Study Table 20.1, page 506 in your text. Then check your understanding of minerals in this matching exercise.

Ca. Calcium	F. Fluorine	K. Potassium
Cl. Chlorine	Fe. Iron	Mg. Magnesium
Co. Cobalt	I. Iodine	Na. Sodium

_____ a. Main anion in extracellular fluid, part of HCl in stomach; component of table salt

_____ b. Main cation in extracellular fluids; in table salt

_____ c. Most abundant mineral in the body; found mostly in bones and teeth; important in clotting and normal nerve and muscle function

_____ d. Important component of hemoglobin and also of cytochromes in the electron transport chain (See Checkpoint B5d).

_____ e. Main cation inside cells; critical in formation of action potentials in nerves and muscles

_____ f. Essential component of thyroid hormone

_____ g. Constituent of vitamin B_{12}, so necessary for red blood cell formation

_____ h. Improves tooth structure and inhibits tooth decay

_____ i. Serves as a catalyst for conversion of ADP to ATP; needed for normal muscle and nerve function

■ **A9.** Vitamins are *(organic? inorganic?)*. Most vitamins *(can? cannot?)* be synthesized in the body. In general, what are the functions of vitamins?

■ **A10.** Contrast the two principal groups of vitamins, and list the main vitamins in each group.

a. Fat-soluble _____

b. Water-soluble _____

■ **A11.** Select the vitamin that fits each description.

A	B$_1$	B$_2$	B$_{12}$	C	D	E	K

_____ a. This serves as a coenzyme that is essential for blood clotting, so it is called the antihemorrhagic vitamin; synthesized by intestinal bacteria.

_____ b. Its formation depends on sunlight on skin, as well as healthy kidneys and liver; necessary for calcium absorption.

_____ c. Riboflavin is another name for it; necessary for normal integrity of skin, mucosa, and eyes.

_____ d. This vitamin acts as an important coenzyme in carbohydrate metabolism and is necessary for forming acetylcholine; deficiency leads to beriberi or polyneuritis.

_____ e. Formed from carotene, it is necessary for normal bones and teeth; it prevents night blindness.

_____ f. This substance is also called ascorbic acid; deficiency causes anemia, poor wound healing, and scurvy.

_____ g. This coenzyme, the only B vitamin not found in vegetables, is necessary for normal erythropoiesis; absorption from GI tract depends on intrinsic factor.

_____ h. Also known as tocopherol, it is necessary for normal red blood cell membranes; deficiency is associated with sterility in some animals.

■ **A12.** *A clinical challenge.* Match names of vitamins in the box with descriptions below.

Biotin	Folic acid	Niacin	Pantothenic acid

a. Amy, who hopes to become pregnant, is taking this chemical as a prenatal vitamin, since it does reduce the risk for spina bifida and other neural

tube defects; also essential for normal hemopoiesis: _____ .

b. Miji is experiencing depression, fatigue, and nausea which are symptoms of deficiency of this vitamin:

_____ .

c. Sabrina is taking supplements of this vitamin to reduce her blood cholesterol level, however it does cause her to

experience some tingling of the skin and occasional diarrhea: _____ .

d. Malcolm's fatigue is associated with a deficiency of this vitamin which his body needs to form coenzyme A, a

key molecule needed in metabolism of foods and energy production: _____ .

A13. List three vitamins that serve as antioxidants. _____ _____ _____
Explain the functions of antioxidants.

B. Metabolism (pages 507–515)

■ **B1.** Complete this table comparing catabolism with anabolism.

Process	Definition	Releases or Uses Energy	Examples
a. Catabolism			
b. Anabolism			

■ **B2.** Metabolism utilizes enzymes and coenzymes. Describe them briefly here.

a. Metabolic reactions are accelerated by _____ .

b. For normal function many enzymes require ions such as _____ . (*Hint*: Refer to Table 20.1.)

c. Some enzymes require coenzymes derived from vitamins; name several of these vitamins. (*Hint*: Refer to Table 20.2.)

■ **B3.** Answer these questions about carbohydrate metabolism.

a. The story of carbohydrate metabolism is really the story of _____ metabolism because this is the most common carbohydrate (and in fact the most common energy source) in the human diet.

b. In which organ are other carbohydrates converted to glucose? _____

c. Just after a meal, the level of glucose in the blood (*increases? decreases?*). Cells use some of this glucose; by

_____ glucose, they release energy.

d. List two or more mechanisms by which excess glucose is used.

e. In order for cells to use glucose, glucose must move into those cells. This occurs by the process of (*active transport? facilitated diffusion?*), which is enhanced by the hormone _____ .

■ **B4.** Which molecule is a product of metabolism that contains greater potential energy? *(ATP? ADP?).* Which molecule receives some of the energy that is released during glucose catabolism? *(NAD? NADH + H$^+$?)*

Step 1 (_____): Glucose ⟶ 2 pyruvic acids + little ATP

Step 2 (_____): ⤷ acetyl CoA + CO_2

Step 3 (_____): Acetyl CoA ⟶ energized coenzymes + ATP + CO_2

Step 4 (_____): Energized coenzymes + ADP + O_2 ⟶ ATP + H_2O

Figure LG 20.1   Summary of glucose catabolism, described in four steps. Refer to Checkpoint B5, and write names of each step on lines provided.

B5. Refer to Figure 20.3, page 510 of your text, and Figure LG 20.1 above. Then summarize glucose catabolism in this Checkpoint.

a. Step 1 is the process known as _____ . Products include two molecules of

_____ and a small amout of the energy-storing molecule _____ .
Step 1 happens in *(cytosol? mitochondria?)* of cells. This process *(does? does not?)* require oxygen, so it is called *(aerobic? anaerobic?)*

b. In step 2, the products of step 1 (pyruvic acids) must give up CO_2 and convert to molecules of

_____ .

c. Step 3 consists of a series of reactions known as the _____ cycle. These occur in the *(cytosol? mitochondria?).* Three types of products result: (1) *(O_2? CO_2?)* which passes out into blood to lungs; (2) energized coenzymes, namely *(FAD and NAD? FADH$_2$ and NADH + H$^+$?)*; and (3)*(ADP? ATP?).*

d. What happens in step 4? This is a series of reactions along a chain of chemicals known as the _____ chain. Here, energy that had been stored in energized coenzymes (FADH$_2$ and NADH + H$^+$) is tapped off and trapped in *(ADP? ATP?).* This step *(does? does not?)* require O_2 molecules. The process finally leads to the formation

of _____ .

e. Fill in blanks in Figure LG 20.1.

■ **B6.** Complete this exercise about glucose metabolism.

a. You learned earlier that excess glucose may be stored as glycogen. In other words, glycogen consists of long

chains of _____ . Name the process of glycogen formation:

_____ . Where is most (75 percent) of glycogen in the body stored?

_____

b. Between meals, when glucose is needed, glycogen can be broken down again to release glucose.

Name two hormones that stimulate glycogenolysis: _____ _____ .
   (Hint: Refer to Figure LG 13.3, page 198 of the Learning Guide.)

c. Define *gluconeogenesis*.

Name two hormones that stimulate gluconeogenesis. _____ _____

■ **B7.**   Contrast the sources of energy and types of training that are the best preparation for different types of athletic events. *Hint*: It may help to refer to Checkpoint C8, Chapter 8, LG page 112. Choose from the following answers:

> A. Aerobic training of moderate to vigorous intensity that increases size and number of mitochondria
> B. Anaerobic, glycolytic pathways by interval training (high-intensity workout alternated with rest periods)
> C. Creatine phosphate and ATP via power-lifting of very heavy weights

_____ a. For short-term events such as a 100-meter sprint

_____ b. For high intensity events up to about 90 seconds such as a 400-meter run or 100-meter swim

_____ c. For long distance runs or games lasting more than a few minutes

■ **B8.**   Do this exercise about fat metabolism.

a. The initial step in catabolism of triglycerides is their breakdown into _____ and

_____ . Glycerol can then be converted into _____ ; this molecule can enter glycolytic pathways. If the cell does not need to generate ATP, then glyceraldehyde 3-phosphate can be

converted into _____ . This anabolic step is an example of gluco-

_____ .

b. Recall that fatty acids are long chains of carbons, with attached hydrogens and a few oxygens. Two-carbon pieces

are "snipped off" fatty acids; the resulting molecules, named _____ , may enter the Krebs cycle.

c. Ketogenesis occurs when cells are forced to turn to fat catabolism. State one reason why cells might carry out excessive fat catabolism leading to ketogenesis.

d. The presence of excessive amounts of keto acids in blood *(raises? lowers?)* pH, leading to acidosis. This condition is *(rarely serious? potentially lethal?)*.

■ **B9.**   Edward eats 4000 Calories a day in a high-carbohydrate, high-protein, low-fat diet. Explain why he is putting on weight.

Besides formation of adipose tissue, for what other purposes can the Edward's triglycerides be used?

■ **B10.** Most lipids are water-*(soluble? insoluble?)*. For transport in blood, lipids are "coated" with chemicals

(largely protein) to form _____ . Match the four classes of lipoproteins in the box with the related description below.

C. Chylomicrons	LDLs. Low-density lipoproteins
HDLs. High-density lipoproteins	VLDLs. Very low-density lipoproteins

_____ a. Formed by cells lining the small intestine; transport lipids from foods to fat tissue for storage there

_____ b. Called the "good" cholesterol because these lipoproteins remove excess cholesterol from body cells and transport it to the liver for elimination

_____ c. Called the "bad" cholesterol because excessive amounts of these lipoproteins help to form atherosclerotic plaques in arteries

_____ d. Transport triglycerides and eventually form LDLs

■ **B11.** Discuss cholesterol levels that can maximize cardiovascular health. *Hint*: It may help to refer to Checkpoint F2, Chapter 15, LG page 234.

a. Total blood cholesterol should be less than _____ mg/dL. Because *(HDLs? LDLs?)* make up most of the total cholesterol, it is important to keep these lipoproteins at a low level.

b. "Total cholesterol to HDL" ratio should be *(less than? greater than?) 3:1*. Two ways to improve (lower) this ratio

are to _____-crease total cholesterol level and to _____-crease HDL level. Regular aerobic exercise can help achieve both of those goals.

B12. Throughout your study of systems of the body so far, you have learned about a variety of roles of proteins. List at least six functions of proteins in the body, using one or two words for each function. Include some structural and some regulatory roles.

_____   _____

_____   _____

_____   _____

■ **B13.** Refer to Figure LG 20.2 and describe the uses of protein.

Two sources of proteins are shown: (a) _____ and (b) _____ . Amino

acids from these sources may undergo the process known as (c) _____ to form new body proteins. If other energy sources are used up, amino acids may undergo catabolism. The first step is (d)

_____ , in which an amino group (NH_2) is removed in the liver. The remaining portion of the

amino acid may enter (e) _____ pathways at a number of points (Figure 20.5, page 513 in

your text). In this way, proteins can lead to formation of (f) _____ . Amino acids may also be

converted to (g) _____ by the process of gluconeogenesis or to (h)

_____ .

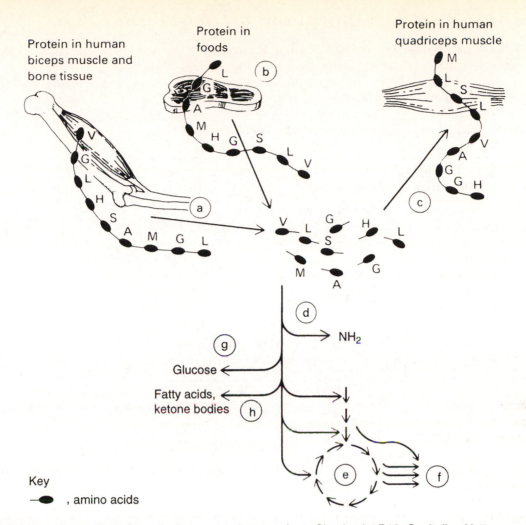

Protein in human biceps muscle and bone tissue

Protein in foods

Protein in human quadriceps muscle

Key

—●— , amino acids

Figured LG 20.2 Metabolism of protein. Lowercase letters refer to Checkpoint B13. Capitalized letters represent different amino acids.

■ **B14.** List hormones that stimulate protein:

a. Anabolism

b. Catabolism

B15. Contrast *essential amino acids* with *nonessential amino acids*.

How many amino acids are considered essential to humans? _____

B16. Explain why newborns are screened for PKU.

C. Metabolism and body heat (pages 515–518)

■ **C1.** Catabolism leads to *(release? absorption?)* of heat. If you ingest, digest, absorb, and catabolize a slice of bread, the energy released from the bread equals about 80 to 100 *(calories? Calories?)*.

C2. Define *basal metabolic rate (BMR)*. BMR amounts to about _____ – _____ Calories per day.

Explain why you will probably require more Calories today than the BMR.

C3. Write an example of each of these types of heat loss. One is done for you.

a. Radiation

b. Convection: cooling by a draft while taking a shower

c. Conduction

d. Evaporation

■ **C4.** During an hour of active exercise Bill produces 1 liter of sweat. On a humid day, *(more? less?)* evaporation of Bill's sweat occurs, so Bill would be cooled *(more? less?)* on such a day.

C5. Lil is outside on a snowy day without a warm coat. Her skin is pale and chilled and she begins to shiver. Explain how these responses are attempts of her body to maintain homeostasis of temperature. What other responses might raise her body temperature? Include these words in your answer: *sympathetic, metabolism, adrenal medulla, thyroid,* and *coat*.

■ **C6.** The body's "thermostat," which controls body temperature, is located in the _____ . Which of the following are mechanisms that result in heat loss (rather than heat production)?

A. Sweating   C. Dilation of blood vessels in skin

B. Shivering   D. Increase of metabolic rate by increase in thyroxin

C7. Discuss *hypothermia*:

a. Causes

b. Signs and symptoms

D1.   In what ways is a fever beneficial?

D2.   Define obesity, and state several causes of obesity.

D3.   Contrast terms in each pair:

a. *Heat cramps/heatstroke*

b. *Kwashiorkor/marasmus*

ANSWERS TO SELECTED CHECKPOINTS: CHAPTER 20

A2.   Carbohydrates, proteins, lipids, minerals, vitamins, and water.

A3.   (a) 4, about 1800. (b) 9, about 3500–4000.

A4.   (a1) 1600. (a2) 2800. (a3) 2800. (a4) 2200. (a5) 2200. (b) See Table below.

Type of nutrients	% of 2000 Calories needed each day	Calories/ day	Calories/ gram	Grams/ day
1. Carbohydrate	50–60%	1000–1200	4	250–300
2. Fat	30%	600	9	67
3. Protein	12–15%	240–300	4	60–75

A6.   (a) 6–11. (b) 3–5. (c) 3–5. (d) 2–3. (e) 2–3. (f) Sparingly.

A7.   Inorganic substances; 4, skeleton.

A8.   (a) Cl. (b) Na. (c) Ca. (d) Fe. (e) K. (f) I. (g) Co. (h) F. (i) Mg.

A9.   Organic; cannot; most serve as coenzymes, maintaining growth and metabolism.

A10.   (a) A D E K. (b) B complex and C. See Table 20.2.

A11.   (a) K. (b) D. (c) B_2. (d) B_1. (e) A. (f) C. (g) B_{12}. (h) E.

A12.   (a) Folic acid. (b) Biotin. (c) Niacin. (d) Pantothenic acid.

B1.

Process	Definition	Releases or Uses Energy	Examples
a. Catabolism	Breakdown of complex organic compounds into smaller ones	Energy released as heat or stored in ATP	Glycolysis, Krebs cycle, glycogenolysis
b. Anabolism	Synthesis of complex organic molecules from smaller ones	Uses energy (ATP)	Synthesis of protein, fats, or glycogen

B2.   (a) Enzymes. (b) Calcium, copper, iron, magnesium, zinc. (c) Thiamine (B_1), riboflavin, niacin (B_3), pyroxidine (B_6), cyanocobalamin (B_{12}), biotin, folic acid, C (ascorbic acid), and K.

B3.   (a) Glucose. (b) Liver. (c) Increases; catabolism of. (d) Conversion of glucose to glycogen (glycogenesis) or to triglycerides or fats (lipogenesis). (e) Facilitated diffusion, insulin.

B4.   ATP; NAD.

B5.   (a) Glycolysis; pyruvic acid, ATP; cytosol, does not, anaerobic. (b) Acetyl coenzyme A. (c) Krebs; mitochondria; (c1) CO_2; (c2) $FADH_2$ and NADH + H^+; (c3) ATP. (d) Electron transport; ATP; does; water (H_2O). (e) Step 1: Glycolysis;

Step 2: transition step; Step 3: Krebs cycle; Step 4: electron transport chain.

B6. (a) Glucose; glycogenesis; skeletal muscle cells. (b) Glucagon and epinephrine. (c) Formation of glucose from "new" (=neo) sources, that is, from noncarbohydrate sources of fats and amino acids; cortisol and glucagon.

B7. (a) C. (b) B. (c) A.

B8. (a) Fatty acids, glycerol; glyceraldehyde-3-phosphate; glucose; neogenesis. (b) Acetyl coenzyme A (acetyl CoA). (c) Lack of glucose as energy source in cells due to diabetes mellitus. (d) Lowers; potentially lethal.

B9. Amino acids and glucose can be converted to fatty acids, combined with glycerol, and stored as tryglycerides; phospholipids in plasma membranes, lipoproteins (See Checkpoint B10), thromboplastin (in blood clotting pathways), and myelin sheaths over neurons.

B10. Insoluble; lipoproteins. (a) C. (b) HDLs. (c) LDLs. (d) VLDLs.

B11. (a) 200; LDLs. (b) Less than; de, in.

B13. (a) Worn-out cells as in bone or muscle. (b) Ingested foods. (c) Anabolism. (d) Deamination. (e) Krebs cycle. (f) $CO_2 + H_2O$ + ATP. (g) Glucose. (h) Fatty acids or ketone bodies.

B14. (a) Insulinlike growth factors, insulin, thyroid hormone, estrogen, and testosterone. (b) Cortisol.

C1. Release; Calories (or kilocalories).

C4. Less, less.

C6. Hypothalamus; A C.

CRITICAL THINKING: CHAPTER 20

1. Contrast anabolism and catabolism. Give one example of each process in metabolism of carbohydrates, lipids, and proteins.

2. Explain why red blood cells derive all of their ATP from glycolysis. (Hint: which organelles do they lack?)

3. When you wake up at 7 AM and have not eaten since dinner last night, which hormones are helping you to maintain an adequate blood level of glucose? How do they accomplish this?

4. Explain why "fats are more fattening" than carbohydrates or proteins.

5. Describe guidelines for healthy eating and discuss roles foods may play in disorders such as atherosclerosis.

6. Discuss homeostatic mechanisms of temperature regulation.

7. Today Ruth's dietary intake includes 40 grams of protein and 56 grams of fat. If she keeps her total intake today at 2000 Calories, how many grams of carbohydrates should be in her diet? Does her intake of fat meet the guidelines for healthy eating?

8. Discuss the pro's and con's of taking vitamin or mineral supplements for the following groups: (a) pregnant women; (b) nonpregnant women in their twenties or thirties; (c) adults in their sixties. (d) vegetarians.

9. Amelia has been taking antibiotics that destroy bacteria, including many of the bacteria in the large intestine. For which vitamin deficiencies is she at greatest risk?

MASTERY TEST: ■ CHAPTER 20

Questions 1–6: Circle the letter preceding the one best answer to each question.

1. Which of these processes is anabolic?
   A. Pyruvic acid → $CO_2 + H_2O$ + ATP
   B. Glucose → pyruvic acid + ATP
   C. Protein synthesis
   D. Digestion of starch to maltose
   E. Glycogenolysis

2. Which of the following can be represented by the equation: Glucose → pyruvic acids + small amount ATP?
   A. Glycolysis
   B. Transition step between glycolysis and Krebs cycle
   C. Krebs cycle
   D. Electron transport system

3. The process of forming glucose from amino acids or glycerol is called _____ .
   A. Glycolysis   C. Glycogenesis
   B. Glycogenolysis   D. Gluconeogenesis

4. Choose the *false* statement about vitamins.
   A. They are organic compounds.
   B. They regulate physiological processes.
   C. Most are synthesized by the body.
   D. Many act as parts of enzymes or coenzymes.

5. Choose the *false* statement about temperature regulation.
   A. Some aspects of fever are beneficial.
   B. Fever is believed to be due to a "resetting of the body's thermostat."
   C. Vasoconstriction of blood vessels in skin tends to conserve heat.
   D. Heat-producing mechanisms that occur when you are in a cold environment are primarily parasympathetic.

6. Which of the following vitamins is water-soluble? A, C, D, E, K?

Question 7: Arrange the answers in correct sequence.

_____ _____ _____ 7. Refer to the Food Pyramid and arrange these food groups according to number of servings suggested per day, from greatest to least:
   A. Meat, fish, poultry, dry beans, eggs, and nuts
   B. Vegetables
   C. Bread, cereal, rice, pasta

Questions 8–10: Circle T (true) or F (false). If the statement is false, change the underlined word or phrase so that the statement is correct.

T   F   8. The electron transport chain is located in the <u>mitochondria</u>.

T   F   9. Thiamine, riboflavin, and pyroxidine are all <u>B</u> vitamins.

T   F   10. <u>Carbohydrates, proteins, and vitamins</u> provide energy and serve as building materials.

Questions 11–15: Fill-ins. Complete each sentence with the word or phrase that best fits.

_____ 11. Catabolism of each gram of carbohydrate or protein results in release of about

_____ kcal, whereas each gram of fat leads to about _____ kcal.

_____ 12. Write a chemical equation showing the overall reaction for aerobic respiration.
   Glucose + 6 oxygen

_____ 13. _____ is the mineral that is most common in extracellular fluid (ECF); it is also important in osmosis, buffer systems, and in nerve impulse conduction.

_____ 14. Aspirin and acetaminophen (Tylenol) reduce fever by inhibiting synthesis of

_____ so that the body's "thermostat" located in the _____ is reset to a lower temperature.

_____ 15. Name two hormones that stimulate glycogen catabolism.

ANSWERS TO MASTERY TEST: ■ CHAPTER 20

Multiple Choice
1. C
2. A
3. D
4. C
5. D
6. C

Arrange
7. C B A

True or False
8. T
9. T
10. F. Carbohydrates and proteins but not vitamins

Fill-ins
11. 4, 9
12. $6CO_2 + 6H_2O + 36\text{–}38$ ATP
13. Sodium (Na^+)
14. Prostaglandins (PGs), hypothalamus
15. Glucagon and epinephrine

FRAMEWORK 21
Urinary System

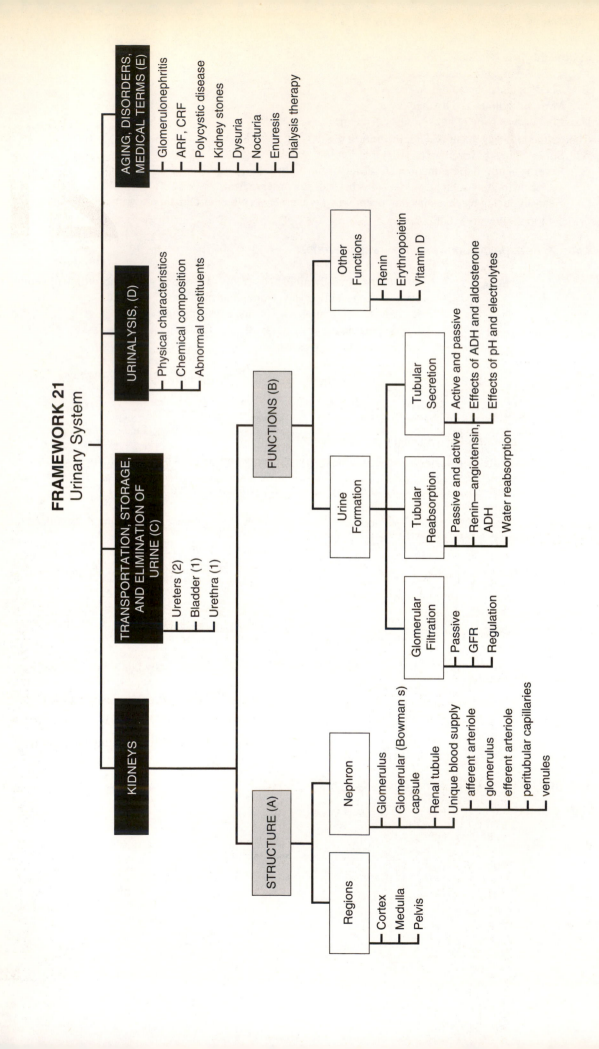

The Urinary System

Wastes such as urea and hydrogen ions (H^+) are products of metabolism that must be removed or they will lead to deleterious effects. The urinary system works with lungs, skin, and the GI tract to eliminate wastes. Kidneys are designed to filter blood and selectively determine which chemicals stay in the blood or exit in the urine. The process of urine formation in the kidney requires an intricate balance of factors such as blood pressure and hormones (ADH and aldosterone). Examination of urine (urinalysis) can provide information about the status of the blood and possible kidney malfunction. A healthy urinary system requires an exit route for urine through ureters, bladder, and urethra.

Start your study of the urinary system with a look at the Chapter 21 Framework and familiarize yourself with the key concepts and terms. As you begin this chapter, carefully examine the Chapter 21 Topic Outline and check off each objective after you meet it.

TOPIC OUTLINE AND OBJECTIVES

A. Structure of kidneys

☐ 1. List the components of the urinary system and their general functions.

☐ 2. Describe the structure and the blood supply of the kidneys.

B. Functions of kidneys

☐ 3. Identify the three basic functions performed by nephrons and collecting ducts and indicate where each occurs.

C. Transportation, storage, and elimination of urine

☐ 4. Describe the structure and functions of the ureters, urinary bladder, and urethra.

D. Urinalysis

E. Aging, homeostasis, common disorders, and medical terminology

☐ 5. Describe the effects of aging on the urinary system.

WORDBYTES

Study each wordbyte, its meaning, and an example of its use in a term. Check your understanding by jotting meanings of wordbytes in margins. Identify other examples of terms that contain these wordbytes as you continue through the text and *Learning Guide*.

Wordbyte	Meaning	Example(s)	Wordbyte	Meaning	Example(s)
af-	toward	*af*ferent	insipid	without taste	diabetes *insipid*us
anti-	against	*anti*diuretic	juxta-	next to	*juxta*glomerular
azot-	nitrogen-containing	*azot*emia	macula	spot	*macula* densa
calyx	cup	major *calyx*	myo-	muscle	*myo*genic
cyst-	bladder	*cyst*ostomy	recta	straight	vasa *recta*
densa	dense	macula *densa*	ren-	kidney	*ren*al vein
ef-	out	*ef*ferent	retro-	behind	*retro*peritoneal
-ferent	carry	af*ferent*	-ulus	small	glomer*ulus*
-genic	producing	myo*genic*	urin-	urinary	*urin*alysis
glomus	ball	*glom*erular			

CHECKPOINTS

A. Structure of kidneys (pages 524–527)

■ **A1.** Complete this checkpoint on kidney functions.

a. List at least seven examples of waste products eliminated by urine.

b. List at least five ions eliminated by kidneys. _____

c. Which ion do kidneys excrete when blood pH is too low? _____

d. Which vitamin is activated by kidneys? _____

e. Kidneys produce _____ , which regulates blood pressure and blood volume, and

_____ , which is vital to normal red blood cell formation.

■ **A2.** Identify the organs that make up the urinary system on Figure LG 21.1 and answer the following questions about them.

a. The kidneys are located at just above (*waist? hip?*) level. Each kidney is about _____ cm (_____ inches) long. Visualize kidney location and size on yourself.

b. The kidneys are in an extreme (*anterior? posterior?*) position in the abdomen. They are described as

_____ because they are posterior to the peritoneum.

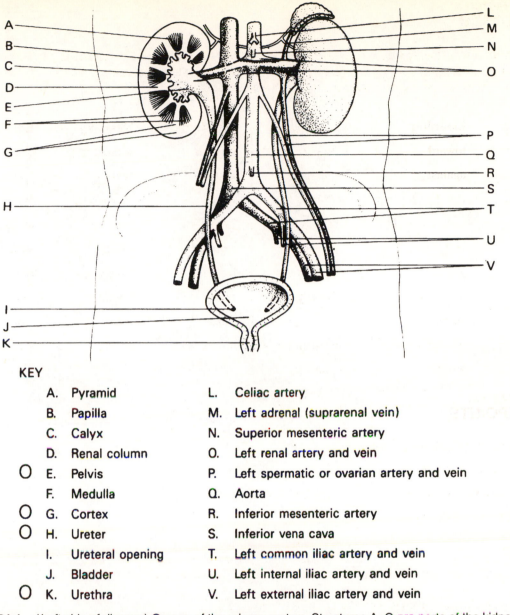

KEY

	A.	Pyramid	L.	Celiac artery
	B.	Papilla	M.	Left adrenal (suprarenal vein)
	C.	Calyx	N.	Superior mesenteric artery
	D.	Renal column	O.	Left renal artery and vein
O	E.	Pelvis	P.	Left spermatic or ovarian artery and vein
	F.	Medulla	Q.	Aorta
O	G.	Cortex	R.	Inferior mesenteric artery
O	H.	Ureter	S.	Inferior vena cava
	I.	Ureteral opening	T.	Left common iliac artery and vein
	J.	Bladder	U.	Left internal iliac artery and vein
O	K.	Urethra	V.	Left external iliac artery and vein

Figure LG 21.1 (*Left side of diagram*) Organs of the urinary system. Structures A–G are parts of the kidney. Identify and color as directed in Checkpoints A2 and C1. (*Right side of diagram*) Blood vessels of the abdomen. Color as directed in Chapter 16, Checkpoint D6, page 247 of the *Learning Guide*.

c. Identify the parts of the internal structure of the kidney on Figure LG 21.1. Then color all structures as indicated by color code ovals.

■ **A3.** About what percentage of cardiac output passes through the kidneys each minute? _____ percent.

This amounts to about _____ mL/min.

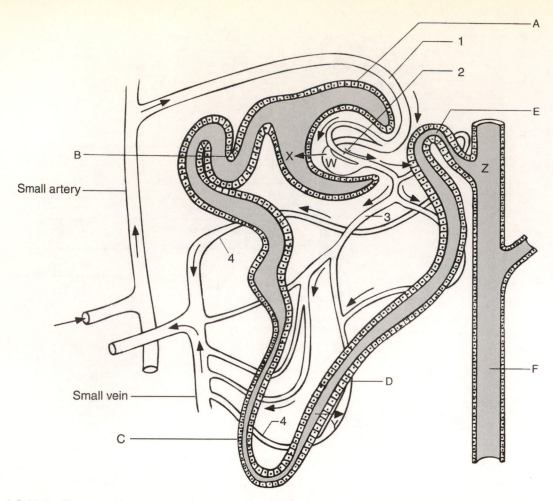

Figure LG 21.2 Diagram of a nephron. Numbers refer to Checkpoint A4. Letters A–F refer to Checkpoint A6, and letters W–Z refer to Checkpoints B2, B5, and B6.

■ **A4.** Follow the pathway that blood takes through kidneys by naming vessels 1–4 on Figure LG 21.2. Some arrows are shown; add more to reinforce your understanding of the pathway of blood.

1. _____   3. _____

2. _____   4. _____

■ **A5.** Explain what is unique about the blood supply of kidneys in this activity.

a. In virtually all tissues of the body, blood in capillaries flows into vessels named _____ .

However, in kidneys, blood in glomerular capillaries flows directly into _____ .

b. Which vessel has a larger diameter? (*Afferent? Efferent?*) arteriole. State one effect of this difference.

■ **A6.** On Figure LG 21.2 label parts A–F of the nephron, with letters arranged according to the flow of urine.

326

B. Functions of kidneys (pages 528–533)

■ **B1.** List the three steps in urine production:

_____ _____ _____

■ **B2.** Refer to Figure LG 21.2. Now describe the first step in urine production in this Checkpoint.

a. Glomerular filtration is a process of *(pushing? pulling?)* fluids and solutes out of _____

and into the fluid known as _____ . The direction of glomerular filtration is shown by the

arrow from W to X on the figure. This movement of substances is aided by presence of spaces between _____ and also by the considerable blood pressure in these capillaries (See B2c below).

b. *(Most? Only a few?)* types of substances are forced out of blood in the process of glomerular filtration. In fact, during filtration all solutes are freely filtered from blood, but two large components of blood

(_____ and _____) do not pass across the glomerular membrane.

c. Blood pressure in glomerular capillaries is *(higher? lower?)* than that in other capillaries of the body. This extra pressure is accounted for by the fact that the diameter of the efferent arteriole is *(larger? smaller?)* than that of the afferent arteriole. Picture three garden hoses connected to each other, representing the afferent arteriole, glomerular capillaries, and efferent arteriole. The third one (efferent arteriole) is extremely narrow; it creates such resistance that pressure builds up in the first two. Fluids are forced out through the highly permeable middle (glomerular) hose.

d. Determine the effect of blood colloid pressure (COP) in glomeruli. (*Hint*: Refer to Checkpoint A4b–c in Chapter 16.) Due to the presence of plasma proteins, blood COP is a *(pushing? pulling?)* pressure that *(mimics? opposes?)* filtration pressure. The direction is form *(W to X? X to W?)* in Figure LG 21.2. Combining these two pressures (along with glomerular capsule pressure), the overall pressure out of glomerular capillaries is known

as _____ *filtration pressure*.

■ **B3.** Define GFR in this exercise.

a. To what do the letters GFR refer? g _____ f _____

r _____

b. Write a normal value for GFR: _____ mL/min. Multiplying this by 1440 min/day (60 min/hour × 24 hours/day),

the amount of fluid moving from glomerular blood to filtrate is _____ mL/day (or _____ L/day).

c. Describe these two causes of decreased GFR. GFR is likely to decrease if blood pressure in kidneys _____-creases.

This can result from sympathetic nerve-induced vaso-_____ of renal vessels during stress. (See details in

Checkpoint B4b.) Either an enlarged prostate gland or kidney _____ lodged in the ureter will cause a backup of urine into renal tubules of the kidneys, opposing glomerular blood pressure, and this will decrease GFR also.

d. What consequences may be expected to accompany an abnormally low GFR?

e. Normal daily urine output is about 1500 mL (1.5 L) per day. A scanty output of 50 to 250 mL/day is the condition

known as _____ . *Anuria* refers to a daily urine output of _____ .

■ **B4.** Do this exercise about two mechanisms that regulate glomerular filtration rate (GFR).

a. ANP is a hormone that is produced by cells of the _____ when the walls of this organ are stretched by extra

blood volume. The acronym ANP stands for _____ _____ _____ . This name

indicates that ANP (acting like a diuretic) promotes *(loss? retention?)* of water and also loss of _____ by

increasing GFR. As a result, blood volume _____-creases.

b. Neural control of GFR is especially prominent during periods of *(high? low?)* stress, for example, during vigorous
exercise or hemorrhage. Sympathetic nerves especially narrow the *(afferent? efferent?)* arterioles. Imagine three
garden hoses linked together, representing the afferent arteriole, glomerular capillaries, and efferent arteriole. If
the first hose is narrowed, then *(more? less?)* blood flows into the second and third garden hoses. As a

result, GFR _____-creases, urinary output _____-creases, and blood volume _____-creases, allowing more blood
flow for stress responses in body tissues.

■ **B5.** Refer to Figure LG 21.2 and complete this Checkpoint about the next step in urine formation.

a. We have already seen that glomerular filtration results in movement of substances from *(blood to filtrate/forming
urine? filtrate/forming urine to blood?)*, as shown by the arrow between W and X. If this were the only step in
urine formation, all the substances in the filtrate (now called *filtration fluid*) would leave the body in urine. Note

from Table 21.1 (page 529 in your text) and Check point B3b that the body would produce _____ liters of urine
each day! And it would contain many valuable substances. Obviously some of these "good" substances must be
drawn back into blood and saved.

b. Recall that blood in most capillaries flows into vessels named _____ , but blood in glom-

erular capillaries flows into _____ and then into _____ . This
unique arrangement permits blood to recapture some of the substances indiscriminately pushed out during filtra-

tion. This occurs during the second step of urine formation, called _____ . This process
moves substances in the *(same? opposite?)* direction as glomerular filtration, as shown by the direction of the
arrow at area Y of Figure LG 21.2.

■ **B6.** Refer to areas Y and Z on Figure LG 21.2 and to the two columns on the right of Table 21.1, page 529 of your
text. Discuss the effects of tubular reabsorption in this learning activity.

a. Which solutes are 100 percent reabsorbed into area Y, so that virtually none remains in urine (area Z)?

b. About what percentage of water that is filtered out of blood is reabsorbed back into blood? _____ percent

c. Which other chemicals are at least 90 percent reabsorbed?

d. Which solute is about 50 percent reabsorbed? _____

e. Most tubular reabsorption occurs in the _____ convoluted tubules.

f. Tubular reabsorption is a *(nonselective? discriminating?)* process that enables the body to save valuable nutrients,
ions, and water. In other words, tubular reabsorption involves *(only passive? both active and passive?)* transport
processes.

■ **B7.** Denise has been diagnosed with diabetes mellitus. Her blood glucose level is often higher than ideal. Explain how this condition can lead to:

a. Glycosuria

b. Polyuria

■ **B8.** The third step in urine production is _____ . It involves movement of substances from *(blood to urine? urine to blood?)*. In other words, tubular secretion is movement of substances in the *(same? opposite?)* direction as movement occurring in filtration. List four substances that are secreted by the process of tubular secretion.

(a) Explain why regulation of K^+ is especially important.

(b) The pH of urine is typically *(acidic? alkaline?)*. Explain why.

■ **B9.** Now identify functions of the four "A" hormones that regulate urinary output. (*Hint*: It may help to refer to Figure LG 13.4, page LG 200.) Choose from answers in the box.

ADH. Antidiuretic hormone	Ang. II. Angiotensin II
Ald. Aldosterone	ANP. Atrial natriuretic peptide

_____ a. A vasoconstrictor that also stimulates release of aldosterone.

_____ b. Produced by the adrenal cortex, it causes collecting ducts to reabsorb more Na^+ and water and increases secretion of K^+.

_____ c. A hormone produced by heart cells, this chemical opposes the actions of renin, aldosterone, and ADH as it *increases* GFR.

_____ d. A hypothalamic hormone, it increases reabsorption and blood volume by increasing permeability of final portions of renal tubules.

■ **B10.** Diuretics have the *(same effect as? opposite effect to?)* ADH. Diuretics _____ -crease urinary output and

_____ -crease blood volume. Therefore they tend to _____ -crease blood pressure. List four commonly ingested fluids that have diuretic effects.

C. Transportation, storage, and elimination of urine (pages 535–537)

■ **C1.** Refer to Figure LG 21.1 as you complete the following exercise.

a. Ureters connect _____ to _____ .

b. The bladder is lined with _____ epithelium. Explain how the presence of this type of epithelium serves the primary function of the urinary bladder.

c. The urinary bladder is located in the *(abdomen? pelvis?)*. Two sphincters lie just inferior to it. The *(internal? external?)* sphincter is under voluntary control.

d. Urine leaves the bladder through the _____ . This tube is shorter in *(females? males?)*.

■ **C2.** In the micturition reflex *(sympathetic? parasympathetic?)* nerves stimulate the _____ muscle which forms most of the bladder wall, and causes *(contraction? relaxation?)* of the internal sphincter. What stimulus initiates the micturition reflex?

■ **C3.** Define *incontinence*.

This condition is *(normal? abnormal?)* in 1-year-old infants. This condition is *(normal? abnormal?)* in 21-year-old adults. List three causes of *stress incontinence*.

■ **C4.** Do this exercise on urinary tract infections (UTIs).

a. UTIs typically occur more often in *(women? men?)*. State a rationale for this fact.

b. Which bacteria most often cause UTIs? _____ Which microbe is most often the cause of "yeast infections"

of the vagina or urethra? _____

c. List common signs and symptoms of UTIs.

d. List three health practices that can reduce risk of UTIs.

■ **D1.** The typical volume of urine eliminated is about _____ mL/day.

D2. Describe the following characteristics of urine. Suggest causes of variations from the normal in each case.

a. Volume: _____ mL/day

b. Color: _____

c. Turbidity: _____

d. Odor: _____

e. pH: _____

f. Specific gravity: _____

D3. The following substances are not normally found in urine. Explain what the presence of each might indicate.

a. Albumin

e. Red blood cells

b. Glucose

f. Casts

c. Bilirubin

g. Ketone bodies

d. Protein

■ **D4.** Write arrows (↑ or ↓) to indicate the effect of different hormones on level of the following ions in urine.

a. Parathyroid hormone _____Ca^{2+} in urine

b. Parathyroid hormone _____Mg^{2+} in urine

c. Parathyroid hormone _____HPO_4^{2-} in urine

d. Aldosterone _____K^+ in urine

E. Aging, homeostasis, common disorders, and medical terminology (pages 537–539)

■ **E1.** Describe typical aging changes of the urinary system by filling in the blanks with terms in the box. Use each term once.

Decrease	Nephritis
Dysuria	Nocturia
Hematuria	Pyuria
Increase	Renal calculi

a. With aging, kidney mass, renal blood flow, GFR, and urinary output all tend to _____ .

b. Frequency of kidney inflammations known as _____ and kidney stones

(_____) are greater in older adults. These conditions are two possible causes of blood

in the urine (_____).

c. Frequency of UTIs is likely to _____ related to prostate enlargement in men or decreased amounts of urine "flushing" the urinary tract in either gender. UTIs are often associated with painful urination

(_____) and urinary frequency. White blood cells in urine (_____) can also indicate UTIs.

d. Weakness of urethral sphincters (and prostate enlargement in men) may contribute to the need to urinate

frequently throughout the night (_____).

■ **E2.** Complete this Checkpoint by identifying body systems on which the urinary system exerts its impact.

a. Kidneys play a critical role in regulating blood levels of calcium and phosphates that affect the

_____ system(s).

b. The male urethra functions in two systems: one is the urinary system, the other is the

_____ .

c. Kidneys synthesize renin and erythropoietin, and also regulate blood volume levels that impact the

_____ system.

d. As discussed in Chapter 22, blood pH is regulated largely by the urinary and the _____ systems.

■ **E3.** Fill in blanks in this exercise with terms in the box. Use each term once.

Anuria	Intravenous pyelogram
Dialysis	Oliguria
Diuresis	Polycystic renal disease
End-stage renal failure	Polyuria
Enuresis	Uremia

a. When kidneys fail, wastes such as urea and creatinine remain in blood, and may reach a toxic level, the condition

known as _____ . This is a sign of _____ . A procedure known as

_____ cleanses blood of toxic wastes, and can prolong life of persons in renal failure.

b. One cause of end-stage renal failure is _____ , a quite common hereditary disease in which kidney tissue is replaced by useless fluid-filled cysts.

c. Failing kidneys may lead to a decreased urine volume (_____) or eventually virtually

no urine (_____).

d. A large urinary output may be described by two terms: _____ or

_____ .

e. Bed-wetting, as occurs in about 15% of 5-year-old children, is known as _____

f. An x-ray of kidneys after venous injection of dye, for example to test for kidney stones, is called a(n)

_____ .

E4. Define and state two or more possible causes of *urinary retention*.

■ **E5.** Which condition is more likely to be fatal?

ARF. Acute renal failure	CRF. Chronic renal failure

ANSWERS TO SELECTED CHECKPOINTS: CHAPTER 21

A1. (a) Nitrogen-containing chemicals such as ammonia, urea, uric acid, creatinine, bilirubin, drugs and toxins, ions, and excessive water. (b) H^+, Na^+, K^+, Ca^{2+}, Cl^-, HPO_4^{2-}. (c) H^+ which will lower urine pH but raise blood pH. (d) Calcitriol (active vitamin D). (e) Renin, erythropoietin.

A2. (a) Waist; 10–12 (4–5). (b) Posterior; retroperitoneal. (c) See KEY to Figure LG 21.1, p. 325.

A3. 25; 1200.

A4. 1, Afferent arteriole; 2, glomerulus; 3, efferent arteriole; 4, peritubular capillaries.

A5. (a) Venules; efferent arterioles. (b) Afferent; increases glomerular blood pressure. (See Checkpoint B2c.)

A6. A, glomerular (Bowman's) capsule; B, proximal convoluted tubule; C, descending limb of the loop of Henle; D, ascending limb of the loop of Henle; E, distal convoluted tubule; F, collecting duct.

B1. Glomerular filtration, tubular reabsorption, tubular secretion.

B2. (a) Pushing, blood, filtrate; podocytes. (b) Most; proteins and blood cells. (c) Higher; smaller. (d) Pulling, opposes, X to W; net.

B3. (a) Glomerular filtration rate. (b) 105–120 mL/min; 150,000–180,000 mL/day (150–180 L/day). (c) De; constriction; stones. (d) Additional water and waste products (such as H^+ and urea) normally eliminated in urine are retained in blood, possibly resulting in hypertension and acidosis. (e) Oliguria; less than 50 mL.

B4. (a) Atria of the heart; atrial natriuretic peptide; loss, Na^+; de. (b) High, afferent; less; de, de, in.

B5. (a) Blood to filtrate/forming urine; 180. (b) Venules; efferent arterioles, capillaries; tubular reabsorption; opposite.

B6. (a) Glucose and bicarbonate (HCO_3^-). (b) 99. (c) Na^+, Cl^-, K^+, and uric acid. (d) Urea. (e) Proximal. (f) Discriminating; both active and passive.

B7. (a) The amount of glucose in renal blood and in tubular fluid exceeds the amount that can be reabsorbed, so some glucose remains in urine. (b) Glucose exerts an osmotic effect, pulling more water into urine and increasing urine volume.

B8. Tubular secretion; blood to urine; the same; K^+, H^+, urea, ammonia (NH_3), creatinine, and penicillin. (a) Cardiac irregularities (which may

be lethal) are associated with K$^+$ imbalances.
(b) Acidic because high-protein diets create the need for kidneys to secrete H$^+$ into urine.

B9. (a) Ang. II. (b) Ald. (c) ANP. (d) ADH.

B10. Opposite effect to; in, de; de; caffeine drinks such as coffee, tea, and cola, as well as alcohol since it inhibits ADH production.

C1. (a) Kidneys, urinary bladder. (b) Transitional; this type of epithelium stretches and becomes thinner as the bladder fills. (c) Pelvis; external. (d) Urethra; females.

C2. Parasympathetic, detrusor, contraction; stretching the walls of the bladder by more than 200–400 mL (about 0.8–1.5 cups) of urine.

C3. Lack of voluntary control over micturition; normal; abnormal; any factor (sneezing, coughing, laughing) that causes abdominal pressure to surpass the pressure of the external urethral sphincter (which can be weakened by factors such as multiple pregnancies).

C4. (a)Women; the urethra is much shorter, so microbes travel a shorter distance from the outside of the body into the bladder. (b) *Escherichia coli; Candida.* (c) Burning or painful urination, urinary urgency and frequency, cloudy or blood-tinged urine, urethral discharge, fever, chills, nausea, and back pain (if infection reaches kidneys). (d) Drink plenty of fluids, wipe from front to back (so fecal or vaginal microbes do not move into urethra), and take precautions against STDs.

D1. 1000 to 2000 mL/day (1 to 2 L/day)

D4. (a) ↓ . (b) ↓ . (c) ↑ . (d) ↑ .

E1. (a) Decrease. (b) Nephritis, renal calculi; hematuria. (c) Increase; dysuria; pyuria. (d) Nocturia.

E2. (a) Skeletal (and others such as muscular, nervous, and cardiovascular). (b) Reproductive. (c) Cardiovascular. (d) Respiratory.

E3. (a) Uremia; end-stage renal failure; dialysis. (b) Polycystic renal disease. (c) Oliguria, anuria. (d) Diuresis, polyuria. (e) Enuresis. (f) Intravenous pyelogram (IVP).

E5. CRF.

CRITICAL THINKING: CHAPTER 21

1. Discuss the significance of the structure and arrangement of blood vessels within a nephron.

2. Describe how blood pressure is increased by mechanisms initiated by renin and aided by angiotensin I and II, aldosterone, and ADH.

3. Tamika, an accident victim, was admitted to the hospital 3 hours ago. Her chart indicates that she had been hemorrhaging at the scene of the accident. Tamika's nurse monitors her urinary output and notes that it is currently 12 mL/hr. (Normal output is 30–60 mL/hr.) Explain why a person in such severe stress is likely to have decreased urinary output (oliguria).

4. Who is more likely to be at higher risk for urinary tract infections (UTIs)? Explain why in each case. (a) 30-year-old women or men. (b) 30-year-old men or 70-year-old men.

5. Discuss signs and symptoms that are likely to be present in a person who has chronic renal failure (CRF). Compare the likely outcome (prognosis) for a person with a diagnosis of CRF compared to a diagnosis of ARF.

MASTERY TEST: ■ CHAPTER 21

Questions 1–3: Arrange the answers in correct sequence.

_____ _____ _____   1. From superior to inferior:
A. Ureter
B. Bladder
C. Urethra

_____ _____ _____ _____   2. From most superficial to deepest:
A. Renal capsule
B. Renal medulla
C. Renal cortex
D. Renal pelvis

_____ _____ _____ _____ _____ 3. Pathway of glomerular filtrate:
        A. Ascending limb of loop
        B. Descending limb of loop
        C. Collecting tubule
        D. Distal convoluted tubule
        E. Proximal convoluted tubule

Questions 4 and 5: Circle the letter preceding the one best answer to each question.

4. Which one of the hormones listed here causes increased urinary output and natriuresis?
   A. ANP   C. Aldosterone
   B. ADH   D. Angiotensin II

5. Which of these is a normal constituent of urine?
   A. Albumin   D. Casts
   B. Urea   E. Acetone
   C. Glucose

Questions 6–10: Circle T (true) or F (false). If the statement is false, change the underlined word or phrase so that the statement is correct.

T F   6. During stressful situations, urinary output is likely to <u>decrease</u>, partly due to <u>increase in sympathetic</u> nerve impulses leading to <u>vasoconstriction</u> of renal vessels.

T F   7. Aldosterone <u>increases</u> permeability of part of the distal convoluted tubule and the collecting duct by <u>inserting transporter proteins</u> into the wall of these sections of tubule.

T F   8. Blood in glomerular capillaries flows into <u>arterioles</u>, not into <u>venules</u>.

T F   9. A person taking diuretics is likely to urinate <u>less</u> than a person taking no medications.

T F 10. An efferent arteriole normally has a <u>larger</u> diameter than an afferent arteriole.

Question 11–15: Fill-ins. Complete each sentence with the word or phrase that best fits.

_____ 11. Most water reabsorption normally takes place across the _____ tubule of a nephron.

_____ 12. A _____ consists of a glomerulus and a renal tubule.

_____ 13. A toxic level of urea in blood is a condition known as _____ .

_____ 14. Sympathetic impulses cause greater constriction of _____ arterioles with resultant _____ -crease in GFR and _____ -crease in urinary output.

_____ 15. A person with chronic renal failure is likely to be in acidosis because kidneys fail to excrete _____ , may be anemic because kidneys do not produce _____ , and may have symptoms of hypocalcemia because kidneys do not activate _____ .

ANSWERS TO MASTERY TEST: ■ CHAPTER 21

Arrange
1. A B C
2. A C B D
3. E B A D C

Multiple Choice
4. A
5. B

True or False
6. T
7. T
8. T
9. F. More
10. F. Smaller

Fill-ins
11. Proximal convoluted
12. Nephron
13. Uremia
14. Afferent, de, de
15. H^+, erythropoietin, vitamin D

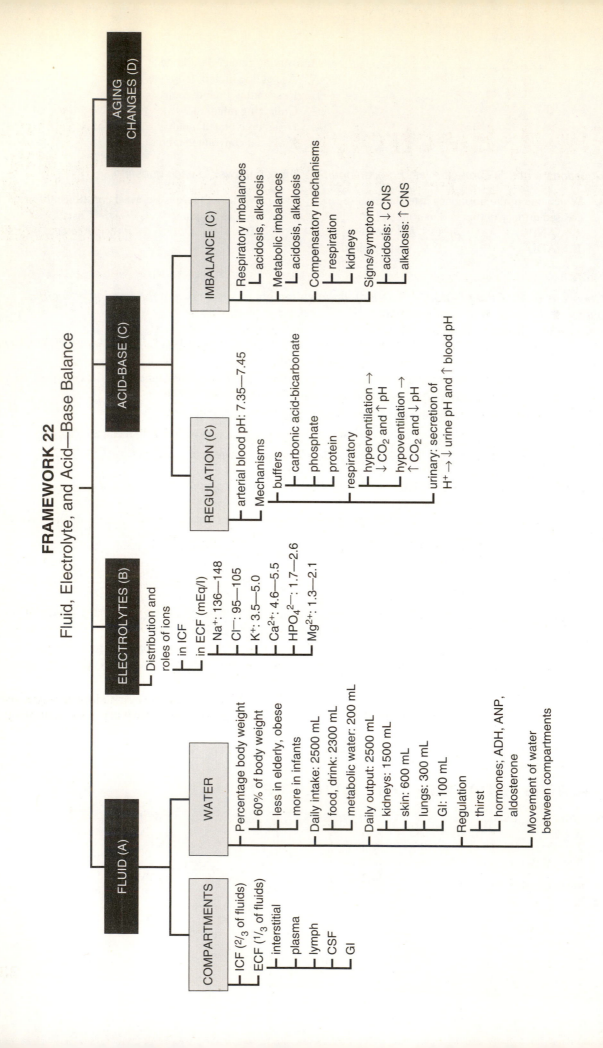

FRAMEWORK 22

Fluid, Electrolyte, and Acid—Base Balance

FLUID (A)

COMPARTMENTS
- ICF (2/3 of fluids)
- ECF (1/3 of fluids)
  - interstitial
  - plasma
  - lymph
  - CSF
  - GI

WATER
- Percentage body weight
  - 60% of body weight
  - less in elderly, obese
  - more in infants
- Daily intake: 2500 mL
  - food, drink: 2300 mL
  - metabolic water: 200 mL
- Daily output: 2500 mL
  - kidneys: 1500 mL
  - skin: 600 mL
  - lungs: 300 mL
  - GI: 100 mL
- Regulation
  - thirst
  - hormones; ADH, ANP, aldosterone
- Movement of water between compartments

ELECTROLYTES (B)

- Distribution and roles of ions
  - in ICF
  - in ECF (mEq/l)
    - Na^+: 136—148
    - Cl^-: 95—105
    - K^+: 3.5—5.0
    - Ca^{2+}: 4.6—5.5
    - $HPO_4{}^{2-}$: 1.7—2.6
    - Mg^{2+}: 1.3—2.1

ACID-BASE (C)

REGULATION (C)
- arterial blood pH: 7.35—7.45
- Mechanisms
  - buffers
    - carbonic acid-bicarbonate
    - phosphate
    - protein
  - respiratory
    - hyperventilation → ↓ CO_2 and ↑ pH
    - hypoventilation → ↑ CO_2 and ↓ pH
  - urinary: secretion of H^+ → ↓ urine pH and ↑ blood pH

IMBALANCE (C)
- Respiratory imbalances
  - acidosis, alkalosis
- Metabolic imbalances
  - acidosis, alkalosis
- Compensatory mechanisms
  - respiration
  - kidneys
- Signs/symptoms
  - acidosis: ↓ CNS
  - alkalosis: ↑ CNS

AGING CHANGES (D)

Fluid, Electrolyte, and Acid–Base Balance

CHAPTER
22

For the maintenance of homeostasis, levels of fluid, ions or electrolytes, and acids and bases must be kept within acceptable limits. Without careful regulation of pH, for example, nerves become overactive (in alkalosis) or underfunction, leading to coma (in acidosis). Alterations in blood levels of electrolytes can be lethal: excess K^+ can cause cardiac arrest, and decrease in blood Ca^{2+} can result in tetany of the diaphragm and respiratory failure. Systems you studied in Chapter 18 (respiratory) and Chapter 21 (urinary), along with buffers in the blood, normally achieve fluid, electrolyte, and acid–base balance.

Look first at the Chapter 22 Framework and the essential terms. As you begin this chapter, carefully examine the Chapter 22 Topic Outline and check off each objective after you meet it.

TOPIC OUTLINE AND OBJECTIVES

A. Fluid compartments and fluid balance

☐ 1. Compare the locations of intracellular fluid (ICF) and extracellular fluid (ECF), and describe the various fluid compartments of the body.

☐ 2. Describe the sources of water and solute gain and loss, and explain how each is regulated.

B. Electrolytes and body fluids

☐ 3. Compare the electrolyte composition of the three major fluid compartments: plasma, interstitial fluid, and intracellular fluid.

☐ 4. Discuss the functions of sodium, chloride, potassium, and calcium ions, and explain how their concentrations are regulated.

C. Acid–base balance

☐ 5. Compare the roles of buffers, exhalation of carbon dioxide, and kidney excretion of H^+ in maintaining pH of body fluids.

☐ 6. Define acid–base imbalances, describe their effects on the body, and explain how they are treated.

D. Aging changes

☐ 7. Describe the changes in fluid, electrolyte, and acid–base balance that may occur with aging.

WORDBYTES

Study each wordbyte, its meaning, and an example of its use in a term. Check your understanding by jotting meanings of wordbytes in margins. Identify other examples of terms that contain these wordbytes as you continue through the text and *Learning Guide*.

Wordbyte	Meaning	Example(s)	Wordbyte	Meaning	Example(s)
extra-	outside	*extra*cellular	intra-	within	*intra*cellular
hyper-	above	*hyper*ventilation	-osis	condition of	alkal*osis*
hypo-	below	*hypo*tonic			

CHECKPOINTS

A. Fluid compartments and fluid balance   (pages 544–547)

■ **A1.** Do this exercise about Tyler, age 13, who weighs 100 pounds.

a. Tyler's body weight is likely to consist of _____ lbs of fluid. His remaining _____ lbs of body weight consists of solids such as protein and fat.

b. Intracellular fluid (ICF) accounts for about _____ lbs of his body weight, whereas extracellular fluid (ECF)

accounts for about _____ lbs of his body weight.

c. Because (*80%? 20%?*) of Tyler's ECF is blood plasma, plasma accounts for about _____ lbs of his weight.

■ **A2.** Circle all the fluids that are considered *extracellular fluids* (ECF).

Blood plasma	Lymph	Pericardial fluid
Fluid within liver cells	Synovial fluid	Aqueous humor and vitreous body of eyes
Fluid immediately surrounding liver cells	Cerebrospinal fluid (CSF)	Endolymph and perilymph of ears

■ **A3.** *A clinical challenge.* Is a 45-year-old woman who has a water content of 55 percent likely to be in fluid balance?

Explain how her fluid level is related to her electrolyte level.

■ **A4.** Daily intake and output (I and O) of fluid usually both equal _____ mL (_____ liters). Of the intake,

about _____ mL are ingested in food and drink, and _____ mL are a product of catabolism of foods (discussed in Chapter 20).

■ **A5.** Now list four systems of the body that eliminate water. Indicate the amounts lost by each system on an average day. One is done for you.

a. Integument (skin): 600 mL/day

b. _____ : _____ mL/day

c. _____ : _____ mL/day

d. _____ : _____ mL/day

Explain how fluid output is likely to change as a result of heavy exercise.

A6. Think of the last time you felt very thirsty. What led you to drink fluids? List three mechanisms.

All three of these mechanisms ultimately signal the thirst center located in your (*hypothalamus? medulla? pharynx?*). Your thirst dissipates due to presence of fluids in your mouth and pharynx.

■ **A7.** Do this exercise on hormones that regulate water and solutes. *Hint*: Refer back to Chapter 21, Checkpoint B9, page LG 329.

a. Review mechanisms of angiotensin and aldosterone production (Figure LG 13.4 page 200 of the *Learning Guide*).

Recall that low blood volume and low BP trigger a(n) _____ -crease in renin, angiotensin II, and aldosterone production; aldosterone then promotes (*retention? excretion?*) of Na^+, Cl^-, and water by kidneys. As a result, the

renin-angiotensin-aldosterone (RAA) mechanisms _____ -crease blood volume and blood pressure.

b. Homeostatic mechanisms also come into play when blood volume is too *high* and must be *lowered*. High blood volume cause atria of the heart to stretch (*more? less?*), stimulating this tissue to release the hormone

_____ , which promotes (*retention? excretion?*) of Na^+, Cl^-, and water by kidneys. In other words, ANP causes natriuresis, which means a (*large? small?*) volume of urine that is (*high? low?*) in Na^+.

c. A second effect of elevated blood volume is a(n) _____ -crease in renin production from kidneys. A *decrease*

in renin leads to a(n) _____ -crease in angiotensin II. Since the role of angiotensin II is to promote aldosterone production, now (*more? less?*) aldosterone is produced, leading to (*retention? excretion?*) of Na^+, Cl^-, and water by kidneys. As a result, blood volume and blood

pressure _____ -crease. In other words, aldosterone and ANP have (*similar? opposite?*) effects.

d. Another effect of *reduced* angiotensin II levels of blood is (*constriction? dilation?*) of blood vessels, including renal vessels. As a result, more blood flows into kidneys, GFR and urinary output _____ -crease, causing blood volume and blood pressure to _____ -crease.

e. The name ADH indicates that this hormone (*promotes? opposes?*) diuresis, which is defined as a (*large? small?*) volume of urine. On a hot summer day when you exercise vigorously and sweat profusely, your hypothalamus will _____ -crease ADH production to _____ -crease urine volume and _____ -crease blood volume.

Then after blood volume increases, your ADH production will _____ -crease.

■ **A8.** *A clinical challenge*. Address these cases about fluid or electrolyte imbalances.

a. Explain why IV fluids administered to dehydrated clients typically contain solutes.

b. Explain how abnormal aldosterone level can cause *hypovolemia*.

c. Describe how sodium imbalance can lead to *edema*.

B. Electrolytes and body fluids (pages 548–550)

B1. List four general functions of electrolytes.

■ **B2.** Write two or more examples of each of the following types of ions. (*Hint*: Refer to Figure 22.5, page 548 of the text.)

a. Cations

b. Anions

■ **B3.** On Figure LG 22.1 write the chemical symbols for electrolytes that are found in greatest concentrations in each of the three compartments. Write one symbol on each line provided. Use the following list of symbols. (*Hint:* Refer to Figure 22.5 page 548 of the text.)

Cl^-	Chloride	Mg^{2+}	Magnesium
HCO_3^-	Bicarbonate	Na^+	Sodium
HPO_4^{2-}	Phosphate	$Protein^-$	Protein
K^+	Potassium		

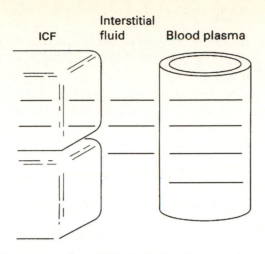

Interstitial
ICF   fluid   Blood plasma

Figure LG 22.1 Diagrams of fluid compartments. Write in electrolytes according to Checkpoint B3.

■ **B4.** Analyze the information in Figure LG 22.1 by answering these questions.

a. Circle the two compartments that are most similar in electrolyte concentration.
   ICF   plasma   interstitial fluid (IF)

b. List one major difference in electrolyte content of these two similar compartments.

c. Most body protein is located in which compartment? _____

d. Which cation is most concentrated in intracellular fluid (ICF)? _____ Which cation is most concentrated in

   extracellular fluid (ECF)? _____

e. Concentrations of electrolytes in body fluids are usually measured in _____ /liter.

B5. Check your understanding of the functions and disorders related to major electrolytes by completing
Table LG 22.1.

■ **B6.** A *clinical challenge.* Norine has had her serum electrolytes analyzed. Complete this exercise about her results.
(*Hint*: it may help to refer to Table 22.2, page 549 of the text.)

a. Her Na^+ level is 128. Norine's value indicates that she is more likely to be in a state of (*hyper? hypo?*)-natremia.
   Write two or more signs or symptoms of this electrolyte imbalance.

b. A diagnosis of hypochloremia would indicate that Norine's blood level of _____ ion is lower than normal.

   A normal range for serum chloride (Cl^-) level is _____ mEq/liter. One cause of low chloride is excessive

   _____ .

c. A normal range for K^+ in serum is 3.5–5.0 mEq/liter. This range is considerably (*higher? lower?*) than that for
   Na^+. This is reasonable because K^+ is the main cation in (*ECF? ICF?*), yet lab analysis of serum electrolytes is
   examining K^+ in (*ECF? ICF?*).

Electrolyte (Name and Symbol)	Normal Range (Serum Level), in mEq/liter	Principal Functions	Signs and Symptons of Imbalances:	
			Hyper-	**Hypo-**
a.	136–148			
b.				Hypochloremia: muscle spasms, alkalosis, ↓respirations, coma
c.		Most abundant cation in ICF; helps maintain fluid volume of cells; functions in nerve transmission		
d.			Hypercalcemia: weakness, lethargy, confusion, stupor, and coma	

d. Norine's potassium (K$^+$) level is 5.6 mEq/liter, indicating that she is in a state of hyper-

_____ . Hyperkalemia and hyponatremia (as described in *a*) may be caused by (*high? low?*) levels of aldosterone. Explain why.

If Norine's serum K$^+$ continues to increase, for example, to 8.0 mEq/liter, her most serious risk, possibly leading

to death, is likely to be _____ .

e. Norine's Ca^{2+} level is 3.9 mEq/liter. A normal range is 4.5–5.5 mEq/liter. This electrolyte imbalance, known as

_____ , is most closely linked to _____ -creased levels of parathyroid hormone (PTH)

and _____ -creased secretion of the thyroid hormone _____ . Low blood levels of both Ca^{2+} and Mg^{2+} cause overstimulation of the central nervous system and muscles. List two symptoms of these electrolyte imbalances.

f. When blood levels of calcium decrease (as in *e* above), blood levels of phosphate are likely to _____ -crease. In other words, there is a(n) (*direct? opposite or inverse?*) relationship between these two electrolytes. Which is a normal level of phosphate? (*2.0? 4.0? 10? 100?*) mEq/liter.

B7. Do this activity on dehydration.

a. Explain how vigorous activity with profuse sweating can lead to dehydration.

b. Write several signs or symptoms of dehydration.

c. List health practices that can help athletes to avoid dehydration.

C. Acid–base balance (pages 550–553)

■ **C1.** Complete this Checkpoint about acid–base regulation.

a. The pH of blood and other ECF should be maintained between _____ and _____ .

b. Name the three major mechanisms that work together to maintain acid–base balance.

c. List three important buffer systems in the body.

■ **C2.** Finish the chemical reaction to show formation and dissociation of carbonic acid.

$CO_2 + H_2O \rightarrow$ _____ $\rightarrow$ _____
(Also refer to Chapter 18, Checkpoint C3d, page 281 of the *Learning Guide*.)

■ **C3.** Refer to Figure LG 22.2 and complete this exercise about buffers.

a. In Figure LG 22.2 you can see why buffers are sometimes called "chemical sponges." They consist of an anion (in this case bicarbonate, or HCO_3^-) that is combined with a cation, which in this buffer system may be either H^+

or Na^+. When H^+ is drawn to the HCO_3^- "sponge," the weak acid _____ is formed.

When Na^+ attaches to HCO_3^-, the weak base _____ is present. As you color H^+ and Na^+, imagine these cations continually trading places as they are "drawn in or squeezed out of the sponge."

b. Buffers are chemicals that help the body to cope with (*strong? weak?*) acids or bases that are easily ionized and cause harm to the body. When a strong acid, such as hydrochloric acid (HCl), is added to body fluids, buffering occurs, as shown in Figure LG 22.2(b). The easily ionized hydrogen ion (H^+) from HCl is "absorbed by the sponge" as Na^+ is released from the buffer/sponge to combine with Cl^-. The two products,

_____ and _____ , are weak (less easily ionized).

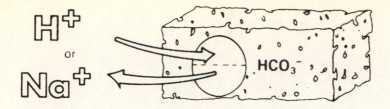

a. H$^+$ and Na$^+$: Only one of these ions can enter the "sponge" (buffer) at any time.

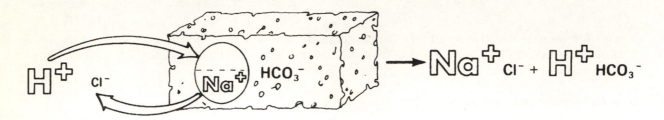

b. Na$^+$ is squeezed out as H$^+$ is drawn in.

○ H$^+$ and arrows showing movement of H$^+$

○ Na$^+$ and arrows showing movement of Na$^+$

Figure LG 22.2   Buffering action of carbonic acid–sodium bicarbonate system. Complete as directed in Checkpoint C3.

c. When the body continues to take in or produce excess strong acid, the concentration of the

_____ member of this buffer pair will decrease as it is used up in the attempt to maintain homeostasis of pH.

d. Although buffer systems provide rapid response to acid–base imbalance, they are limited because one member of the buffer pair can be used up. And they can convert only strong acids or bases to weak ones; they cannot eliminate them. Two systems of the body that can actually eliminate acidic or basic substances are

_____ and _____ .

■ **C4.**   Phosphates are more concentrated in *(ECF? ICF?)*. (For help, see Figure 22.5, page 548 in the text.) Name

one type of cell in which the phosphate buffer system is most important. _____

■ **C5.**   Complete this Checkpoint about respiration and kidneys related to pH.

a. Hyperventilation tends to *(raise? lower?)* blood pH because as the person exhales CO$_2$, less CO$_2$ is available for

formation of _____ acid and free hydrogen ion.

b. A slight decrease in blood pH tends to *(stimulate? inhibit?)* the respiratory center and so *(increases? decreases?)* respirations.

c. The kidneys regulate acid–base balance by altering their tubular secretion of _____ or

elimination of _____ ion.

■ **C6.** *A clinical challenge.* Which sign is more characteristic of acidosis?

A. Coma (indicating serious depression of the central nervous system)

B. Spasm (indicating overexcitability of the central nervous system)

■ **C7.** Complete this Checkpoint on compensatory mechanisms for acid–base imbalances.

a. Mrs. Davis, a diabetic, has a blood pH of 7.2 that indicates *(acidosis? alkalosis?)*, possibly due to ketoacidosis. Mrs. Davis is likely to *(hyper? hypo?)*-ventilate since exhaling extra CO_2 is likely to move her pH towards normal. (For help, refer to Checkpoint C2 in this chapter on page 343 of the *Learning Guide*.)

b. Mrs. McLaughlin has an advanced case of emphysema: her lungs have lost their elasticity and cannot properly exhale CO_2. As a result, she is likely to develop the acid–base imbalance of respiratory *(acidosis? alkalosis?)*. Which compensatory mechanism is she likely to use to try to return her blood pH to normal? *(Hyperventilation? Elimination of H^+ from kidneys?)*

c. Mrs. Davis's arterial blood moves from 7.20 to 7.42; in other words, her respiratory compensation is *(complete? partial?)*. Mrs. McLaughlin's compensatory mechanisms bring her arterial blood pH from 7.18 to 7.28; her renal compensation is *(complete? partial?)*.

D. Aging changes (page 553)

D1. Older adults are at higher risk for fluid–electrolyte balance and acid–base balance. Explain how changes in the following systems may increase such risk.

a. Integumentary (sweat glands)

b. Sensory (thirst receptors)

c. Muscular (muscle mass)

d. Urinary (kidney function)

ANSWERS TO SELECTED CHECKPOINTS: CHAPTER 22

A1. (a) 60; 40. (b) 40 (= ⅔ of 60 lbs fluid), 20 (⅓ of 60 lbs). (c) 20, 4 (= ⅕ of 20 lbs).

A2. All answers except fluid within liver cells.

A3. The percentage is within the normal range; however, more information is needed because the fluid must be distributed correctly among the three major compartments: cells (ICF) and ECF in plasma and interstitial fluid. For example, excessive distribution of fluid to interstitial space occurs in the fluid imbalance known as edema. Osmosis is the major mechanism of movement of water across membranes, and osmosis is regulated by electrolyte concentrations.

A4. 2500 (2.5); 2300, 200.

A5. (Any order) (b) Kidney, 1500. (c) Lungs, 300. (d) GI, 100. Output of fluid increases greatly via skin and slightly via respirations. To compensate, urine output decreases and intake of fluids increases.

A7. (a) In, retention; in. (b) More, ANP, excretion; large, high. (c) De; de; less, excretion; de; opposite. (d) Dilation; in, de. (e) Opposes, large; in, de, in; de.

A8. (a) Excessive intake of pure water can over-hydrate cells (water intoxication). (b) Deficient aldosterone level causes excessive urinary elimination of Na^+ and H_2O, so lower blood volume and blood pressure. (c) Excessive aldosterone causes retention of Na^+ and H_2O; also high dietary Na^+ intake.

B2. (a) Cl^-, HCO_3^-, HPO_4^{2-}, SO_4^{2-}, proteins. (b) Na^+, K^+, Ca^{2+}, Mg^{2+}.

B3.

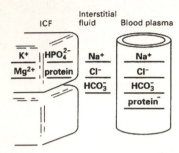

Figure LG 22.1A   Diagram of fluid compartments.

B4. (a) Plasma and interstitial fluid (IF). (b) Plasma contains more plasma protein. (c) Intracellular (ICF). (d) K^+; Na^+. (e) mEq.

B6. (a) Hypo; examples: headache, low blood pressure, tachycardia, and weakness, possibly leading to confusion, stupor, and coma if her serum Na^+ continues to drop. (b) Cl^-; 95–105; vomiting, dehydration, or use of certain diuretics. (c) Lower; ICF, ECF. (d) Kalemia; low; aldosterone promotes reabsorption of Na^+ and H_2O and also Mg^{2+} into blood and kidneys, and secretion of K^+ into urine; ventricular fibrillation resulting in cardiac arrest. (e) Hypocalcemia, de, in, calcitonin; tetany, convulsions. (f) In; opposite or inverse; 2.0.

C1. (a) 7.35, 7.45. (b) Buffers, respirations, and kidney excretion. (c) Carbonic acid–bicarbonate, phosphate, protein.

C2. $H_2CO_3 \rightarrow HCO_3^- + H^+$

C3. (a) Carbonic acid, or H_2CO_3; sodium bicarbonate, or $NaHCO_3$. (b) Strong; the salt sodium chloride (NaCl) and the weak acid H_2CO_3 ($H \cdot HCO_3$). (c) $NaHCO_3$ or weak base. (d) Respiratory (lungs), urinary (kidneys).

C4. ICF; kidney cells.

C5. (a) Raise, carbonic. (b) Stimulate, increases. (c) H^+, HCO_3^-.

C6. A.

C7. (a) Acidosis; hyper. (b) Acidosis; elimination of H^+ from kidneys. (c) Complete; partial.

CRITICAL THINKING: CHAPTER 22

1. Contrast the chemical composition of intracellular fluid with that of plasma and interstitial fluid.
2. Explain why changes in fluids result in changes in concentration of electrolytes in interstitial and intracellular fluids.
3. Describe how you might be able to distinguish two people with acid–base imbalances: one in acidosis and one in alkalosis.
4. Describe effects of ADH, ANP, and aldosterone on fluid and electrolyte balance.

MASTERY TEST: ■ CHAPTER 22

Questions 1–3: Circle the letter preceding the one best answer to each question.

1. On an average day the greatest volume of fluid output is via:
   A. Feces
   B. Sweat
   C. Lungs
   D. Urine

2. All of the following factors will tend to increase fluid output *except*:
   A. ANP
   B. ADH
   C. Diuretics
   D. Increased glomerular filtration rate

3. Which two electrolytes are in greatest concentration in ICF?
   A. Na^+ and Cl^-
   B. Na^+ and K^+
   C. K^+ and Cl^-
   D. K^+ and HPO_4^{2-}
   E. Na^+ and HPO_4^{2-}
   F. Ca^{2+} and Mg^{2+}

Questions 4–10: Circle T (true) or F (false). If the statement is false, change the underlined word or phrase so that the statement is correct.

T F 4. Hyperventilation tends to <u>raise</u> blood pH.

T F 5. During edema there is increased movement of fluid <u>out of plasma and into interstitial fluid.</u>

T F 6. Under normal circumstances fluid intake each day <u>is greater than</u> fluid output.

T F 7. Parathyroid hormone (PTH) causes a <u>decrease</u> in the blood level of calcium.

T F 8. When aldosterone is in <u>high</u> concentrations, sodium is conserved (in blood) and potassium is excreted (in urine).

T F 9. About <u>two-thirds</u> of body fluids are found in intracellular fluid (ICF) and <u>one-third</u> of body fluids are found in extracellular fluid (ECF).

T F 10. Sodium plays a much <u>greater</u> role than magnesium in osmotic balance of ECF because <u>more</u> mEq/liter of sodium are in ECF.

Questions 11–15: Fill-ins. Complete each sentence with the word or phrase that best fits.

_____ 11. List four functions of electrolytes.

_____ 12. The thirst control center is located in the _____.

_____ 13. _____ is the most abundant buffer system in the body.

_____ 14. To compensate for acidosis, kidneys increase secretion of _____ , and in alkalosis kidneys eliminate more _____ .

_____ 15. Normally blood serum contains about _____ mEq/liter of Na^+ and _____ mEq/liter of K^+.

ANSWERS TO MASTERY TEST: ■ CHAPTER 22

Multiple Choice
1. D
2. B
3. D

True or False
4. T
5. T
6. F. Equals
7. F. Increase
8. T
9. T
10. T

Fill-ins
11. Serve as essential minerals, help maintain osmotic and acid–base balance, and carry electrical current
12. Hypothalamus
13. Protein
14. H^+, HCO_3^-
15. 136–148, 3.5–5.0.

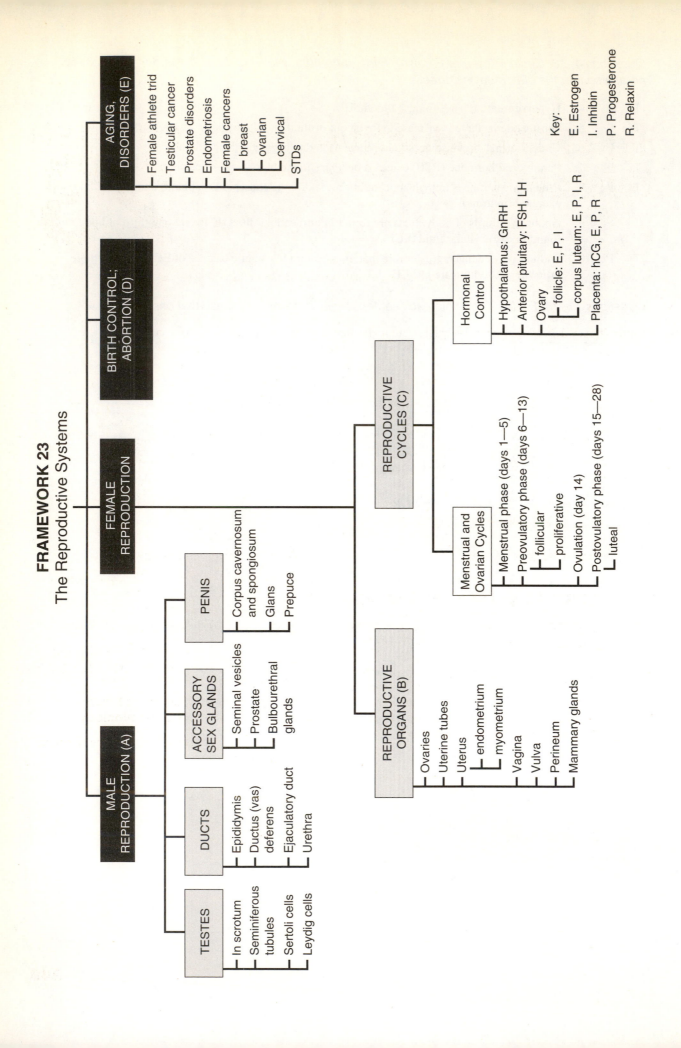

FRAMEWORK 23
The Reproductive Systems

The Reproductive Systems

Chapter 23 introduces the systems in females and males that provide for continuity of the human species, not simply homeostasis of individuals. The reproductive system produces sperm and ova that may unite and develop to form offspring. In order for sperm and ova to reach a site for potential union, they must each pass through duct systems and be nourished and protected by secretions. External genitalia include organs responsive to stimuli and capable of sexual activity. Hormones play significant roles in the development and maintenance of the reproductive systems. The variations of hormones in ovarian and uterine (menstrual) cycles are considered in detail, along with methods of birth control. Start your study with the Chapter 23 Framework and become familiar with the terms presented there. Then carefully examine the Chapter 23 Topic Outline and check off each objective after you meet it.

TOPIC OUTLINE AND OBJECTIVES

A. Male reproductive system

☐ 1. Describe the location, structure, and functions of the organs of the male reproductive system.
☐ 2. Describe how sperm cells are produced.
☐ 3. Explain the roles of hormones in regulating male reproductive functions.

B. Female reproductive system

☐ 4. Describe the location, structure, and functions of the organs of the female reproductive system.
☐ 5. Describe how oocytes are produced.

C. Female reproductive cycle

☐ 6. Describe the major events of the ovarian and uterine cycles.

D. Birth control and abortion

☐ 7. Compare the various types of birth control methods and outline the effectiveness of each.

E. Aging and reproduction; common disorders and medical terminology

☐ 8. Describe the effects of aging on the reproductive system.

WORDBYTES

Study each wordbyte, its meaning, and an example of its use in a term. Check your understanding by jotting meanings of wordbytes in margins. Identify other examples of terms that contain these wordbytes as you continue through the text and *Learning Guide*.

Wordbyte	Meaning	Example(s)	Wordbyte	Meaning	Example(s)
a-	without	*a*menorrhea	mast-	breast	*mast*ectomy
andro-	man	*andro*gen	meio-	smaller	*meio*sis
-arche	beginning	men*arche*	men-	month	*men*opause, *men*ses
cervix-	neck	*cervix* of uterus	-metrium	uterus	endo*metrium*
corp-	body	*corp*us cavernosum	mito-	thread	*mito*sis
crypt-	hidden	*crypt*orchidism	oo-, ov-	egg	*oo*genesis, *ov*um
dys-	apart	*dys*menorrhea	orchi-	testis	*orchi*dectomy
-genesis	formation	spermato*genesis*	rete	network	*rete* testis
gyn-	woman	*gyn*ecologist	-rrhea	a flow	dysmeno*rrhea*
hyster(o)-	uterus	*hyster*ectomy	salping-	tube, trumpet	*salping*itis
labia	lips	*labia* majora	troph-	nutrition	*troph*oblast
mamm-	breast	*mamm*ogram			

CHECKPOINTS

A. Male reproductive system (pages 557–563)

A1. Briefly state functions of each category of reproductive organs.

a. Gonad

c. Accessory sex glands

b. Ducts

d. Supporting structures

■ **A2.** As you go through the chapter, note which structures of the male reproductive system are included in each of the above categories.

■ **A3.** Describe the scrotum and testes in this exercise.

a. The temperature in the scrotum is a little *(higher? lower?)* than the temperature in the abdominal cavity. Of what significance is this?

b. When the scrotum is exposed to *(cold? warm?)* temperature or during sexual arousal, *(smooth? skeletal?)* muscles contract to elevate testes closer to the warmth of the pelvis.

c. During fetal life, testes develop in the *(scrotum? posterior of the abdomen?)*.

Undescended testes is a condition known as _____ . This is more likely to occur in *(premature? full term?)* males. Explain why.

d. Cryptorchid testes *(are? are not?)* likely to descend on their own into the scrotum during the first year of life. If they do not, why is surgical correction performed?

■ **A4.** Discuss spermatogenesis in this exercise.

a. Spermatozoa are one type of gamete; name the other type: _____ . Gametes are *(haploid? diploid?)*. Human gametes contain _____ chromosomes.

b. Fusion of gametes produces a cell with *(haploid? diploid?)* chromosome number. This cell and all those descending from it (except in ovaries or testes) are said to be *(n? 2n?)*. These chromosomes, received from both parents,

are said to exist in _____ pairs.

c. Gametogenesis ensures production of haploid gametes from otherwise diploid humans. In males, this process,

called _____-genesis, occurs in the _____ .

■ **A5.** Complete this exercise about the process of sperm formation.

a. Use answers in the box to indicate types of cells involved in different phases of sperm formation. The first one is done for you.

1°Sc. Primary spermatocyte	St. Spermatid
2°Sc. Secondary spermatocyte	Sz. Spermatozoon (sperm)
Sg. Spermatogonium	

Sg ⟶ _____ ⟶ _____ ⟶ _____ ⟶ _____

         Meiosis I   Meiosis II   Spermio-
                                     genesis

b. Crossing-over occurs in meiotic division *(I? II?)*. What advantage is provided by crossing-over?

c. The entire process of meiosis in seminiferous tubules requires about two to three *(hours? days? months?)*. Each original diploid spermatogonium develops into *(1? 2? 4?)* mature sperm.

■ **A6.** Describe spermatozoa (sperm) in this Checkpoint.

a. Number produced in a normal male: _____ /day

b. Life expectancy of sperm once inside female reproductive tract: _____ hours

c. Portion of sperm containing nucleus: _____

d. Part of sperm containing mitochondria: _____

e. Function of acrosome:

f. Function of tail: _____

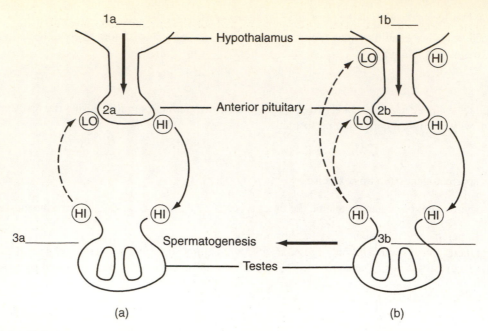

Figure LG 23.1 Male hormones and their regulation. Complete as directed in Checkpoint A7. HI = high levels of the hormone; LO = low levels of the hormone; ⟶, Direct effect; – – –→, Inverse (opposite or negative feedback) effect.

■ **A7.** Refer to Figure LG 23.1 and do this exercise on male hormones.

a. The *(hypothalamus? anterior pituitary?)* secretes the hormone called GnRH, or gonadotropin-

_____ hormone. Write the name of this hormone on lines 1a and 1b of Figure LG 23.1.

b. GnRH causes the anterior pituitary to release a hormone named _____ , which stimulates spermatogenesis in testes. Write the name of this hormone on line 2a in the figure. On line 3a write the name of a hormone made by testes that inhibits FSH.

c. GnRH causes the anterior pituitary to release another hormone named LH (write on line 2b in Figure LG 23.1).

LH stimulates production of the hormone _____ (line 3b).

d. Figure LG 23.1(b) demonstrates that a high level of GnRH stimulates a *(high? low?)* level of LH, and then the high level of LH leads to a *(high? low?)* level of testosterone. However, a rise in testosterone levels results in a *(positive? negative?)* feedback mechanism, which *(raises? lowers?)* both GnRH and LH and consequently *(raises? lowers?)* testosterone.

e. What other hormone besides FSH spurs on spermatogenesis? *(Hint:* refer to Figure LG 23.1.)

■ **A8.** Select hormones in the box that fit descriptions below.

DHT. Dihydrotestosterone	I. Inhibib	T. Testosterone

_____ a. Before birth, stimulates development of internal genitalia

_____ b. Before birth, stimulates development of external genitalia

_____ c. At puberty, facilitates development of male sexual organs and secondary sex characteristics (two answers)

_____ d. Inhibits GnRH and LH

_____ e. Produced by Leydig cells of testes

_____ f. Produced by Sertoli cells of testes

_____ g. Made within prostate and seminal vesicles

_____ h. Inhibits FSH, therefore slows down sperm production

A9. List five or more secondary sex characteristics promoted at puberty by testosterone and DHT.

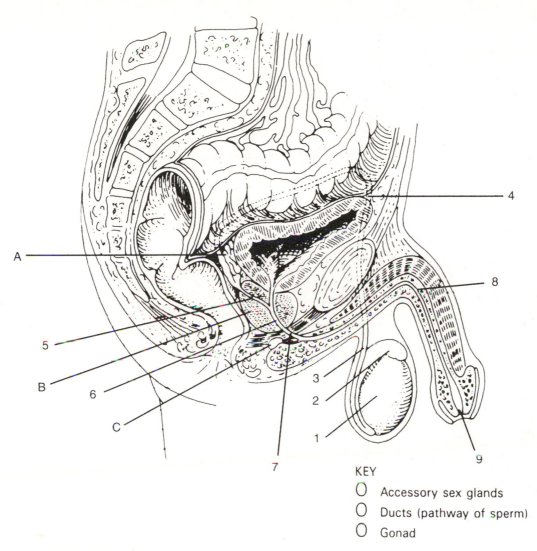

KEY

O Accessory sex glands
O Ducts (pathway of sperm)
O Gonad

Figure LG 23.2 Male organs of reproduction seen in sagittal section. Structures are numbered in order along the pathway taken by sperm. Color and label as directed in Checkpoints A10, A11, A14, and A17.

■ **A10.** Refer to Figure LG 23.2. Structures in the male reproductive system are numbered in order along the pathway that sperm take from their site of origin to the point where they exit from the body. Match the structures in the figure (using numbers 1 to 9) with descriptions below.

_____ a. Ejaculatory duct

_____ b. Epididymis; site where sperm mature and become motile over a period of up to two weeks; sperm are also stored here

_____ c. Urethra (portion within prostate)

_____ d. Urethra (portion within penis)

_____ e. Urethra (portion within urogenital diaphragm)

_____ f. Ductus deferens (or vas) where sperm can be stored and propelled forward (portion within abdomen)

_____ g. Ductus deferens (or vas) where sperm can be stored and propelled forward (portion within scrotum)

_____ h. External urethral orifice

_____ i. Testis

■ **A11.** Which structures numbered 1–9 in Figure LG 23.2 are paired? Circle their numbers:

1   2   3   4   5   6   7   8   9

Which structures are single? Write their numbers: _____

■ **A12.** Answer these questions about the ductus deferens.
a. This duct is also known as the *(epididymis? urethra? vas?)*.

b. It is located *(entirely? partially?)* in the abdomen. It enters the abdomen via the _____ .

c. A vasectomy is performed at a point along the section of the vas that is within the _____ .
Must the peritoneal cavity be entered during this procedure? *(Yes? No?)* Will the procedure affect testosterone level and sexual desire of the male? *(Yes? Not normally?)* Why? *(Hint:* see page 574 of the text.)

d. What other structures besides the vas contribute to the formation of the *spermatic cord?* _____

■ **A13.** Write *BUG* (bulbourethal glands), *P* (prostate), or *SV* seminal vesicles next to related descriptions below.

_____a. Doughnut-shaped gland that surrounds the urethra

_____b. Secretes slightly acidic fluid containing nutrients for sperm, enzymes, and PSA

_____c. Pea-sized glands that secrete alkaline fluid that protects sperm

_____d. Pouchlike structures that lie at the base of the urinary bladder

_____e. Secrete alkaline fluid containing, clotting proteins, prostaglandans, and fructose; this fluid makes up 60% of semen

■ **A14.** Color parts of the male reproductive system on Figure LG 23.2 according to color code ovals. Label the *seminal vesicle, prostate gland,* and *bulbourethral gland* on the figure.

■ **A15.** Complete the following exercise about semen.

a. The average amount per ejaculation is _____ mL.

b. The average range of number of sperm is _____ /mL. When the count falls below

_____ /mL, the male is likely to be sterile. Only one sperm fertilizes the ovum. Explain why such a high sperm count is required for fertility.

c. The pH of semen is slightly *(acid? alkaline?)*. State the advantage of this fact.

d. Write a term that mean "blood in semen." _____

■ **A16.** Complete this exercise about male external genitalia.

a. The _____ , or foreskin, is a covering over the _____ of the penis.

Surgical removal of the foreskin is known as _____ .

b. The names *corpus*_____ and *corpus* _____ indicate that these

354

bodies of tissue in the penis contain spaces that distend in the presence of large amounts of
_____ . This temporary state is known as the *(erect? flaccid?)* state and results from *(sympathetic? parasympathetic?)* nerve reflexes.

c. The urethra passes through the corpus _____ . The urethra functions in the transport of

_____ and _____ . During ejaculation, what prevents sperm from entering the urinary bladder and urine from entering the urethra?

d. Define erectile dysfunction (ED).

Explain how medications such as sildenafil (Viagra) may help men with ED.

■ **A17.** Label these parts of the penis on Figure LG 23.2: *prepuce, glans, corpus cavernosum,* and *corpus spongiosum of the penis.*

B. Female reproductive system (pages 564–570)

B1. Refer to Figure LG 23.3 and match the structures numbered 1–5 with their descriptions. Note that structures are numbered in order along the pathway taken by secondary oocytes from the site of formation to the point of exit from the body.

_____ a. Uterus (body)   _____ d. Uterine (fallopian) tube

_____ b. Uterus (cervix)   _____ e. Vagina

_____ c. Ovary

O Rectum (brown)
O Urinary bladder (yellow)
O Uterus (red)

Figure LG 23.3   Female organs of reproduction. Structures are numbered in order in the pathway taken by ova. Color and label as directed in Checkpoints B1, B2, B6, and B7.

■ **B2.** Indicate which of the five structures in Figure LG 23.3 are paired.

Circle the numbers: 1   2   3   4   5

Which structures are singular? Write their numbers: _____

B3. Contrast the structure and function of terms in each pair:

a. *Ovarian cortex/germinal epithelium*

b. *Ovarian follicle/mature (graafian) follicle*

c. *Corpus luteum/corpus albicans*

■ **B4.** Check your understanding of oogenesis and fertilization in this Checkpoint. Select names of cells from those in the box. Then circle *n* or *2n* to indicate whether cells are haploid (n) or diploid (2n).

1PB. First polar body	Og. Oogonium
2PB. Second polar body	OV. Ovum
1°Oc. Primary oocyte	Z. Zygote
2°Oc. Secondary oocyte	

a. The cell type that undergoes mitosis in fetal life is a(n) _____ *(n? 2n?)*.

b. During fetal life, some oogonia develop into large cells named _____ *(n? 2n?)* that begin meiosis I during fetal life but do not complete this reduction division until after puberty.

c. Completion of meiosis division I normally occurs in one follicle each month, resulting in two cells: a large cell

named _____ *(n? 2n?)*and a small cell named _____ *(n? 2n?)*.

d. The cell named _____ *(n? 2n?)* begins meiosis II within a mature (graafian) follicle, but stops the process before its completion. This cell is then ovulated.

e. The second meiotic division is completed only if fertilization of the secondary oocyte occurs. The two resulting

cells are a larger cell named the _____ *(n? 2n?)* and a smaller cell, the _____ *(n? 2n?);* the latter is discarded.

f. The union of the ovum and the nucleus of the sperm form the _____ *(n? 2n?)*.

■ **B5.** Complete this exercise about the uterine tubes.

a. Uterine tubes are also known as _____ tubes.

b. The lateral ends of the uterine tube consist of a fingerlike fringe known as _____ .

c. What structural features of the uterine tube help to draw the secondary oocyte into the tube and propel it toward the uterus?

d. Besides facilitation of movement of the secondary oocyte, what other important function do uterine tubes carry out?

B6. Color the pelvic organs on Figure LG 23.3. Use color code ovals on the figure.

■ **B7.** Refer to Figure LG 23.3 and complete this exercise about the uterus.

a. The organ that lies anterior and inferior to the uterus is the _____ . The _____ lies posterior to the uterus.

b. The fundus of the uterus is normally tipped *(anteriorly? posteriorly?)*. Which part of the uterus is most inferior in location? *(body? cervix? fundus?)* In its normal position, at what angle does the uterus join the

vagina?_____

c. Most of the uterus consists of _____-metrium. This layer is *(smooth muscle? epithelium?)*. It should appear red in the figure (as directed above).

d. The innermost layer of the uterus is the _____-metrium which is a *(mucous? serous?)* membrane. This layer *(is? is not?)* glandular. Write two roles of the endometrium.

■ **B8.** Name the relatively painless procedure in which a small number of cells are removed from the cervical area

and examined microscopically for possible changes indicating cancer. _____ smear.

(*Hint*: Refer to text page 581.) Write the term that means surgical removal of the uterus. _____

■ **B9.** The pH of vaginal mucosa is *(acid? alkaline?)*. What is the clinical significance of this fact?

■ **B10.** Refer to Figure 23.10, page 568 in the text, and answer these questions about female external genitalia or vulva.

a. The mons pubis marks the *(anterior? posterior?)* border of the vulva.

b. Arrange from most anterior to most posterior: _____ _____ _____ _____
   A. Anus   C. Clitoral prepuce   U. Urethral orifice   V. Vaginal orifice

c. Arrange from most lateral to most medial: _____ _____ _____
   LMaj. Labia majora   LMin. Labia minora   V. Vaginal orifice

d. The greater vestibular (Bartholin's) glands are located just inside the orifice of the:
   A. Anus   U. Urethra   V. Vagina

e. The site of an episiotomy is between the _____ and the _____ . (Select two answers)
   A. Anus   C. Clitoral prepuce   U. Urethral orifice   V. Vaginal orifice

■ **B11.** Describe mammary glands in this Checkpoint.

Each breast consists of 15–20 _____ composed of lobules. Milk-secreting glands called (*areola? alveoli?*)

empty into a duct system that terminates at the _____ . Milk production is stimulated by the hormones
(*oxytocin? prolactin?*) as well as estrogens and progesterone. Milk letdown (from glands into ducts) is triggered

by the hormone _____ .

C. Female reproductive cycle (pages 570–574)

C1. Contrast *menstrual cycle* and *ovarian cycle*.

■ **C2.** Select names of hormones produced by each endocrine gland listed below. (*Hint:* Refer to Figure 23.13,
page 573 in your text.)

E. Estrogen	LH. Luteinizing hormone
FSH. Follicle-stimulating hormone	P. Progesterone
GnRH. Gonadotropin-releasing hormone	R. Relaxin
I. Inhibin	

Hypothalamus: _____   Ovary: _____

Anterior pituitary: _____   Placenta: _____

■ **C3.** Now select the hormones listed in Checkpoint C2 that fit descriptions below.

_____ a. Stimulates release of FSH

_____ b. Stimulates release of LH

_____ c. Inhibits release of FSH

_____ d. Inhibits release of LH

_____ e. Initiates development of ovarian
         follicles and their secretion of estrogens

_____ f. Completes development of ovarian
         follicles and enhances secretion of estrogens

_____ g. Triggers ovulation and converts ovarian
         follicles into corpus luteum; stimulates
         it to secrete four hormones

_____ h. Stimulates development and maintenance
         of uterus, breasts, and secondary sex
         characteristics; enhances protein
         anabolism

_____ i. Prepares endometrium for implantation
         by fertilized ovum

_____ j. Dilates cervix for labor and delivery

_____ k. Lowers blood cholesterol level

■ **C4.** Refer to Figure 23.12, page 571 in the text and check your understanding of events of the female cycles in this
Checkpoint.

a. The average duration of the menstrual cycle is _____ days. The event that marks day 1 of the menstrual cycle

   is _____ . The menstrual phase usually lasts _____ days.

b. Following the menstrual phase is the _____ phase. During both of these phases,

   _____ develop in the ovary. Usually only _____ matures each month, becoming the

   dominant follicle, and later the _____ (graafian) follicle.

358

c. Follicle development occurs under the influence of the hypothalamic hormone _____ , which regulates the

anterior pituitary hormones _____ and later _____ .

d. What are the two main functions of follicles?

e. How do estrogens affect the endometrium?

For this reason, the preovulatory phase is also known as the _____ phase. Because

follicles reach their peaks during this phase, it is also known as the _____ phase.

f. A moderate increase in estrogen production by the growing follicles initiates a *(positive? negative?)* feedback mechanism that *(increases? decreases?)* FSH level. It is this change in FSH level as well as the

hormone _____ that causes degeneration of the remaining partially developed follicles.

g. The principle of this negative feedback effect is used in oral contraceptives. By taking "pills" consisting of

hormones estrogen and progesterone, a woman will maintain a very low level of the hormones _____ and

_____ . Without these hormones, follicles and ova will not develop, so the woman will not

_____ . (See text page 574.)

h. The second anterior pituitary hormone released during the cycle is _____ . The surge of this hormone results from the *(high? low?)* level of estrogens present just at the end of the preovulatory phase. This is an example of a *(positive? negative?)* feedback mechanism.

i. The LH surge causes the release of the secondary oocyte, the event known as _____ . The LH surge is the basis for a home test for *(ovulation? pregnancy?)*.

j. Following ovulation is the _____ phase. It lasts until day _____ of a 28-day cycle.

Under the influence of the hormone _____ , follicle cells are changed into the corpus

_____ . These cells secrete four hormones, the major one being

_____ , which prepares the endometrium for _____ . What preparatory changes occur?

Because of events during this period, the postovulatory phase is also known as the _____

or _____ phase.

k. What effect do rising levels of target hormones estrogens and progesterone have on GnRH and LH?

In addition, the hormone named _____ produced by the corpus luteum inhibits production of LH somewhat.

l. One function of LH is to form and maintain the corpus luteum. So as LH _____-creases, the corpus luteum

disintegrates, forming a scar on the ovary known as the corpus _____.

m. The corpus luteum had been secreting _____ and _____. With the demise of the corpus luteum, the levels of these hormones rapidly *(increase? decrease?)*. Because these hormones were maintaining the endometrium, endometrial tissue now deteriorates and will be shed during the next

_____ phase.

n. If fertilization occurs, estrogens and progesterone are needed to maintain the endometrial lining. The

_____ around the developing embryo secretes a hormone named

_____ . It functions much like LH in that it maintains the corpus

_____ , even though LH has been inhibited (step k above). The corpus luteum continues to secrete estrogens and progesterone until the placenta itself can secrete sufficient amounts. Incidentally, hCG is

present in the urine of pregnant women only and so is routinely used to detect _____ .

D. Birth control and abortion (pages 574–575 and 577)

■ **D1.** Match the methods of birth control listed in the box with descriptions below.

Con. Condom	Nor. Norplant	Symp. Sympto-thermal method
Depo. Depo-provera	OC. Oral contraceptive	T. Tubal ligation
Dia. Diaphragm	RU 486. RU 486	Vag. Vaginal ring
IUD. Intrauterine device	Sperm. Spermatocide	Vas. Vasectomy

_____ a. Foams, creams, jellies or douches that kill sperm; example: nonoxynol 9

_____ b. "The pill," combination of progesterone and estrogens, causing decrease in GnRH, FSH, and LH so ovulation does not occur

_____ c. Removal of a portion of the ductus (vas) deferens

_____ d. Tying off of the uterine tubes

_____ e. Periodic abstinence based on symptoms and temperature changes that indicate fertility or infertility

_____ f. A mechanical method of birth control in which a dome-shaped structure is placed over the cervix; accompanied by use of a jelly or cream

_____ g. Small object placed in the uterus by physician

_____ h. A nonporous latex sheath covering the penis

_____ i. Slender, progestin-containing capsules that are surgically implanted under the skin of the arm; effects last for five years unless capsules are removed.

_____ j. A ring worn for three weeks each month; the ring releases progestin or estrogen and progestin.

_____ k. Competitive inhibitor of progesterone; a chemical that induces miscarriage.

_____ l. Injections of progestin taken every three months to prevent maturation of ova

■ **D2.** Categorize birth control methods listed in the box above in this Checkpoint.

    a. Which methods are considered hormonal? _____

    b. Which are barrier methods? _____

    c. Which are forms of sterilization? _____

■ **D3.** Which of the forms of birth control listed in the box (in Checkpoint D1) is *least* effective with typical use?

_____Which nonsurgical method listed in the box is *most* effective with typical use?_____

D4. Refer back to Checkpoint C4g and review the mechanism of action or oral contraceptives in preventing pregnancy.

D5. Contrast *spontaneous abortion* with *induced abortion*.

List three methods of induced abortion.

E. Aging and reproduction; female athlete triad; common disorders and medical terminology (pages 576–581)

E1. Contrast these terms: menarche/menopause.

Menopause is considered a *(normal aging change? disorder?)*. List several symptoms of menopause.

E2. Summarize normal aging changes that occur in the reproductive systems of:

a. Females

b. Males

■ **E3.** Check your understanding of the "female athlete triad" in this Checkpoint.
  a. List the three components of this triad.

  b. Explain why ongoing amenorrhea can lead to osteoporosis.

  c. List categories of athletes at increased risk for this triad of disorders.

  d. Defend or dispute this statement. "Exercise increases bone density."

■ **E4.** Benign prostatic hyperplasia (BPH) is *(rare? common?)* among older men. List three signs of BPH and explain why these occur.

■ **E5.** Check your understanding of disorders of the reproductive system in this Checkpoint.
  a. Cancer of the *(prostate? testes?)* is the most common cancer of young adult males. Pain *(is? is not?)* a classic sign of this form of cancer.

  b. Cancer of the _____ is the most common lethal form of reproductive cancer in U.S. males. List two tests for this type of cancer.

  c. PMS refers to a distressing condition in women who are *(pre? post?)*-ovulatory. List three or more signs or symptoms of PMS.

  d. Endometriosis is a *(benign? malignant?)* condition involving excessive growth of the lining of the _____

  under the influence of hormones _____ and _____ . List several sites where this tissue may grow.

  e. Ovarian cancer *(is? is not?)* usually detected before it metastasizes. The occurrence of this type of cancer is greater *(before? after?)* age 50, in *(black? white?)* women, and in those who *(have been? have never been?)* pregnant.
  f. Risk for cervical cancer is linked to the virus that causes *(herpes simplex II? genital warts?)*. It is more common in women who had first intercourse *(early? later?)* in life.

■ **E6.** *A clinical challenge.* Mrs. Rodriguez, a 55-year-old woman, tells the nurse that she is concerned about having breast cancer. Do this exercise about the nurse's conversation with Mrs. Rodriguez.

a. The nurse asks some questions to evaluate Mrs. Rodriguez's risk of breast cancer. Which of the following responses indicate an increased risk of breast cancer?

A. "My older sister had cancer when she was 40 and she had the breast removed."
B. "I've never had any kind of cancer myself."
C. "I have two daughters; they were born when I was 36 and 38."
D. "I don't smoke and I never did."
E. "I know I eat too many high-fat foods."

b. The nurse tells the client that the American Cancer Society recommends three methods of early detection of breast cancer. These are listed below. On the line next to each of these, write the frequency that the nurse is likely to suggest to Mrs. Rodriguez.

1. Self breast exam (SBE) _____

2. Mammogram _____

3. Breast exam by physician _____

E7. Describe the main characteristics of these sexually transmitted diseases (STDs) by completing the table.

Disease	Causative Agent	Main Symptoms
a. Gonorrhea		
b.	*Treponema pallidum*	
c.	Type II herpes simplex virus (HSV)	
d. Chlamydia		

■ **E8.** Check your understanding of the disorders listed in the box by matching them with the correct descriptions.

A. Amenorrhea	D. Dysmenorrhea	M. Menorrhagia
Can. Candidiasis	E. Endometriosis	S. Syphilis
Chl. Chlamydia	G. Gonorrhea	

_____ a. Painful menstruation, partly due to contractions of uterine muscle

_____ b. Possible cause of blindness in newborns if bacteria transmitted to eyes during birth

_____ c. Spreading of uterine lining into abdominopelvic cavity via uterine tubes

_____ d. Caused by bacterium *Treponema pallidum;* may involve many systems in tertiary stage

_____ e. Absence of menstrual periods

_____ f. Yeastlike vaginal infection with itching, pain, and cheesy discharge, experienced by about 75% of females at some time in their lives.

_____ g. The most prevalent STD in the U.S., and most common cause of pelvic inflammatory disease (PID)

_____ h. Excessivley long and heavy menstrual periods

■ **E9.** Oophorectomy refers to surgical removal of a(n) _____ , whereas a salpingectomy is surgical removal of a(n) _____ . The "D" of the procedure known as a "D & C" refers to _____ of the cervix.

ANSWERS TO SELECTED CHECKPOINTS: CHAPTER 23

A2. (a) Testes. (b) Seminiferous tubules, epididymis, vas deferens, ejaculatory duct, urethra. (c) Prostate, seminal vesicles, bulbourethral glands. (d) Scrotum, penis.

A3. (a) Lower; lower temperatures are required for production and survival of sperm. (b) Cold; skeletal. (c) Posterior of the abdomen; cryptochidsim; premature; testes usually descend into the scrotum during the seventh month of fetal life. (d) Are (80% chance); cryptorchid testes are likely to be sterile and have a high risk for malignancy.

A4. (a) Secondary oocytes (potentially mature ova); haploid; 23. (b) Diploid; 2n; homologous. (c) Spermato, seminiferous tubules of testes.

A5. (a) Sg. → 1°Sc. → 2°Sc. → St. → Sz. (b) I; recombination of genes may occur. (c) Months; 4.

A6. (a) 300 million. (b) 48. (c) Head. (d) Middle piece. (e) Releases enzymes that help the sperm to penetrate the secondary oocyte. (f) Propels sperm.

A7. (a) Hypothalamus, releasing; write GnRH on lines 1a and 1b in Figure LG 23.1. (b) FSH; inhibin. (c) Testosterone. (d) High, high; negative, lowers, lowers. (e) Testosterone.

A8. (a) T. (b) DHT. (c) T and DHT. (d) T. (e) T. (f) I. (g) DHT. (h) I.

A10. (a) 5. (b) 2. (c) 6. (d) 8. (e) 7. (f) 4. (g) 3. (h) 9. (i) 1.

A11. Paired: 1–5; singular: 6–9.

A12. (a) Vas. (b) Partially; inguinal canal. (c) Scrotum; no; not normally; hormones exit through testicular veins, which are not cut during a vasectomy. (d) Blood and lymph vessels, autonomic nerves.

A13. (a) P. (b) P. (c) BUG. (d) SV. (e) SV.

A14. Accessory sex glands: A (seminal vesicles), B (prostate), C (bulbourethral glands); ducts, 2–9; gonad, 1.

A15. (a) 2.5–5.0 ml. (b) 50–150 million; 20 million; only a small percentage of sperm ever reach the secondary oocyte. (c) Alkaline; protects sperm against acidity of the male urethra and the female vagina.(d) Hemospermia.

A16. (a) Prepuce, glans; circumcision. (b) Cavernosa, spongiosum, blood; erect, parasympathetic. (c) Spongiosum; urine, semen; sphincter action at base of bladder. (d) Formerly known as impotence, ED is consistent inability to attain or hold an erection long enough for sexual intercourse to take place; enhances release if nitric oxide which dilates penile arteries.

B1. (a) 3. (b) 4. (c) 1. (d) 2. (e) 5.

B2. 1, 2: paired; 3–5, singular.

B4. (a) Og (2n). (b) 1°Oc (2n). (c) 2°Oc (n), 1PB (n). (d) 2°Oc (n). (e) Ov (n), 2PB (n). (f) Z (2n).

B5. (a) Fallopian. (b) Fimbriae. (c) Movements of cilia, fimbriae, and smooth muscle within walls of the tubes. (d) Fertilization normally takes place here.

B7. (a) Urinary bladder; rectum. (b) Anteriorly; cervix; close to a 90° angle. (c) Myo; smooth muscle. (d) Endo; mucous; is; nourishes sperm and zygote and is shed during menstruation.

B8. Papanicolaou (Pap); hysterectomy.

B9. Acid; retards microbial growth, but may harm sperm cells.

B10. (a) Anterior. (b) C U V A. (c) LMaj LMin V. (d) V. (e) V A.

B11. Lobes; alveoli, nipple; prolactin; oxytocin.

C2. Hypothalamus: GnRH; anterior pituitary: FSH, LH; ovary: E, P, I, R; placenta: E, P, R.

C3. (a, b) GnRH. (c, d) I; E and P indirectly because they inhibit GnRH. (e) FSH. (f, g) LH. (h) E. (i) P and E. (j) R. (k) E.

C4. (a) 28; the start of menstrual flow; 5. (b) Preovulatory, secondary follicles; 1, mature. (c) GnRH, FSH, LH. (d) Secretion of estrogens and development of the secondary oocyte. (e) They stimulate thickening of the endometrium to replace the sloughed-off lining; proliferative; follicular. (f) Negative, decreases; inhibin. (g) GnRH, FSH; ovulate. (h) LH; high; positive. (i) Ovulation; ovulation. (j) Postovulatory; 28; LH, luteum; progesterone, implantation; thickening by growth and secretion of glands and increased retention of fluid; luteal, secretory. (k) Inhibition of these hormones; inhibin. (l) De, albicans. (m) Progesterone, estrogens, decrease; menstrual. (n) Chorion layer of the placenta, human chorionic gonadotropin (hCG); luteum; pregnancy.

D1. (a) Sperm. (b) OC. (c) Vas. (d) T. (e) Symp. (f) Dia. (g) IUD. (h) Con. (i) Nor. (j) Vag. (k) RU 486. (l) Depo.

D2. (a) Depo, OC, Nor, Vag. (b) Con, Dia. (c) T, Vas.

D3. (a) Sperm (unless used with Con or Dia); complete abstinence, then Depo.

E3. (a) Disordered eating (anorexia or bulimia), amenorrhea, and osteoporosis. (b) Lack of normal menstrual cycles decreases estrogen production so that bones do not retain calcium. (c) Female runners, gymnasts, figure skaters, and divers. (d) Typically, weight-bearing exercise does increase osteoblast activity and bone density; however, excessive exercise combined with inadequate dietary intake can lead to the female athlete triad.

E4. Common; since it surrounds the urethra (like a doughnut), an enlarged prostate can cause painful and difficult urination (dysuria) and bladder infection (cystitis) due to incomplete urination.

E5. (a) Testes; is not. (b) Prostate; PSA and digital rectal exam. (c) Post; shorter; see text page 579. (d) Benign, uterus, estrogens and progesterone; on the outside of uterus, ovaries, colon, urinary bladder, kidneys, on lymph nodes or abdominal wall. (e) Is not; after, white, have never been. (f) Genital warts; early.

E6. (a) A, C, and E. (b) 1: monthly; 2 and 3: annually or more often based on the decision made by physician and client.

E7. (a) D. (b) G. (c) E. (d) S. (e) A. (f) Can. (g) Chl. (h) M.

E8. Ovary, uterine (fallopian) tube; dilation.

CRITICAL THINKING: CHAPTER 23

1. Describe factors that enhance possibilities for semen to reach the uterine tubes so that fertilization occurs. Include these terms in your description: *cilia, prostaglandins, oxytocin, contractions.*

2. Contrast five common sexually transmitted diseases (STDs). Include causative microorganisms, usual signs and symptoms, and treatments (if any).

3. Contrast the processes of oogenesis and spermatogenesis.

4. Identify the phase that is the "follicular, proliferative, estrogenic phase" of the menstrual cycle.

5. Tell exactly what chemical is tested for in over-the-counter tests for: (a) ovulation; (b) pregnancy. State which specific hormones are found in oral contraceptives and in Depo-provera injections.

6. Identify which form of birth control you think is both most effective and has the least harmful side effects. State your rationale.

MASTERY TEST: ■ CHAPTER 23

Questions 1–6: Arrange the answers in correct sequence.

_______ _______ _______ 1. Thickness of the endometrium, from thinnest to thickest:
A. On day 5 of cycle
B. On day 13 of cycle
C. On day 23 of cycle

_______ _______ _______ 2. From anterior to posterior:
A. Uterus and vagina
B. Bladder and urethra
C. Rectum and anus

_______ _______ _______ _______ 3. Order in which hormone levels begin to increase, from day 1 of menstrual cycle:
A. FSH
B. LH
C. Progesterone
D. Estrogen

_______ _______ _______ _______ 4. Order of events in the female monthly cycle, beginning with day 1:
A. Ovulation
B. Formation of dominant secondary follicle
C. Menstruation
D. Formation of corpus luteum

_______ _______ _______ _______ 5. From anterior to posterior:
A. Anus
B. Vaginal orifice
C. Clitoris
D. Urethral orifice

_______ _______ _______ _______ 6. Pathway of sperm:
A. Testis
B. Urethra
C. Ductus (vas) deferens
D. Epididymis

Question 7: Circle the letter preceding the one best answer to the question.

7. In the normal male, there are two of each of the following structures *except:*
A. Testis
B. Seminal vesicle
C. Prostate
D. Ductus deferens
E. Epididymis
F. Ejaculatory duct

Questions 8–10: Circle letters preceding all correct answers to each question.

8. Which of the following are produced by the testes?
   A. Spermatozoa   D. GnRH
   B. Testosterone   E. FSH
   C. Inhibin   F. LH

9. Which of the following are produced by the ovaries and then *leave* the ovaries?
   A. Follicle   D. Corpus albicans
   B. Secondary oocyte E. Estrogen
   C. Corpus luteum   F. Progesterone

10. Which of the following are functions of LH?
    A. Begin the development of the follicle
    B. Stimulate change of follicle cells into corpus luteum cells
    C. Stimulate release of secondary oocyte (ovulation)
    D. Stimulate corpus luteum cells to produce estrogens and progesterone
    E. Stimulate release of GnRH

Questions 11–15: Fill-ins. Complete each sentence with the word or phrase that best fits.

_____ 11. _____ is a term that means undescended testes.

_____ 12. The most inferior portion of the uterus is called the _____ .

_____ 13. The ovarian cycle is a series of changes in the _____ , whereas the menstrual cycle is a series of changes in the _____ .

_____ 14. Menstrual flow occurs as a result of a sudden _____-crease in the hormones _____ and _____ .

_____ 15. Three major functions of estrogens are development and maintenance of _____ , protein _____ , and lowering of blood _____ level.

ANSWERS TO MASTERY TEST: ■ CHAPTER 23

Arrange
1. A B C
2. B A C
3. A D B C
4. C B A D
5. C D B A
6. A D C B

Multiple Choice
7. C

Multiple Answers
8. A, B, C
9. B, E, F
10. B, C, D

Fill-ins
11. Cryptorchidism
12. Cervix
13. Ovaries, uterus (or endometrium)
14. De, progesterone, estrogens
15. Female reproductive structures and secondary sex characteristics, anabolism, cholesterol

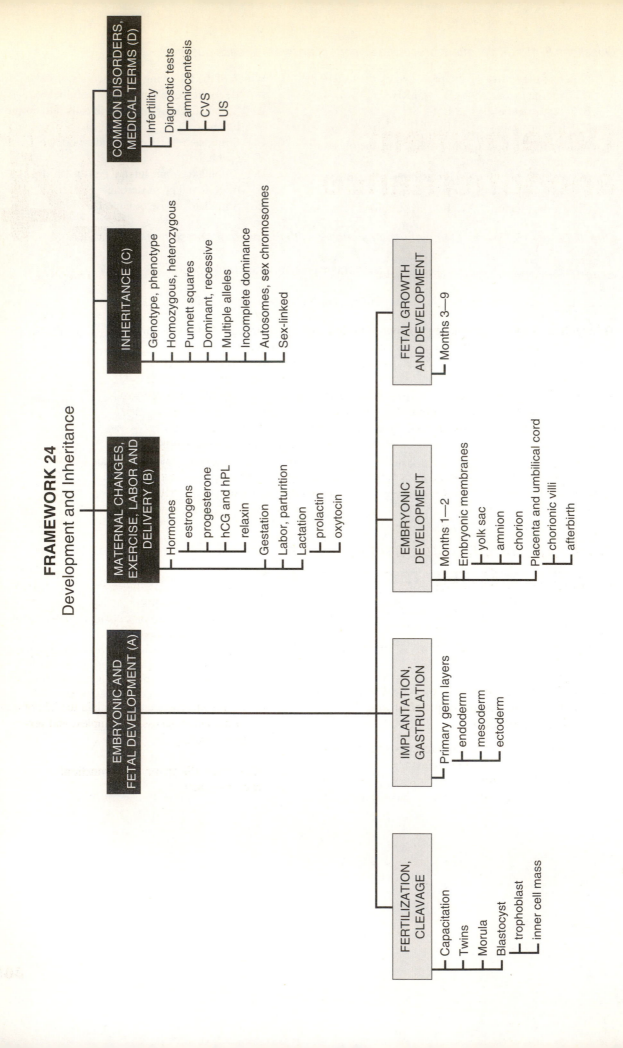

FRAMEWORK 24
Development and Inheritance

Development and Inheritance

Twenty-three chapters ago you began a tour of the human body, starting with the structural framework and varieties of buildings (chemicals and cells) and moving on to the global communication systems (nervous and endocrine), transportation routes (cardiovascular), and refuse removal systems (respiratory, digestive, and urinary). This last chapter synthesizes all the parts of the tour: a new individual, with all those body systems, is conceived and develops during a nine-month pregnancy. Adjustments are made at birth for survival in an independent (outside of mother) world. But parental ties are still evident in inherited features, signs of intrauterine care (mother's avoidance of alcohol or cigarettes), and provision of nourishment, as in lactation.

Look at Framework 24 first. And congratulations on having completed your first tour through the wonders of the human body. As you begin this chapter, carefully examine the Chapter 24 Topic Outline and check off each objective after you meet it.

TOPIC OUTLINE AND OBJECTIVES

A. Embryonic and fetal development

☐ 1. Explain the major developmental events that occur during the embryonic period.
☐ 2. Define the fetal period and outline its major events.

B. Maternal changes during pregnancy; exercise and pregnancy; labor and delivery; lactation

☐ 3. Describe the sources and functions of the hormones secreted during pregnancy.
☐ 4. Describe the hormonal, anatomical, and physiological changes in the mother during pregnancy.

☐ 5. Explain the effects of pregnancy on exercise, and of exercise on pregnancy.
☐ 6. Outline the events associated with the three stages of labor.
☐ 7. Discuss the hormonal control of lactation.

C. Inheritance

☐ 8. Define inheritance, and explain the inheritance of dominant, recessive, complex, and sex-linked traits.

D. Common disorders and medical terminology

WORDBYTES

Study each wordbyte, its meaning, and an example of its use in a term. Check your understanding by jotting meanings of wordbytes in margins. Identify other examples of terms that contain these wordbytes as you continue through the text and *Learning Guide*.

Wordbyte	Meaning	Example(s)	Wordbyte	Meaning	Example(s)
-blast	germ, to form	*blast*omere, *blast*ocyst	-mere	part	blasto*mere*
-cele	hollow	blasto*cele*	meso-	middle	*meso*derm
-centesis	puncture	amnio*centesis*	morula	mulberry	*morula*
ecto-	outside of	*ecto*derm, *ecto*pic	syn-	together	*syn*gamy
endo-	inside of	*endo*derm	-troph	nourish	syncytio*troph*oblast
fertil-	fruitful	*fertil*ization	zygo-	a joining	*zygo*te; di*zygo*tic
lact-	milk	*lact*ation			

CHECKPOINTS

A. Embryonic and fetal development   (pages 587–599)

■ **A1.**   Answer these questions related to human development.

a. Paula is 12 weeks pregnant; she has a(n) *(embryo? fetus?)* in her uterus. Paula's full term pregnancy would take *(38? 40?)* weeks from the day of fertilization.

b. _____ is the division of medicine that deals with pregnancy, labor and delivery, and the

neonatal period (which is the first _____ weeks after birth of the baby).

■ **A2.**   Do this exercise about events associated with fertilization.

a. How many sperm are ordinarily introduced into the vagina during sexual intercourse?

_____ About how many reach the area where the developing ovum is located?

_____

b. How long are sperm usually in the female reproductive tract before they fertilize the female secondary oocyte?

_____ hours. Describe the changes sperm undergo during this time that can enhance the likelihood of fertilization.

These changes in the sperm are known as _____ .

c. Name the usual site of fertilization. _____ What do the corona radiata and zona pellucida have to do with fertilization?

d. How many sperm normally penetrate the developing ovum? _____ What prevents further sperm from entering the female gamete?

e. Fertilization of secondary oocyte by the sperm causes the oocyte to complete meiosis *(I? II?)*. The larger of the two resulting cells is the *(ovum? polar body?)*. Its fusion with the sperm results in formation of the _____ which has *(23? 46?)* chromosomes.

■ **A3.** Contrast types of multiple births by choosing the answer that best fits each description.

| C. Conjoined twins | F. Fraternal twins | I. Identical twins |

_____ a. Also known as dizygotic twins; may not be of the same gender

_____ b. Two infants derived from a single fertilized egg that divides once within about a week of fertilization

_____ c. Two infants derived from a single fertilized egg that divides once more than 8 days after fertilization

_____ d. Two infants derived from two secondary oocytes each fertilized by a different sperm; the infants may be as different as any other siblings

■ **A4.** Complete this checkpoint about the next events following formation of the zygote.

a. Rapid mitotic divisions of the zygote are called _____ . The first of these divisions is completed by about *(6? 30? 48?)* hours after fertilization. By the end of day 3 there are _____ cells, each of which is *(the same size as? progressively smaller than?)* previous cells.

b. A solid sphere is formed called a _____ and is composed of cells known as _____ . This sphere is *(about the same size as? considerably larger than?)* the original zygote. The morula moves through the uterine tube and reaches the uterine cavity by about day _____ .

c. The next stage is the _____ . Draw a diagram of a blastocyst. Label: *blastocyst cavity, inner cell mass*, and *trophoblast*. Color red the cells that will become embryo. Color blue the cells that will become part of the placenta. (Refer to Figure 24.3b, page 589 in the text.)

A5. List two sources of *pluripotent stem cells*.

Describe possible therapeutic uses of pluripotent stem cells.

■ **A6.** Describe the process of implantation in this Checkpoint.

a. The blastocyst remains free in the uterine cavity for about _____ days.

Implantation is likely to occur about _____ days after ovulation (or _____ days after fertilization). Note that the developing baby is implanted in the uterine wall before the mother "misses" the first day of her menstrual period.

b. Typically, the developing individual implants with the inner cell mass *(away from? towards?)* the endometrium of the fundus or body of the uterus. List three or more sites where implantation could occur but which would not

support a full term pregnancy—a condition known as an _____ pregnancy.

c. List several factors that increase risk for ectopic pregnancy.

■ **A7.** Recall from Checkpoint A4c that the blastocyst consists of two portions, the _____ and the _____ . Refer to text Figure 24.5 and describe these parts in the exercise below.

a. About the time of implantation, the trophoblast forms two layers that will both become part of the *(amnion?*

chorion?) that makes up the major portion of the placenta. One layer secretes _____ that will help the blasto-

cyst to implant. The other layer secretes the hormone named _____ . Describe the function of this hormone.

b. Inner cell mass cells also divide into two layers, the _____ and the _____ . Together, these form the

_____ embryonic disc.

c. The epiblast develops the *(amnion? chorion?),* commonly known as the " _____ of _____ ."

The amnion surrounds fluid known as _____ fluid. What is the source of this fluid? _____
List functions of this fluid.

d. Describe the procedure known as an *amniocentesis.*

e. Find cells of the hypoblast in text Figure 24.5. These cells line the primitive gut (_____ tract). These cells from the yolk sac which has several other important functions. List them.

f. Find the *lacunar network* on Figure 24.5 in the text. Explain the function of this network.

g. Find the *extraembryonic mesoderm* on Figure 24.5. This tissue, together with the trophoblast layers, forms

the _____ . (See A7a above)

■ **A8.** The process of formation of primary germ layers is known as _____ ; it is completed by the end of

week _____ following fertilization. During the process, the bilaminar disc is converted into a *(3? 4? 6?)*-layer disc consisting of *(3? 4? 6?)* germ layers.

■ **A9.** Identify which primary germ layer forms structures listed below. Choose from answers in the box.

| Ecto. Ectoderm | Endo. Endoderm | Meso. Mesoderm |

_____ a. Epithelial lining of digestive, respiratory, and genitourinary tracts except near openings to the exterior of the body

_____ b. Epidermis of skin, epithelial lining of entrances to the body (such as mouth, nose, and anus), hair, nails

_____ c. All of the skeletal system (bone, cartilage) and muscles

_____ d. Entire nervous system

■ **A10.** Complete this Checkpoint about additional embryonic development.

a. The notochord is located in the *(anterior? posterior?)* of the developing baby's trunk. The notochord is developed

from _____-derm.

b. Arrange in correct sequence the events in neurulation. Formation of: _____

  NF. Neural fold   NP. Neural plate   NT. Neural tube

c. Which structures fail to develop normally in these conditions?

  Anencephaly   *Neural tube defects (NTDs)*

■ **A11.** Further describe the placenta in this exercise.

a. The placenta has the shape of a _____ embedded in the wall of the uterus. Its fetal por-

tion is the _____ fetal membrane. Fingerlike projections, known as

_____ , grow into the uterine lining. Consequently maternal and fetal blood are *(allowed to mix? brought into close proximity but do not mix?)*. Fetal blood in umbilical vessels is bathed by maternal

blood in _____ spaces.

b. The maternal aspect of the placenta is a portion of the _____-metrium of the uterus.

c. What is the ultimate fate of the placenta?

d. For what purposes can the discarded human placenta be used?

■ **A12.** *A clinical challenge.* Following a birth, vessels of the umbilical cord are carefully examined. A total of

_____ vessels should be found, surrounded by a mucous jellylike connective tissue. The vessels should include *(1? 2?)* umbilical artery(-ies) and *(1? 2?)* umbilical vein(s). Lack of a vessel may indicate a congenital malformation.

373

■ **A13.** Discuss organ formation in this Checkpoint.

a. By the end of week *(3? 5? 8?)*, all major organs are present, at least in rudimentary form.

b. Briefly describe the process of embryonic folding.

As a result, the embryo is *(flat? C-shaped?)* (Figure 24-9c in the text).

c. What does the *otic placode* form (Figure 24.9d)? _____ What about the *lens placode?* _____

d. During which week do limb buds occur? *(4? 6? 8?)* Which limb buds appear first? *(Upper [arm]? Lower [leg]?)* By the end of week 4, a tail *(is? is not?)* found in the human embryo (Figure 24.9c).

e. During weeks five and six of development, the head grows *(larger? smaller?)* in proportion to the trunk, and the

heart has _____ chambers.

f. By the end of the embryonic period (which is the end of week _____), limbs are distinct, digits *(are? are not?)* webbed, and the tail *(is? is not?)* present.

■ **A14.** Study Table 24.1 (page 596 in your text). Then write the number of the month when each of the following events occurs.

_____ a. Heart starts to beat.

_____ b. Heartbeat can be detected.

_____ c. Backbone and vertebral canal form

_____ d. Eyes are almost fully developed; eyelids are still fused.

_____ e. Eyelids separate; eyelashes form.

_____ f. Limb buds develop.

_____ g. Ossification begins.

_____ h. Limb buds become distinct as arms and legs; digits are well formed.

_____ i. Nails develop.

_____ j. Fine (lanugo) hair covers body.

_____ k. Lanugo hair is shed.

_____ l. Fetus is capable of survival.

_____ m. Subcutaneous fat is deposited.

_____ n. Testes descend into scrotum.

B. **Maternal changes during pregnancy; exercise and pregnancy; labor and delivery; lactation** (pages 600–604)

■ **B1.** Complete this exercise about the hormones of pregnancy.

a. During pregnancy, the level of progesterone and estrogens must remain *(high? low?)* in order to support the endometrial lining. During the first two months, these hormones are produced principally by the

_____ located in the _____ , under the influence of the tropic

hormone _____ .

b. Human chorionic gonadotropin (hCG) reaches its peak about the _____ week of preg-

nancy. Because it is present in blood, it is filtered into _____ , where it can readily be de-

tected as an indication of pregnancy as early as the _____ day of pregnancy.

hCG may be the cause of _____ sickness in the first trimester of pregnancy. See page 616 of the text.

c. During the last seven months of the pregnancy, estrogens and progesterone are secreted by the _____ . The name *progesterone* indicates that this hormone is progestation. Explain.

374

d. _____ is a hormone that relaxes pelvic joints and helps dilate the cervix near the time of

birth. This hormone is produced by the _____ and _____ .

e. Name the chorionic hormone helps to prepare the mother's breasts for lactation. _____

f. The newly found chorionic hormone CRH, if secreted in high levels early in pregnancy, is associated with *(premature? late?)* delivery. Is this hormone produced in people who are not pregnant?

■ **B2.** Complete this exercise about maternal adaptations during pregnancy. Describe normal changes that occur.

a. Cardiac output _____-creases to accommodate the _____-creased metabolic rate of mother and _____-creased blood flow to the uterus and placenta.

b. Expiratory reserve volume (ERV) _____-creases related to _____-creased upward pressure of the fetus on the diaphragm.

c. _____-creased pressure on the bladder may result in urinary symptoms such as:

d. Gastrointestinal (GI) motility _____-creases, possibly resulting in _____ . Heartburn

may accompany a delay in _____ emptying.

e. Signs of pregnancy also include _____-creased pigmentation in areas such as the face and the areolae of breasts.

B3. List and describe examples of the following categories of chemicals in breast milk:

a. Immune chemicals

b. Vitamins

B4. Defend or dispute this statement: "All women who are pregnant or lactating should avoid exercise because of its deleterious effects."

B5. Contrast these terms: *labor/parturition*

■ **B6.** Discuss regulation of labor and delivery in this Checkpoint.

a. As discussed earlier, progesterone tends to *(increase? decrease?)* uterine contractions during pregnancy. For birth, known as _____ , to occur, levels of estrogen must *(surpass? become less then?)* levels of progesterone.

b. Explain how oxytocin promotes labor and delivery.

Oxytocin is involved in a *(negative? positive?)* feedback loop.

c. The hormone relaxin assists in birth by *(relaxing? tightening?)* the pubic symphysis. Relaxin, estrogens, and prostaglandins work together to *(soften? tighten?)* the cervix.

■ **B7.** What is the "show" produced at the time of birth?

Is the "show" associated with true labor or false labor?

■ **B8.** Identify descriptions of the three phases of labor (first, second, or third).

_____ a. Stage of expulsion: from complete cervi-   _____ c. Time from onset of labor to complete di-
cal dilation through delivery of the baby   lation of the cervix; the stage of dilation
_____ b. Time after the delivery of the baby until
the placenta ("afterbirth") is expelled; the
placental stage

B9. *A clinical challenge.* Explain more about labor and delivery in this Checkpoint.

a. The average duration of labor is about _____ hours, whereas the duration for first labor and delivery is

about _____ hours.

b. The term _____ means painful or difficult labor. In a breech presentation, which parts of the body

enter the birth canal first? _____ If necessary, a C-section or _____ section will be required for such births. Was this procedure named such because Julius Caesar was born by C-section? *(Yes? No?)* Explain.

c. Defend or dispute this statement: "Longer (such as 45 weeks) is better for the fetus when it comes to length of pregnancy."

d. What is the *puerperium?*

376

■ **B10.** Complete this exercise about lactation.

a. The major hormone promoting lactation is _____ , which is secreted by

the _____ . During pregnancy the level of this hormone increases somewhat, but its effectiveness is limited while estrogens and progesterone are *(high? low?)*.

b. Following birth and the loss of the placenta, a major source of estrogens and progesterone is gone, so effectiveness of prolactin *(increases? decreases?)* dramatically.

c. The sucking action of the newborn facilitates lactation in two ways. What are they?

d. What harmone stimulates milk ejection (let-down)? _____

e. The time period from the start of a baby's suckling at the breast to the delivery of milk to the baby is

about _____ seconds. Although afferent impulses (from breast to hypothalamus) require *(only a fraction of*

a second? 45 seconds?), passage of the oxytocin through the _____ to the breasts requires

about _____ seconds.

B11. Contrast *colostrum* with *maternal milk* with respect to time of appearance, nutritional content, and presence of material antibodies.

B12. Defend or dispute this statement: "Breast-feeding is an effective, reliable form of birth control."

B13. List five or more benefits of breast-feeding.

C. Inheritance (pages 604–606)

C1. Define the following terms:

a. Inheritance

b. Genetic counseling

c. Homologues (homologous chromosomes)

d. Allele

■ **C2.** Answer these questions about genes controlling PKU. (See Figure 24.10, page 605 in the text.)

a. The dominant gene for PKU is represented by *(P? p?)*. The dominant gene is for the *(normal? PKU?)* condition.

b. The letters PP are an example of a genetic makeup or *(genotype? phenotype?)*. A person with such a genotype is said to be *(homozygous dominant? homozygous recessive? heterozygous?)*. The phenotype of that individual would *(be normal? be normal, but serve as a "carrier" of the PKU gene? have PKU?)*.

c. The genotype for a heterozygous individual is _____ . The phenotype is

_____ .

d. Use a Punnett square to determine possible genotypes of offspring when both parents are Pp.

C3. In Huntington's disease (HD), which gene is dominant? The gene for *(being normal? having the disorder?)*. Briefly describe this condition.

■ **C4.** Do this exercise on variations in dominant–recessive inheritance.

a. List the three possible alleles for inheritance of the ABO group. _____ _____ _____ Which of these is/are

dominant? _____ List the possible genotypes of persons who produce A antigens on red blood cells.

_____ Describe the phenotype of persons with the *ii* genotype.

b. ABO blood group inheritance is known as *(multiple allele? polygenic?)* inheritance.

c. Carriers for sickle-cell disease are said to have the sickle-cell _____ . They may have some signs or symptoms of the condition. Explain why.

C5. Contrast these two terms: *autosome, sex chromosome*

■ **C6.** Answer these questions about sex inheritance.

a. All *(sperm? ova?)* contain the X chromosome. About half of the sperm produced by a male contain the _____

chromosome, and half contain the _____ .

b. Every cell (except those forming ova) in a normal female contains *(XX? XY?)* genes on sex chromosomes. All male cells (except those forming sperm) are said to be *(XX? XY?)*.

c. Who determines the sex of the child? *(Mother? Father?)* Show this by using a Punnett square.

■ **C7.** Complete this exercise about X-linked inheritance.

a. Sex chromosomes contain other genes besides those determining sex of an individual. Such traits are called

_____ traits. Y chromosomes are shorter than X chromosomes and lack some genes. One

of these is the gene controlling ability to _____ red from green colors. Thus this ability is

controlled entirely by the _____ chromosome.

b. Write the genotype for females of each type: color-blind, _____ ; carrier, _____ ; normal, _____ .

c. Now write possible genotypes for males: color-blind, _____ ; normal, _____ .

d. Determine the results of a mating between a color-blind male and a normal female.

D. Common disorders and medical terminology   (pages 607–608)

■ **D1.** Fill in the term: *(impotence* or *infertility)* that better fits each description. Then list several causes of each condition:

a. Inability of a man to have his sperm fertilize a secondary oocyte: _____

b. Inability to attain and maintain an erection: _____

D2. Describe how conditions of being underweight or obese may affect reproductive hormones and fertility.

D3. Describe the following fertility-expanding methods:

a. GIFT

b. Embryo transfer

c. Intracytoplasmic sperm infection (ICSI)

D4. Down syndrome is a disorder that involves an extra chromosome number _____ . Risk for this chromosomal disorder is greater in *(younger? older?)* mothers. Briefly describe manifestations of Down syndrome.

■ **D5.** Match the medical term with the description below.

Con. Conceptus	Pre. Preeclampsia
EG. Emesis gravidarum	SRY. SRY region
FAS. Fetal alcohol syndrome	Ter. Teratogen

_____ a. Morning sickness (nausea) which may be related to high hCG and progesterone level early in pregnancy

_____ b. The gene on the Y chromosome that leads to the male pattern of development

_____ c. All structures that develop from the zygote, including the embryo and fetal membranes

_____ d. Sudden hypertension of pregnancy accompanied by protein in urine and edema

_____ e. Any agent that can cause developmental defects to an embryo

_____ f. A condition in which the newborn is small for gestational age, has characteristic facial features, a pattern of defective organs, and behavioral problems

■ **D6.** Complete this exercise about prenatal diagnostic tests.

a. Amniocentesis involves withdrawal of _____ fluid. This fluid is constantly recycled through the fetus (by drinking and excretion); it *(does? does not?)* contain fetal cells and products of fetal metabolism. For what purposes is amniotic fluid collected and examined?

For _____

b. *(Amniocentesis? Ultrasound?)* is by far the most common test for determining true fetal age if the date of conception is uncertain. The technique *(is also? is not?)* used for diagnostic purposes.

c. CVS (meaning c _____ v _____ s _____)
is a diagnostic test that can be performed as early as *(4? 8? 16? 22?)* weeks of pregnancy. It *(does? does not?)*

involve penetration of the uterine cavity. Cells of the _____ layer of the placenta embedded in the uterus are studied. These *(are? are not?)* derived from the same sperm and egg that united to form the

fetus. Therefore this procedure, like amniocentesis, can be used to diagnose _____ .

A1. (a) Fetus; 38. (b) Obstetrics, four.

A2. (a) Approximately 200 million; about 200. (b) At least seven hours and up to 48 hours; development of flagella and changes in the plasma membrane; capacitation. (c) Uterine tube; they are coverings over the oocyte that sperm must penetrate, and the zona pellucida triggers the acrosome to release enzymes. (d) One; sperm penetration of the secondary oocyte causes changes that prevent entry of other sperm. (e) II; ovum; zygote, 46.

A3. (a) F. (b) I. (c) C. (d) F.

A4. (a) Cleavage; 30; 16, progressively smaller than. (b) Morula, blastomeres; about the same size as; 4 or 5. (c) Blastocyst; see Figure 24.3.

A6. (a) 2; 7. (6). (b) Towards; ectopic; most often the uterine tube, but possibly in the cervix on an ovary, or on other abdominopelvic structures. (c) Tubal (or other) scarring from previous infection; abnormal tubal anatomy.

A7. Trophoblast and inner cell mass. (a) Chorion; enzymes; HCG; much like LH, HCG maintains the corpus luteum until placental secretion of estrogens and progesterone eliminates the need for production of these hormones by the corpus luteum; and HCG is the chemical tested for in home pregnancy tests. (b) Hypoblast (primitive endoderm) and epiblast (primitive ectoderm); bilaminar. (c) Amnion, bag (of) waters; amniotic; filtrate of mother's blood (and later, fetal urine); shock absorber for developing baby; see also page 590. (d) See text page 607. (e) Gastrointestinal; see text page 590. (f) Provide nourishment via maternal blood and secretions and serve as a disposal site for fetal wastes. (g) Chorion.

A8. Gastrulation; 3; 3, 3.

A9. (a) Endo. (b) Ecto. (c) Meso. (d) Ecto.

A10. (a) Posterior (noto = "back"); meso. (b) NP NF NT. (c) Anencephaly: brain; NTDs: vertebral column, affecting spinal nerves, cord, and meninges.

A11. (a) Flattened cake (or thick pancake); chorion; chorionic villi; brought into close proximity but do not mix; intervillous. (b) Endo. (c) It is discharged from the mother at birth as the "afterbirth." (d) Sources of hormones, drugs, and blood; for burn coverage; and to contribute placental coverage for women with placenta previa. The umbilical vessels can be used for vessel grafts; and cord blood is a source of pluripotent stem cells.

A12. 3; 2, 1.

A13. (a) 8. (b) Conversion of the flat, two-dimensional disc to a three-dimensional cylinder; C-shaped.

(c) Ear; eye. (d) 4; upper [arm]; is. (e) Larger, four. (f) 8, are not, is not.

A14. (a) 1. (b) 3. (c) 1. (d) 3. (e) 6. (f) 1. (g) 2. (h) 2. (i) 3. (j) 5. (k) 9. (l) 7. (m) 8. (n) 7.

B1. (a) High; corpus luteum, ovary, hCG. (b) Ninth; urine; eighth; morning. (c) Placenta (chorion); this hormone relaxes the myometrium and keeps the cervix tightly closed. (d) Relaxin; placenta, ovaries. (e) Human placental lactogen (hPL). (f) Premature; yes, made by the hypothalamus, it is the releasing hormone for ACTH.

B2. (a) In, in, in. (b) De, in. (c) In, urinary frequency, urgency, and stress incontinence. (d) De, constipation; gastric. (e) In.

B6. (a) Decrease; parturition, surpass. (b) Oxytocin stimulates uterine contractions which force the baby into the cervix. Positive because cervical stretching triggers nerve impulses to the hypothalamus where more oxytocin is released. (c) Relaxing; soften.

B7. A discharge of blood-containing mucus that is present in the cervical canal during labor; true labor.

B8. (a) Second. (b) Third. (c) First.

B10. (a) Prolactin (PRL), anterior pituitary; high. (b) Increases. (c) Stimulates release of prolactin and also release of oxytocin. (d) Oxytocin. (e) 45; only a fraction of a second, bloodstream, 45.

C2. (a) P; normal. (b) Genotype; homozygous dominant; be normal. (c) Pp; normal but a carrier of PKU gene. (d) 25% PP, 50% Pp, and 25% pp.

C4. (a) I^A, I^B, and i; I^A and I^B; $I^A I^A$ or $I^A i$; produces neither A nor B antigens on red blood cells (Type O). (b) Multiple allele; (c) Trait; the dominant gene for sickle-cell is incompletely dominant over the recessive gene.

C6. (a) Ova; X, Y. (b) XX; XY. (c) Father, as shown in the Punnett square (Figure 24.11(b), page 606 in the text) because only the male can contribute the Y chromosome.

C7. (a) X-linked; differentiate; X. (b) $X^c X^c$; $X^c X^c$; $X^c X^C$. (c) $X^c Y$; $X^C Y$. (d) All females are carriers and all males are normal.

D1. (a) Infertility. (b) Impotence.

D5. (a) EG. (b) SRY. (c) Con. (d) Pre. (e) FAS.

D6. (a) Amniotic; does; for detecting suspected genetic abnormalities (14–16 weeks); assessing fetal maturity (35 weeks). (b) Ultrasound; is also; (c) Chorionic villus sampling, 8; does not; chorion; are; genetic and chromosomal defects.

CRITICAL THINKING: CHAPTER 24

1. Summarize the three major events in the embryonic period following implantation.
2. Describe the layers that form the placenta and outline its protective functions.
3. Contrast the three primary germ layers and the types of tissues and organs each will form.
4. Contrast use of amniocentesis and chorionic villi sampling (CVS). Describe when during gestation each procedure is likely to be performed, what is sampled, and how each test gives information about potential genetic anomalies.
5. Describe effects of the following maternal or fetal hormones upon labor and parturition: progesterone, estrogens, oxytocin, and relaxin.
6. Discuss effects of embryonic or fetal environment upon genetic traits. Include effects of cigarettes, alcohol and other drugs, and radiation.
7. Describe the hormones that make lactation possible.

MASTERY TEST: ■ CHAPTER 24

Questions 1–4: Circle the letter preceding the one best answer to each question.

1. These fetal characteristics—weight of 8 ounces to 1 pound, length of 10 to 12 inches, lanugo covers skin, and brown fat is forming—are associated with which month of fetal life?
   A. Third month   C. Fifth month
   B. Fourth month   D. Sixth month

2. All of the following structures are developed from mesoderm *except:*
   A. Aorta   D. Humerus
   B. Biceps muscle   E. Sciatic nerve
   C. Heart   F. Neutrophil

3. Implantation of a developing individual (blastocyst stage) usually occurs about _____ after fertilization.
   A. three weeks   D. seven hours
   B. one week   E. seven minutes
   C. one day

4. All of the following events occur during the first month of embryonic development *except:*
   A. Endoderm, mesoderm, and ectoderm are formed.
   B. Amnion and chorion are formed.
   C. Limb buds develop.
   D. Ossification (bone formation) begins.
   E. The heart begins to beat.

Questions 5–10: Circle T (true) or F (false). If the statement is false, change the underlined word or phrase so that the statement is correct.

T F 5. Maternal pulse rate, blood volume, and tidal volume <u>all normally increase</u> during pregnancy.

T F 6. A <u>phenotype</u> is a chemical or other agent that causes physical defects in a developing embryo.

T F 7. Blood in the umbilical artery is normally <u>more highly</u> oxygenated than blood in the umbilical vein.

T F 8. Maternal blood <u>mixes with fetal blood</u> in the placenta.

T F 9. <u>Amniocentesis</u> is a procedure in which amniotic fluid is withdrawn for examination at about <u>8–10</u> weeks during the gestation period.

T F 10. The <u>expulsion</u> stage of labor refers to the period during which the "afterbirth" is expelled.

Questions 11–15: Fill-ins. Write the word or phrase that best fits the description.

_____ 11. The outer covering of cells of the blastocyst is known as the _____ .

_____ 12. As a prerequisite for fertilization, sperm must remain in the female reproductive tract for 12–24 hours so that _____ can occur.

_____ 13. hCG is a hormone made by the _____ ; its level peaks at about the end of the _____ month of pregnancy, and it can be used to detect _____ .

_____ 14. Determine the probable genotypes of children of a couple in which the man is color-blind and the woman is normal. _____

_____ 15. This is the _____ mastery test question in this book.

ANSWERS TO MASTERY TEST: ■ CHAPTER 24

Multiple Choice
1. C
2. E
3. B
4. D

True or False
5. T
6. F. Teratogen
7. F. Less highly
8. F. Does not mix with but comes close to fetal blood
9. F. Amniocentesis, 14–16
10. F. Placental

Fill-ins
11. Trophoblast
12. Capacitation (dissolving of the covering over the ovum by secretion of acrosomal enzymes of sperm)
13. Chorion of the placenta, second, pregnancy
14. All male children normal; all female children carriers
15. Last or final. Congratulations! Hope that the knowledge you gained here will serve as a foundation for a lifetime of learning.

NOTES

NOTES

NOTES

NOTES

NOTES

NOTES

NOTES

NOTES

NOTES

NOTES

NOTES

NOTES

NOTES

NOTES